IN THE LENA DELTA

A NARRATIVE OF THE

SEARCH FOR LIEUT.-COMMANDER DeLONG AND HIS COMPANIONS

FOLLOWED BY

AN ACCOUNT OF THE GREELY RELIEF EXPEDITION

AND

A PROPOSED METHOD OF REACHING THE NORTH POLE

BY

GEORGE W. MELVILLE

CHIEF ENGINEER U. S N.

EDITED BY

MELVILLE PHILIPS

WITH MAPS AND ILLUSTRATIONS

BOSTON

HOUGHTON, MIFFLIN AND COMPANY

New York: 11 East Seventeenth Street

The Riverside Press, Cambridge

1885

EDITOR'S PREFACE.

IF it be true — and Emerson affirms as much — that great deeds deserve a fit and permanent record, then assuredly there can be no need of explanation, much less of apology, for the appearance of this work. "We need books of this tart cathartic virtue," wrote our New England sage; and so the editor flatters himself for having contributed to the birth of this one.

The world knows the story of the lost Jeannette, the luckless cruise and tedious drift of many months, the amazing march and terrible tribulation, the heroic endeavor and sad ending, — the world is aware of all this, to be sure, because in all the world's history it has no parallel; and no matter the why or the wherefore, there is yet in this story a human sympathy that cannot be disguised, an abiding interest that overlooks the question of utility.

Perhaps there will be readers of this volume who, already acquainted with the prominent part played by our author in the many adventures attending the long ice-blockade of the Jeannette, may have expected a fuller account of that interesting period than will be found herein; and will consequently regard our brief narrative of it as insufficient and unsatisfactory. Certainly it does appear, considering the few pages devoted

to this prolonged and remarkable drift, that we have
treated it too lightly and displayed an undue haste in
transporting the reader to the shores of Siberia.

Not all, it is true, of the Jeannette's experiences in
the ice have been chronicled, but then more than enough
have been published to enable the reader to gain an adequate idea of the wonderful voyage and retreat, and in
the melancholy " ice-journals "[1] of Commander De Long
these may be found embodied in their most permanent,
authentic, and interesting form.   Hence it was deemed
advisable for this work, after affording the reader a running survey of the cruise and march, to begin its more
elaborate discourse at the date of the separation of the
three boats in the gale of September 12, 1881.   The
events, indeed, which followed can have no competent
historian save Chief Engineer Melville.   He it was who
directly gave rise to them, and was the prime mover and
central figure in all the exploits " In the Lena Delta."

The greater portion of the author's manuscript was
prepared between the months of January and April of
this year (1884), and the final chapters were written at
sea ; for, undaunted by his previous experiences, he sailed
again in Arctic waters to the relief of Lieutenant Greely ;
and, moreover, he herein proposes, upon an original plan,
to attain the goal that has baffled the daring of Parry
and of Franklin.

Cui bono? asks the utilitarian.   There are numerous
and well-known advantages that would result from the
success of such a venture.   Aside from the many useful
facts that would be established concerning the laws of
storms and wind-waves, the flattening of the earth at the

---

[1] *Voyage of the Jeannette*, etc., Houghton, Mifflin & Co.

pole would be measured, and geographical science be plainly benefited. Additional information, too, would be gained in astronomy, meteorology, ocean physics, and natural history, a more thorough knowledge of which would certainly add directly or indirectly to the comfort and safety of mankind.

As to the rest, I refer the reader to the theory itself, merely observing that — " Prejudice, which man pretends to hate, is," according to Carlyle, " his absolute lawgiver. . . . Thus, let but a rising of the sun, let but a creation of the world, happen *twice*, and it ceases to be marvelous, to be noteworthy or noticeable."

In other words, let but Chief Engineer Melville reach the North Pole, and besides the scientific benefits issuing from the event will doubtless be another and perhaps more important one to the world at large — his success, in his own words, "may prevent other fools from going there."

MELVILLE PHILIPS.

October 16, 1884.

CONTENTS.

CHAPTER VI.

ON THE LENA DELTA.

CHAPTER VII.

UP THE LENA.

CHAPTER VIII.

AT JAMAVELOCH.

CHAPTER IX.

SIBERIAN LIFE.

CHAPTER X.

KUSMA TO OUR RESCUE.

CHAPTER XI.

A STEP FORWARD.

CHAPTER XII.

AT BELUN.

CHAPTER XVIII.

FROM BELUN TO VERKERANSK.

CHAPTER XIX.

FROM VERKERANSK TO YAKUTSK.

CHAPTER XX.

AT YAKUTSK.

CHAPTER XXI.

NORTH AGAIN.

CHAPTER XXII.

STORM-BOUND.

CHAPTER XXIII.

FINDING THE BODIES.

CHAPTER XXI.

THE BURIAL.

CHAPTER XXV.

SEARCHING FOR CHIPP.

CHAPTER XXVI.

MY FINAL SEARCH TO THE JANA RIVER.

CHAPTER XXVII.

THROUGH SIBERIA.

CHAPTER XXVIII.

HOMEWARD BOUND.

THE GREELY RELIEF EXPEDITION.

LIST OF ILLUSTRATIONS.

MAPS.

THE JEANNETTE LEAVING SAN FRANCISCO.

IN THE LENA DELTA.

CHAPTER I.

OFF FOR THE POLE.

The Jeannette Expedition. — Our Departure. — Unalaska. — St. Michael's. — The Tchuchees. — Nordenskjold. — Frozen in. — Herald Island.

THE *Kuro-Shiwo* (the black current of Japan) runs around the Japanese Islands, threads its way among the Kurile Islands, passes eastward to Kamschatka, and thence northward to Behring Strait, where it separates into two branches. The one branch seeks the west coast of North America, and then runs south, tempering the atmosphere as it goes, until it is lost in the warm water at the equator. The other branch passes into the Arctic Ocean, streaming up into the northeast, and was lately regarded as one of the thermometric gateways to the Pole.

Previous to the Jeannette's voyage no polar expedition had ever set out by way of Behring Strait, although one had indeed been projected by a French lieutenant, but was prevented by the breaking out of the Franco-Prussian War; and a high latitude, it is true, had been gained by the search ships of the English squadron which hoped to intercept Sir John Franklin in case he was successful in making the Northwest Passage.

The object of the Jeannette expedition was thus to

1

reach the North Pole by following up the *Kuro-Shiwo;* but it should be remembered that our first cruise (which unhappily proved both first and last) was only designed at the time to be an experimental voyage. So, resting upon this brief review of the motive of our luckless enterprise, I shall now proceed to chronicle it.

All things being in readiness, on the 8th of July, 1879, the sun shining clear and strong on the beautiful bay, we cleared from the Golden Gate, accompanied by the San Francisco Yacht Club.

Never was departure more auspicious.   It was a gala day for the good people of 'Frisco ; the harbor was alive with their pleasure craft, 'and right royal was the farewell they tendered our adventurous ship.   Cheers rang out from the crowded wharves ; the masts and decks of the myriad vessels on the bay teemed with jolly tars, huzzaing and firing guns with a deafening effect ; and as we steamed abreast of the Presidio, a heavy salute boomed forth from the fortification that seemed a solemn amen to the godspeeds of the people.

Once clear of the Gate, we headed about and left a straight wake for the island of Unalaska.   Unlike the buoyant hearts of her company, the vessel was weighted down below the ordinary line of immersion, and consequently our progress was slow.   The fair weather that attended our departure continued in the Pacific until the Aleutian Group was approached, when fogs set in, and, by the time the Aqueton Pass was reached, grew so dense that it became necessary to anchor and await their lifting.   Even then the breakers had been dangerously neared, for though the islands were invisible, the sea-birds could be distinctly heard cawing on the rocks.

At last the fog lifted from around the Jeannette, and, after a pleasant voyage of twenty-five days, the island of Unalaska hove in sight.

Here, through the kindness of the Alaska Fur Company, a large number of deer-skins, seal-skin blankets, and other furs, to be made up for use during the expedition, were added to our cargo.

With a fresh supply of coal we then departed from Unalaska, and, crossing the shallow sea of Behring, arrived safely at St. Michael's, in Norton Sound, — an old Russian trading-post, with a dilapidated block-house, and several ancient cast-iron guns, which were fired in honor of the Jeannette's arrival. It was here that the schooner Fanny A. Hyde was to meet the ship with our last supply of coal; but, as she was considerably overdue, the provisions were restowed, and the undressed skins sent on shore to be made into clothing by the natives.

Here, too, Alexia, our faithful hunter, and his companion, the womanly-looking Iniguin, joined the ship. Poor Alexia, after a grand palaver with the head man of the village, shipped as hunter and dog-driver. The evening of our departure, the natives thronged on board to bid their friends farewell. Alexia, dressed in " store clothes " furnished by the Alaska Company's agent, with a tile-topped Russian hat encircled by a broad red band, was accompanied by his wife, small, shy, and pretty, and their little boy. Clinging together hand in hand, they wandered and wondered with all the curiosity of children about the ship, until at last with many doubts and fears they affectionately parted, and forever.

Attended by our convoy, we now started across Norton Sound, a large sheet of water making westward into the eastern coast of Siberia. During the passage a heavy gale was encountered, affording us an opportunity to observe the vessel's strength and action. The shallow ocean was troubled and choppy, and at times great seas would roll completely over the deeply-freighted ship as over a sunken rock; but Gibraltar itself was not more firm. The day following the gale, we drew near to Lutke, a beautiful harbor just to the southward of East

Cape, where American whalemen often resort to secure recruits from the native Tchuchees before proceeding on the cruise north of Behring Strait, or after " the catch " to " try out " their oil. As the ship stood in towards the mouth of the harbor, huts were descried on the hillside; and shortly after two *bidaras* — large walrus-skin boats, somewhat similar to the *oomiaks* of the Greenlanders — were seen putting off from the shore and approaching us at a rapid rate. They contained a dozen or more natives, tall, greasy, brawny fellows, none weighing less than a hundred and eighty pounds.

The two *bidaras* pulled alongside, and the natives boarding the ship, inquired, in broken English learned from the whalemen, if we came to hunt the walrus or the whale; if so, they wished to engage themselves for the cruise. To prove their ability, they named the various whaling captains with whom they had served, and concluded with the more important information that Nordenskjöld's ship, the Vega, had been there in the bay of St. Lawrence, and had wintered beyond the peninsula in Kiolutian Bay, where they had visited her.

And now the time was come to push off from the last vestige of civilization. The remainder of the coal and supplies was transferred from the schooner to the ship, and on the evening of the 27th day of August the two vessels stood out of the bay. Once clear of the shoals, we parted, — the Fanny A. Hyde sailing to the south, the Jeannette to the north, — and the cruise had begun in earnest.

During the night we passed through the strait of Behring, — between East Cape (the most easterly cape of Siberia) and the Diomed Islands, three rocky little islets, the stepping-stones between two continents, — perhaps the foundations for that future bridge over which may run the engines of an all-rail route from Cape Horn to the Cape of Good Hope. A sharp lookout was kept for

natives as the ship headed westward along the coast of
Siberia, towards Cape Serdzekamen ("Heart of Stone").
Ice-hills and snow-gorged valleys were no longer a novelty; but very cheerless and inhospitable looked the
bleak black rocks, with no living thing in sight save a few
sea-birds and an occasional walrus or seal. At length a
collection of huts made their appearance, and the ship
was run as close to the shore as possible. Captain De
Long, with Alexia as interpreter and Mr. Dunbar as
pilot, attempted a landing; but the heavy sea rolling
and breaking over the ice-foot compelled him to return.
The natives, knowing quieter water, then came to the
ship in their skin boats. But Alexia was unable to understand them, save that they wanted biscuits, molasses,
and rum, especially rum; so no additional information
of Nordenskjold was obtained. Further on, another village was sighted, and Lieutenant Chipp sent on shore.
He effected a landing, and through an old woman who
formerly lived on King's Island in Norton Sound, and
spoke the same tongue as Alexia, learned that the Vega
had wintered in a bay still to the west, but had passed
safely out of the ice toward the east, in the spring.

Sailing westward, the seal and walrus became more
plentiful. Numerous natives approached the ship, using
with considerable dexterity bladders or floats to prevent
their light skin boats from capsizing, or as fenders to
ward them off from the ship's side, and as buoys when
overladen with the spoils of the chase. In the vicinity
of the Vega's winter-quarters, a large village was observed, and a party was sent on shore in command of
Lieutenant Chipp. After a few hours the boat returned,
bringing a number of articles, such as tin cans, Swedish
money, and coat-buttons, which the natives — who mentioned the names of certain of the Vega's officers, exhibiting presents they had received from them, and who particularly prized the refuse tin cans — said were relics of

Nordenskjöld's stay there. A letter sewed up in canvas was given to the chief, with the request that he deliver it to the first vessel that passed that way.

Having now executed the instructions of the Navy Department as to Nordenskjöld and his party, we pushed northward to Kiolutian Bay, keeping constantly under way to prevent the formation of ice around the ship, and skirting along the edge of the pack, the density of which gradually forced us to the east. Working in and out of the leads of water for several days, and preserving as northerly a course as possible, Herald Island was at length sighted on the evening of September 4th. Every opening in the pack with the least northing to it had now been tried, with the same fruitless result of being crowded off to the eastward ; so that there was positively nothing else to be done than to push boldly in towards Wrangel Land (long regarded as a large continent extending to the Pole), and seek a harbor during the fall, winter, and spring The absurd question has often been asked, "Why did the Jeannette enter the ice-pack?" The answer is this : she was an Arctic ship bound on a polar voyage, and could not be expected to attain the Pole without encountering ice. The best authorities pointed out a continent connecting Wrangel Land with Greenland ; the currents setting among the islands to the Atlantic Ocean were well known ; the rotation of the earth should carry all things from west to east ; and it was fair to presume that, if caught in the ice north of Herald Island, the ship would drift on the coast of Wrangel Land, or to the northeast toward Prince Patrick Land.

But "the best laid schemes o' mice and men gang aft a-gley." As we crowded to the west, the ice set in behind and effectually cut off all chances of retreat, unless the autumn gales and rolling seas might break up the floe. A whale-ship was observed looking for game along

the edge of the pack, and it was afterward regretted
that she had not been spoken for the mail. But we
pushed slowly ahead, with the aid of the hoisting-engine,
making a warp, until the 6th of September; when the
hummocks and masses of ice became solid during the
night, and the ship was frozen in with a list to starboard
at an angle of ten or twelve degrees, rendering motion
on the decks or sleeping in the berths very uncomfortable. Only once was she freed, and then but for an
hour or two, until the final crush in June, 1881.

The crew now dispersed themselves in merry squads
upon the floe, and the dogs were turned loose. Each
man was armed with a pike or staff with which to steady
himself when leaping from hummock to hummock, or as
a protection when slipping into ice-holes — a frequent oc-
currence. Bears were seen, but they kept at a safe distance during the daytime, and would make off in hot
haste when pursued by the " whoop, hurrah! " of the
crew. At night, the *ice* — as they are better named by
Norwegian and Dane, instead of *polar* — bears would cir-
cumambulate the vessel many times, critically examining
the stout hawser that held her to a large ice-anchor, and
regularly escaping the notice of the watch, and of forty-
one dogs, who were shy and timid until they had tasted
the good things that followed a successful bear hunt.
And a most exciting sport is this. The pack of dogs run
yelping and snapping at the heels of poor Ursus, who
dashes across the floe in the direction of rougher country or open water with the ungracefulness of a cow but
the speed of a deer, causing the snow to fly like feathers
in a gale, and leaving dogs and hunters far in the rear.
Often his curiosity will get the better of his judgment,
and he will stop to inspect these strange creatures that
dare put him to flight; for he is monarch of the polar
regions, seal and walrus falling an easy prey to one blow
of his powerful fore-paw. Raising himself on his hind-

legs, he surveys the howling pack half in wonder, half
in contempt, until the fleetest hunter may come within
range, when Ursus, perceiving a new and more formida-
ble enemy, drops on all-fours, and is off again. Selecting
a level amphitheatre surrounded by high hummocks,
where he imagines himself secure, he makes a final stand,
and woe betide the dogs; for he is a wonderful boxer,
and every blow is a fatal one. But the Remington
breech-loader comes into play; the slaughter of the canine
foe is checked, and poor Ursus is dead. In a hunt such
as this, Mr. Dunbar killed a bear weighing about one
thousand pounds, and having a coat of snowy whiteness,
with a single shot; the only time that feat was accomplished during the cruise.

Meanwhile the Jeannette had been steadily drifting
toward Herald Island, and as it was quite certain that she
would be carried by it to the northwest, it became important that a landing should be attempted, in order to erect
a cairn and leave records there. For this purpose, a
party consisting of Lieutenant Chipp, Mr. Dunbar, and
myself, with Alexia as dog-driver, was fitted out, and provisioned for a week. It was thought that the ice closed in
on Herald Island; but when we came within a few miles
of the land, long lanes of open water were found, completely cutting off further progress. It would have been
sheer folly to await the freezing of the water and then
make a dash for the foot of the precipitous rocks, without
the aid of a boat. The ship was rapidly moving past;
no food could possibly have been found on the island,
and in case of a separation from the Jeannette we must
inevitably starve. Taking this view of the situation, we
reluctantly turned back.

And still the floe continued to drift toward Wrangel
Land. Soundings were taken daily, and observations
daily and nightly. A "drift lead" was kept constantly in
place to indicate the impulse of the ice; and the bottom

of the sea was dragged for samples of its natural history. Each officer had his special duty, and the whole ship's company worked together as a unit. At times, the choice library on board would be ransacked for authorities, when the officers, who had been students and observers in all climes, and could be cheerful under all conditions of life, would engage in friendly scientific discussion in the little cabin. Thus, without the appearance of the expected gales, October passed pleasantly away. The ice had been comparatively quiet; now and then low rumbling sounds being transmitted through the floe from distant disturbances. Toward the end of the month Wrangel Land was in plain sight, and it was quite evident that the much-boasted continent was a small island, high and mountainous. When probably at the most northeasterly point to the land, the ship was found shoaling water at sixteen fathoms, and as the floe crowded by the island, it was cast up in great ridges; cracks ran across it in all directions, and the grinding and crushing of the tortured masses sounded like the roar of distant artillery. The ship became entirely surrounded by the towering, telescoping hills of ice. Huge floe-bergs as large as churches bobbed up and down like whales. The situation was now most perilous; for even could the vessel sustain the enormous pressure thus brought to bear upon her, there was imminent danger of the hummocks and "bits" weighing from twenty to fifty tons toppling over, as they were on all sides, and crushing or burying her. In view of the impending disaster, preparations were made for abandoning the ship — a hopeless prospect; when suddenly the floe split along the port side parallel with the keel, leaving a long lane of open water, with the starboard side still imbedded in the ice as in a mould.

This was the moment of extreme danger. Should the floe-pieces come together again and overlap or underrun, the Jeannette would be crushed like an egg-shell. They

separated nearly a thousand yards, and then slowly approached. The poor ship began to creak and groan with
the immense strain; but fortunately the ice, ranging
ahead, acted as a protection. The decks bulged upward;
the oakum and pitch were squeezed out of the seams;
and a bucket, almost full of water, standing on the quar-
ter-deck, was half emptied by the agitation. There was
little sleep obtained; those who turned in doing so with
their clothes on. Yet the discipline of the ship's company was perfect. The men sang and joked with apparent *sang froid*, while they cleared the decks of ice or
pushed away the overhanging masses that were crushing
in the light bulwarks. The powerful trusses fitted in at
Mare Island sturdily withstood the pressure; until at
length the floe gave way somewhere beyond, and, as it
went thundering by, an underrunning piece pushed the
ship out from her bed into the open lead of water. Once
again she was nearly caught stem and stern, but as darkness set in, the young ice began forming, and shortly afterward she was completely frozen in, never to be released
again until a day or two before the fatal 12th of June,
1881.

CHAPTER II.

DRIFTING.

Shipboard Economy — A Frozen Wave. — Lead-Poisoning — My Visit to Henrietta Island.

WINTER set in sharply. Excepting a few minor accidents in the shape of sprains and contusions, we enjoyed entire freedom from disease, and were in constant good humor, — all save Lieutenant Danenhower, who suffered under his terrible affliction from December of the first year until the end of the cruise. As the holidays drew nigh, the crew prepared for the usual theatricals; and at Christmas all hands were summoned to the deck-house to witness a performance replete with jokes at the expense of officers and men.

Early in the experience of the expedition, a strange violation of an accepted physical law presented itself. In accordance with the laboratory teaching of our youth, we had presumed that sea-water passed through the process of freezing yielded perfectly fresh ice. Floe ice was known to be salty, but it was confidently expected that fresh-water snow would be found; and yet it was not. Those who were conversant with the histories of previous Arctic voyages — and nothing of the kind written in the English language had escaped the reading of many — were aware that heretofore no difficulty had been encountered in procuring fresh water for potable purposes from bergs or land snow. There are, however, no bergs proper in this ocean, except those which drop from the small islands, and they are so rare that the only ones met with were seen by my party when we landed at Henrietta Island.

So a distilling apparatus, capable of yielding forty gallons a day, was constructed, and a supply of melted snow kept constantly in tanks on the cabin and forecastle stoves; one large tank being fitted behind the galley to absorb the radiant heat.

And the old year left us busied in this wise with the multiform details of shipboard economy.

Throughout the month of January, 1880, the ice was restless, and the ship experienced many jars and strains. Each gale was followed by the jamming up of the floes; and it was observed that during the continuance of the wind the whole body of ice moved evenly before it, but when it subsided, the mass that had been put in motion crowded and tumbled upon the far-off floes at rest, piling tumultuously upward in a manner terrific to behold.

It was in one of these oppressive intervals succeeding a gale, when the roar and crash of the distant masses could be distinctly heard, that the floe in which the Jeannette was imbedded began splitting in all directions. The placid and almost level surface of ice suddenly heaved and swelled into great hills, buzzing and wheezing dolefully. Giant blocks pitched and rolled as though controlled by invisible hands, and the vast compressing bodies shrieked a shrill and horrible song that curdled the blood. On came the frozen waves, nearer and nearer. Seams ran and rattled across them with a thundering boom, while silent and awestruck we watched their terrible progress. Sunk in an amphitheatre, about five eighths of a mile in diameter, lay the ship, the great bank of moving ice, puffed in places to a height of fifty feet, gradually inclosing her on all sides. Preparations were made for her abandonment, but, — what then? If the mighty circle continued to decrease, escape was hope-

less, death inevitable. To think of scrambling up the
slippery sides of the rolling mass would be of equal folly
with an attempt to scale the falling waters of Niagara.

"The ice is approaching at the rate of one yard per
minute. It is three hundred paces distant; so in three
hundred minutes we shall pass over to the Great Beyond."

Thus one of the crew announced his computation of
the time, distance, and calamity. Certain it is that had
the Jeannette been two hundred yards in any direction
out of the exact spot she then occupied on the floe, she
would have been overwhelmed and destroyed by the
grinding masses, as readily as a "sojer crab" on the
beach is buried beneath the roll of the surf. But her
time had not yet come. The terrible circle slowly contracted to within a few hundred feet, and then stopped
— stopping our prayers; and all was quiet, save the roar
of the underrunning floes at the bottom of the ship.

With light hearts the men dispersed themselves upon
the ice, climbing the slopes of the marble - like basin,
leaping from block to block, clambering up pinnacles and
tumbling down with laughter, calling each other's attention to the marvelous shapes and positions of the confused heaps, speculating upon the chances of escape
had such an one toppled over on the vessel, all hailing
and shouting in boyish glee, — when, suddenly, the dread
cry of "Man the pumps!" put a check to their short-
lived sport, and sent every one scudding back.

Hand-pumps were at once rigged and operated with
all the vigor of the stoutest men; another gang removed
the provisions from the fore-hold, while a third filled
the boiler with water, ice, snow, and slush from the
bilges. The temperature at this time was about 40° Fahrenheit below zero, and as the water rushed into the
hold it almost instantly froze. Pouring steadily in, it
crept above the fire-room floor, and fears were enter-

tained that it might reach the boiler furnaces before the steam-pumps could be started. To prevent this, and keep the after-hold dry, the water was hoisted out of the hatch by means of a barrel. Time meant life or death. The flood was rapidly rising in the bilges; so the furnace was fired long before the boiler had received its regulation supply, and soon the steam giant was casting out the water at such a rate as to win from Jack Cole the admiring observation, — "No 'Spell O!' (the relief call at the pump brake) for that chap."

Winter passed swiftly by, and the bright spring sun, melting the snow, opened up a fresh field of labor by revealing the hideous results of forty dogs and thirty-three men living in one spot for six months. Nor was it without considerable anxiety that the approach of warmer weather was observed. The ship, indeed, would be free; but was she not leaky, and the supply of coal fast failing? How long would the spars, masts, and upper works of the hull hold out after all other fuel had been consumed? These were questions which we could not consider without alarm.

Northwest winds prevailed in the early spring and drove the ship within sight of Herald Island, or Wrangel Land, and about fifty miles from the point where we had entered the ice the preceding fall. During this time, we had drifted nearly five hundred miles in a zigzag course, and so persistently from northwest to southeast that I conceived the existence of two banks or shoals between which we had been moving, or perhaps two great packs of ice, the Polar pack to the north and the Siberian pack to the south, which latter shifted on and off the coast of Siberia with the changing winds and seasons. A canal covered with broken ice was thus formed, and hemmed in between the impenetrable floes and floebergs.

The winter of 1880–81 passed by without much inci-

dent. The novelty of life on the ice had worn off. Our
supplies of jokes and stories were completely exhausted,
and their points had long ago been dulled by much handling. The ship's company, fore and aft, had found their
affinities; and congenial spirits began to walk, talk, and
hunt together in couples. In the cabin there was more
reading and less conversation, and the senior officers
seemed daily bound by a closer band. Stricter attention
was paid to all the sanitary regulations of the ship, particularly to the distillation of water, the preparation of
food, the ship's ventilation, and the healthful exercise of
the men.

This was our second winter in the ice; and in the history of all previous expeditions, scurvy, the bane of the
Arctic voyager, had made its dread appearance long ere
such an interval had elapsed. Why were we exempt?
How long would we thus remain blessed above all other
crews? Like vegetables grown in the dark, we were
bleached to an unnatural pallor; and as spring approached
all exhibited signs of debility. Sleep was fortunately
peaceful and undisturbed, by reason of the floe's solid-
ity; but certain members of the mess were attacked with
fits of indigestion; Mr. Dunbar became very ill; and an
ugly ulcer appeared on Alexia's leg accompanied by
other symptoms which raised suspicions of the presence
of scurvy.

At length, an epidemic seemed to break out among
the whole company. Dr. Ambler was diligent in his
search for the cause. There were no evidences of scurvy
save in Alexia's case, and his was extremely doubtful.
Finally the patients showed symptoms of lead poisoning,
and the question at once arose, whence came the lead?
A few grains of shot found in the bodies of birds (guillemots) served for dinner one day sufficed to direct the
conversation to the subject that was uppermost in the
minds of all; and at the same time, some one chancing

upon several pellets of solder in the canned tomatoes, it was jocularly asked, "Who shot the tomatoes?" which resulted in bringing to light the *raison d'être* of the poisoning. Knowing how deadly wine might become by the dissolution of a single grain of shot left in the bottle from its cleansing, it was easy to understand how the acid fruits and vegetables had absorbed their noxious properties from the many drops of solder, composed of equal parts of lead and tin. And this cumulative poisoning had been in progress for months! Nor, as investigation proved, were these pellets the only source of the malady. Aware of the manufacturer's practice of covering certain qualities of sheet tin with solder, the cans were inspected, and many found to be coated with black oxide of lead. Scraping this off and analyzing it, Dr. Ambler became altogether satisfied as to the origin of the "epidemic."

It was at the beginning of this dark period in the history of our cruise that the cheering cry of "Land ho!" rang out from the crow's-nest. The ice had been slowly disintegrating for weeks, and Mr. Dunbar, our ancient mariner and Arctic authority, had declared a week before that something to the leeward was obstructing and breaking up the floe. Now, a faint line on the horizon with a stationary cloud above it indicated the presence of land. At once, all the younger prophets, — who had for months been seeing vast continents in the shape of various clouds, which they assiduously plotted on charts and named, only to be as regularly laughed at, — turned out, aloft, below, and on the high hummocks, to scan with glasses, or without, the discovered country. There it was, sure enough; and all were as elated as though a second Goshen, or still better our own peerless land of peers, had sprung into view. Speculation was rife as to its distance, size, and inhabitability; sketches were as plentiful as ticks in a southern forest; some of the far-seeing enthusiasts distinctly descried reindeer moving about; and others of still greater ken could plainly distinguish the buck from the doe.

Meanwhile, Mr. Dunbar, with that keenness of vision that comes from forty years' experience at sea, had espied another and separate land beyond, much smaller and lower than the first. As the whole floe was in a swirl, and the Jeannette was drifting rapidly to the northwest, the question arose, was it possible to visit the strange island, and return in safety? General opinion was adverse to the success of the undertaking, albeit there was no scarcity of volunteers. Messrs. Chipp, Danenhower, and Newcomb were prostrated in the cabin; so it was decided that I should go, accompanied by Mr. Dunbar, and a picked four of the crew: namely, Nindemann, Bartlett, Ericksen, and Sharvell. We were supplied with provisions for ten days, and a small boat mounted on a sled drawn by fifteen dogs.

Early in the morning, followed by the cheers and good-wishes of our shipmates, we were off, making a straight line for the island. The condition of the ice, grinding, crashing, and telescoping, sometimes pitching and rolling in such a manner as to render foothold impossible, made our enterprise a particularly perilous one.

Difficulties beset us at the very start. Not five hundred yards from the ship, we came to a lead of water, and dismounting the boat, ferried over the sled and supplies; but nothing could induce the dogs to follow suit. They howled and fought, all resisting with might and main, and a few breaking or slipping from their harness and scampering back to the ship. The thermometer registered many degrees below freezing point; the boat was covered with ice, our clothes were wet, and our hands frost-bitten. The deserters were at length captured and returned by the men on board ship, and again fastened in harness. A rope, tied to their traces, was then stretched across the lead (scarcely twenty yards in width), the whole team pushed into the water, and thus pulled and urged across. It was cruel, I know, but there was no alternative; and once over, and rehitched to the sleds, the poor shivering brutes were soon warming themselves in the hard work ahead of them. We were all equipped with "ruy ruddies" (canvas harness) to assist the dogs in hauling; and as the snow was waist-deep they were almost buried at times. Mr. Dunbar ran on before, leading the way among the hummocks, the rest of us steadying and pushing the sled, two on each side and one behind. Now and then, the team would come to a halt, and everything —a matter of 1,900 pounds — must be unloaded, since it is absolutely impossible to induce or compel a dog-team to pull in concert until the sled is first put in motion.

There is no greater violence done the eternal cause of truth than in those pictures where the Esquimaux are represented as calmly sitting in shoe-shaped sleds, with the lashes of their long whips trailing gracefully behind, while the dogs dash in full cry and perfect unison across smooth expanses of snow. If depicted "true to nature" the scene changes its aspect considerably; it is quite as full of action, but not of progress. A pandemonium of horrors! Dogs yelling, barking, snapping, and fighting; the leaders in the rear, and the wheelers (?) in the middle, all tied in a knot and as hopelessly tangled up as a basketful of eels.

Thus retarded, we toiled on for twelve hours, making roads, filling up chasms with "hummocky bits," and jumping the team across them; four times the boat was launched, and when evening came on we had traveled but four miles from the ship, and made no appreciable gain on the island. Nevertheless we erected our tent under the lee of a large hummock, supped, fed the dogs, and encasing ourselves in sleeping-bags lay down on the snow, partially warmed by the dogs, which were curled on the flaps of the tent, and well pleased with our first day's progress. At six o'clock next morning we were up and active. Sharvell prepared a breakfast of pigs' feet and mutton broth heated together in a can, along with a cup of tea, while the rest of us stored the tent and hitched the team; and by seven we were off again.

After journeying forward in this way for three days, the island at length loomed up before us in all its cloud-crowned majesty. The black serrated rocks, rising precipitously four hundred feet at the coast and towering inland to four times that height, bore at a distance the appearance of a vast heap of scoriæ discharged from some great blast furnace and streaked with veins of iron. They were grown over with moss and lichens, the tops capped with snow and ice and the highest peaks lost in the clouds. As we drew nearer we could distinguish glaciers making down the gorges, and bold headlands standing, as they had been for ages, like sentinels, grimly challenging our strange advent. The silence was awful, was confounding, and the loneliness of our situation indescribably depressive. Before us, like a black monster, arose the lofty island, protected, to a certain degree, from the endless grinding of the floes by an ice-foot, which extended in some places a half mile from the base. Here we stood lost in the contemplation of the wild tumult and rout before us. Millions of tons of blocks were piled up, as though they were the ghastly heaps of slain from the battle that was forever raging among the broken masses; and great bodies of ice were incessantly fleeing, it seemed, from the mad pursuit of those behind; now hurling themselves on top, and now borne down and buried by others. And it was through this chaos of ice that we must force our way to the island.

A glance at the situation convinced me of the utter impossibility of accomplishing a passage by means of the boat; and as we were fast drifting by, I determined to abandon it, together with the gear and most of the provisions, and make a dash for the land across the broken ice, jumping from bit to bit. It was a hazardous expedient, the success of which must be greatly a matter of luck; and still more so, our escape from the island and recapture of the drifting boat and provisions. However, we left these latter on a secure and elevated floepiece; on the tallest hummock of which, as a guide for our retreat, we raised an oar with a black flag lashed to it, and Ericksen's old felt hat on top as a liberty cap. Then with the tent, guns, instruments, and one day's provisions mounted on the sled, we started in a gallop for the island. The dogs were trained to follow a leader; so one of us ran on ahead, relieved in turn by the others, who jogged along with the sled and occasionally rested upon it. But when we reached the broken ice, the team stopped and refused to follow the leader. Poor brutes, they knew full well what it was to be dragged through the water, and hauled out coated with a sheet of ice, more dead than alive. So with the floe bits rolling under their feet they turned round, yelping in an agony of dread, and darted in all directions, the men shouting and belaboring them in vain; man and dog now splashing in the water, and now clambering out; raised at times high up in the air by the pressure of the underrunning floes, only to plunge down again or roll over. Mr. Dunbar had become snow-blind, and was now perched on the sled, greatly to the old gentleman's disgust. It was the first time in his life that he had ever broken down, and it grieved him sorely. He begged in the most distressing manner to be left on the ice rather than retard our progress; but directing him to hold fast, I finally seized the head dog by the neck with my "ruy ruddy," and, followed by the others, sprang forward, dragging team and all after me. Then we waded and struggled through the posh and water, the sled wholly immersed, with Mr. Dunbar still clinging to the cross-bars and Ericksen performing herculean feats of strength. More than once, when the sled stuck fast, did he place his brawny shoulders under the boot and lift it bodily out. Indeed, we all toiled so hard that when the ridge at the edge of the ice-foot was reached, we were barely able to crawl over it and drag Dunbar from the sea like some great seal.

A brief rest, with supper, and I then proceeded to take formal possession of the island. Marching over the ice-foot, without observing any regular order of procession, I, as a commissioned officer and proper representative of the Government, landed first ; and, having claimed the island as the territory of the United States, invited my companions on shore, Hans Ericksen carrying the colors. The ground was then named Henrietta, in honor of Mr. Bennett's mother, and baptized with a few — a very few — drops of corn extract from a small but precious wicker bottle that had been placed in the boat-box for medicinal purposes. After which ceremony, a greater number (and yet too few) of drops from the same vessel being allotted to each member of the party, Mr. Dunbar and myself kept camp while the rest rambled a short distance inland.

The sun at this time was above the horizon the whole twenty-four hours, although it had not been visible, by reason of the foggy weather, since we left the ship. Snow-storms prevailed to such an extent that the island had been completely cut off from view two hours before our landing; we having traveled a compass course. So when I awoke the next morning at ten o'clock, I at once supposed that we had overslept ourselves ; my orders having been to remain no longer than twenty-four hours

on the island, and here we had wasted one-half of that
allowance in the arms of Morpheus. Hastily calling the
men, who yawned and turned out saying they were too
tired to rest well, I directed the performance of the day's
labors. A cairn was built on a bold, high headland,
named by Mr. Dunbar "Melville's Head," but afterwards changed on the chart to "Bald Head;" and in
this we buried a zinc case containing papers, and a copper cylinder containing a record written by Captain De
Long.

I then made a running survey of the island by compass,
Ericksen and Bartlett reading the instrument while I
sketched and recorded. The others ran over the largest
portion of the eastern end of the land, naming many of
its prominent features; and Sharvell shot a few peteu-
larkies and guillemots, which nestled among the rocks in
great numbers. These were the only birds seen; indeed,
we saw no other living thing upon Henrietta Island.

Flushed now with the success of our undertaking, we
once more restowed the sled and set out for the ship,
halting briefly when a mile from the shore, while I took
the bearings of the principal promontories and mountain
peaks from which to plot a map of the island. The retreat was a more difficult task than the landing. We
had drifted far to the northwest; the ice was moving
more rapidly, driving and grinding with greater force;
and the ship, though plainly in sight from the high
grounds of the island, could not be seen on the floe.
The boat was nowhere in view, and the ice, growing more
and more broken at every step, seemed alive. Mr. Dunbar was totally blind for the time being, and as the dogs
were running briskly, it was necessary for him, in order
to keep up with us, to ride on the sled. At one time,
forcing our way through a stream of posh, we had no
more than gained the rounded surface of a small floe-
piece shaped like a whale's back, than it began rolling

to and fro, after the manner of Sindbad the Sailor's adventure. Every one, dogs included, crouched down and awaited events, knowing the floe-piece must soon turn over one way or the other. This it finally did in the very direction we wanted to go, spilling us safely, and the most of us dryly, on the edge of the main floe. But not so the dogs, among whom there was unfortunately a diversity of opinion as to the proper course to pursue; so the majority of them went yelping overboard, and dragged the sled, with Mr. Dunbar sprawled out on top, bodily through the slush and water to the firm ice, while we roared with laughter.

Shaking out our soaked sled-load of animate and inanimate freight, we again started forward toward the ship in the direction of the spot where we had abandoned the boat. I now entertained fears of our not being able to find it, since we failed to encounter any of our previous tracks in the snow, and the whole aspect of the floe appeared to have undergone a change. At length, attaining a large floe-piece, and the weather clearing slightly, Ericksen espied from the top of a high hummock the flag-staff which we had raised in the vicinity of our boat, — a fortunate discovery, since we had consumed the one day's rations taken with us to the island.

From this time on, until we reached the ship, the weather was miserable; and guided altogether by compass we marched forward in the face of a cruel snow and wind storm, constantly impeded by open lanes and leads of water. When we camped, the second night after leaving the island, the storm was at its height. Notwithstanding the broken state of the ice I felt easy in mind, knowing that we must now be close upon the ship; yet several of the party were suffering severely: Nindemann from an attack of cramps, and Ericksen, who since the failure of Mr. Dunbar's eyes had piloted the dogs and kept the sled in a compass course, from snow-blindness.

Poor Nindemann, drawn and doubled up, was enduring
the agonies of the lost, caused, no doubt, by lead poisoning ; so directly after supper, and before I crawled
into my sleeping-bag, I drew forth the medicine box
furnished by Ambler and proceeded to "doctor" him.
There could be no mistake. Inside the box were written
instructions, and Nindemann wanted a cramp antidote ;
Tinc. capsicum, cognac, etc. But my fingers were cold
and sore ; so Ericksen, who must have some sweet-oil to
rub upon his damaged nose (big nose) and chafed body,
would draw the corks. He drew them with a reckless
abandon, spilling the tincture of capsicum (cayenne pepper raised to the nth power) over his cracked and blistered hands. Then, losing his head completely, he applied
the sweet-oil by means of his fiery fingers to the afflicted
portions of his body. The result was at once a surprise
to him and a delight to us. He rolled and squirmed about
in the snow like an eel. Little Sharvell sensibly augmented the animation of the victim by suggesting that he
disrobe and sit down in the snow to cool off ; but then,
fearing that he might melt his way through the floe,
further advised that he station himself on top of a lofty
hummock. This brought about a new and delightful
state of feeling in the tent. Nindemann laughed his
cramps away, and Dunbar found time between his groans
to shout out, —

"Ericksen, are you hot enough to make the snow hiss ?
If you are, the chief can extinguish the fire in the forecastle and use you for a heater."

The next morning, when the mists had lifted, we sighted
the ship. Hoping to reach her before dinner-time, we
pushed on over a course that grew more and more rugged,
coming within a mile of her without attracting the attention of any one on board. Finally, a running stream of
ground ice checked our progress, and after vainly trying
to avoid it by many detours, I decided to launch the boat ;

1 Jeannette men exercising. 2. In the crow's-nest. 3. The Jeannette crushed.
4 A hunting party in luck.

but at this juncture one of the sled's runners gave way, and although we repaired it as best we could, it was yet too frail to sustain all the baggage. Anxious to get rid of the dogs and to place our blinded pilot on board ship at once, I started the sled laden with most of the equipment across the moving mass of hummocks, Mr. Dunbar lying at full length on top and stoutly protesting his ability to walk. It was no time for sentiment, so off they went, men and dogs yelling lustily; Sharvell and I staying with the boat on the far side of the lead until relief could reach us from the ship. Soon we saw Jack Cole, the boatswain, accompanied by a party of men, hastening in our direction. Following my orders, they abandoned the sled, and picked their way across the lead; then by means of a long painter or tow-line, with a man on each side to support the boat, it was shortly hauled over.

Captain De Long, his head bandaged up because of a bout with the windmill, and Dr. Ambler, came out to meet us; and I cannot say which were more pleased, the greeters or the greeted. As for me, all toils and aches were amply compensated for by that welcome, "Well done, old fellow; I am glad to see you back." And the doctor, generous soul that he was, inquiring first after the health of the party, said in his hearty way, "Old man, I am glad you have had the opportunity of first unfurling our flag with honor." Not a demonstrative man was the doctor, but our hug was a close one and heartfelt.

Before boarding the ship I was greatly surprised at discovering the hour of day to be only nine A. M. Sharvell and I, when left alone, had regaled ourselves with some hot broth, supposing it to be noontime. I found, upon comparing my own with the ship's chronometer, that it had not deviated, and so reached the conclusion that we had been prematurely awakened, when on the island, by the unusual brightness of the sun clearing away the

clouds and fog and shining in upon us. Hence we must
have started at three A. M., instead of six; which accounts for our close, unnoticed approach to the ship with
colors flying, before eight A. M., when the crew began to
stir around.

CHAPTER III.

CAST UPON THE ICE.

Life in the Upper Cabin. — Our Situation — The Jeannette goes down. — Camping. — Marching.

WE now had new matter for discussion in the little cabin. Indeed there had never been a stagnation of argument there, where all exchanged ideas freely, and courted criticism. Some of the opinions promulgated therein were no less interesting than original. For instance, one of the mess, ever happy and contented, considered it a very fortunate thing indeed that the ship leaked, inasmuch as the men were thereby "trained and exercised;" and it was so cheerful to lie awake in his berth at night listening to the merry "chug" of the pump!

But now we devoted our time to the consideration of the serious circumstances which so thickly beset us. We were all persuaded that the chances of the ship holding together, in the present state of the ice, were not one in a thousand. Yet she might; but what then? This was the supreme question which constantly presented itself to the minds of all: whether it would not be wiser to abandon the ship at once, and make for the nearest land (New Siberian Islands), instead of tarrying for the fall travel. De Long naturally wished to stay by the ship until the end, or so long as the provisions lasted, proposing that we remain until they had dwindled down to an allowance of ninety days for our retreat. Had a vote been taken of those who gave the matter their undivided thought, there is scarcely any doubt in my mind but that a majority would have decided to abandon the vessel about the middle of June.

However, we had no discretion whatever in the matter. She left us, after sheltering us for so many dreary months; delivering us, Cæsar-like, upon the floe, amid the crashing of her poor old ribs.

On the evening of June 10th the motion of the ice became more violent, the floes far and near cracking and grinding continually. In the silence of the night, when most of the company had retired, the ice started to split around us with fearful frequency; each successive shock being transmitted to the ship as to a centre, and resounding with awful distinctness upon her sides like death strokes. That night it was my tour from nine to twelve P. M., and as officer of the watch it was part of my duty to record the readings of the instruments placed on the ice. Just before the bell struck eight for the midnight hour, and while I was yet on the gang plank making my way towards the observatory, a sharp report like that of a gun rang out on the air, starting the company from their bunks. The floe had split fore and aft on a line with our keel, and the ship, oscillating for a few minutes, came at last to a rest with her starboard side close to the ice, the other floe-piece, on which were the dogs, observatory, and a few small articles, moving off to a distance of a hundred yards or more.

Our situation was now full of peril.

"Well," said De Long in cheery tones to Dunbar, "what do you think of it?"

"She will either be under the floe or on top of it before to-morrow night," replied he.

And so it was.

After the ship had been hauled ahead and fastened within a little cove affording a slight protection, all hands save the watch turned in. Before seven o'clock the next morning the detatched floe-piece, cruelly prolonging our fate, had approached alongside of us, and backed off again. Breakfast over, certain of the men, according to custom, started off hunting, leaving the rest of us to ponder our predicament. Once more the ice drew near, this time closing with the ship and squeezing her gently, as though to test her mettle. The poor Jeannette groaned, and the attacking floe, apparently satisfied, eased off. Meanwhile, there were no signs of trepidation among officers or men. The usual signal was given for the return of the hunters, and they came straggling in as if ignorant of the impending disaster. Yet all were aware of it, and fully appreciated how imminent it was. Preparations had been made for such a catastrophe ever since we entered the ice; every officer and man had his appointed duty to perform, and hence there was neither noise nor confusion when it did occur.

About three o'clock in the afternoon the ice was quiet, the sun shining brightly, and the position of the vessel so strikingly picturesque that De Long told me to bring out the camera and photograph her. I had been acting as photographer during the voyage, and had taken a number of fine views, — all of which, however, were lost with the ship. While developing my plate in the dark room, word was passed for all hands to abandon ship, calling every one except the sick to his post. Under Captain De Long's direction, the colors were hoisted to the masthead, the boats lowered, and, together with the sleds, tents, provisions, and general equipment, placed on the ice about five hundred yards back from the edge. Dr. Ambler took charge of the sick, and with the aid of several men rescued his medical stores. Mr. Chipp was the only patient who really required assistance, and this there were many hands to tender, he being a favorite with all. Everything was conducted quietly but vigorously, and superintended by De Long, who stood coolly smoking his pipe on the ship's bridge.

As the ice continued crowding in, the ship heeled over more and more, until it became impossible to stand on deck without clinging to something. The forecastle watch had supped, and the others were about to follow suit, when the water suddenly began to rise, and so swiftly that many could not escape by the ladder and companion-way, but were forced to leap through the deck ventilator. So those of us at work on floe and deck lost the last evening meal.

Every one at length having left the vessel, De Long jumped on the floe, and waving his cap cried, " Good by, old ship ! " then commanding that thereafter no one should venture on board of her.

We now set about preparing our camp, tenting, as had been arranged months before, by boat crews, in command of the officers originally detailed, except Lieutenant Chipp; whose tent, by reason of his sickness, was given in charge of Mr. Dunbar. Our boats consisted of the first cutter (to have two tents), the second cutter, the first whale-boat, and the second whale-boat; but considering the long march ahead of us before we might meet with open water, if, indeed, we came up with any at all, Captain De Long very wisely concluded to reserve but three boats; so the second whale-boat, being the most unwieldy, was left hanging at the davits. The tents erected, the coffee made, and supper eaten, we finally turned in.

And here we were, cast out upon the ice five hundred miles from the mouth of the Lena River, our nearest hope of succor; with a sick list, and a limited supply of food. Yet, although the seriousness of our situation was appreciated by all, none were despondent, many merry, and shortly after the boatswain "piped down," the whole camp was lost in slumber.

And thankful were we to make our beds on snow instead of beneath the sea, where honest Jack so often finds his endless rest. Honest Jack! Proverbial for his growling, when the day is fair and life is rosy; for his cheerfulness, in times of danger and distress.

We had slept but a few hours when a loud report like that of a cannon awoke us. The floe had split in every direction, one crack making directly into our camp through the centre of De Long's tent; and had it not been for the weight of the sleepers on either end of the rubber blanket, those in the middle must inevitably have dropped into the sea. As it was, they were rescued with great difficulty; and in an instant the camp was alive again. Although the boats, sleds, and provisions had been placed close to the tents to avoid separation by just such a happening as this, we now found ourselves drifting slowly away from them. Boards were at once thrown across the crack, nimble feet sped back and forth, the sleds and boats were successfully jumped over, and when the gap had widened beyond the length of the planks, a way was discovered around it. The provisions recovered, our tents were quickly shifted farther back from the edge of the floe, and we were soon dozing again in our sleeping-bags. During the early hours of the morning Kuhne, the watch, had attentively observed the ship, as she swayed to and fro, creaking and groaning with the movements of the ice. Towards four o'clock, the hour for him to summon relief, he suddenly announced, in addition to his stage whisper to Bartlett, "Turn out, if you want to see the last of the Jeannette. There she goes! There she goes!"

Most of us had barely time to arise and look out, when, amid the rattling and banging of her timbers and iron-work, the ship righted and stood almost upright; the floes that had come in and crushed her slowly backed off; and as she sank with slightly accelerated velocity, the yard-arms were stripped and broken upward parallel to the masts; and so, like a great, gaunt skeleton clapping its hands above its head, she plunged out of sight. Those of us who saw her go down, did so with mingled feelings of sadness and relief. We were now utterly isolated, beyond any rational hope of aid; with our proper means of escape, to which so many pleasant associations attached, destroyed before our eyes; and hence it was no wonder we felt lonely, and in a sense that few can appreciate. But we were satisfied, since we knew full well that the ship's usefulness had long ago passed away, and we could now start at once, the sooner the better, on our long march to the south.

It was nearly a week before we were ready to take up our march, and during this time a thorough organization of the crew was effected. No matter what the issue might be, we were all overjoyed when the day of departure at last arrived. Certainly, judging from the marching experiences of all previous Arctic expeditions, we had a most dismal prospect ahead of us. The crew of the Thegetoff, it is true, all escaped; but they had been so fortunate as to encounter open water less than one degree of latitude from where they abandoned their ship. And only by a similar good fortune could we hope to make good our retreat; for all these marches were as mere bagatelles compared with the one before us.

Previous to the loss of the ship, Captain De Long had taken accurate observations for position almost daily, and after we were cast out upon the ice they were secured whenever the weather would allow. Our route had long been a subject for discussion among the officers. We had been drifting so rapidly toward the west during the last few months, that the New Siberian Islands were pitched upon as a resting-place on our way to the Lena River, which we had selected for our point of destination, knowing it to be navigated by steamboats, and its banks thickly inhabited. Hence if we could succeed in entering it before winter set in, our difficulties would thereafter be few.

Accordingly, the line of retreat was laid due south, and, at first, " true," — De Long and Dunbar performing this part of the work with a series of black flags. On the evening of June 16th orders were issued changing our working hours; so that we slept during the day and labored at night. This was done for various reasons, chief of which were that by such an arrangement we avoided snow-blindness from the sun's glare, and could sleep sounder and warmer, while our wet clothes were drying on the boats and tent-tops. Again, it is decidedly less fatiguing to march and haul in the crisp air of night, or when the sun is low, than when it is high and strong. The temperature during the day in summer-time usually runs up to the melting-point of ice, — sometimes as far as forty degrees, — whereas it always freezes at night, even in midsummer, when the sun has been most powerful; and I have often observed the ice melting on the sunny side of the ship while water was freezing on the shady side.

Before turning in on the morning of the 17th, I conveyed, by De Long's orders, a dog-sled load of provisions for our next day's dinner, to what I supposed was the farthest flag; but unfortunately it had fallen down, and the depot I made was nearly half a mile short of it. Our division of labor was as follows: Captain De Long and Mr. Dunbar, as mentioned before, laid out the course and selected the roads; Dr. Ambler had charge of the sick, and with the aid of a dog-team attended to their transportation, as well as that of the medical stores, tent, etc., having also the direction of the road-making, bridging, and rafting; for throughout the entire march we were forced to make our roads, never coming, except once, upon a straight floe-piece more than half a mile long

3

where a horse could be driven without imminent danger
of breaking its legs. Owing to the sickness of Chipp
and Danenhower, I commanded the working gang. Our
first day's work was a hard one, and disastrous to the
sledges. It had been imagined that each party could advance its sledge, and then all return in a body for the
boats; but upon trial this was found to be utterly impossible, and as De Long thought it best to first haul
forward the boats, in order to have at the front the
tents, cooking utensils, and sleeping-bags, which were
stowed in them, I proceeded to advance the first cutter.
Probably two thirds of the working force were equipped
with harness, called "ruy ruddies," or double bands of
stitched canvas about two inches wide and long enough
to pass over one shoulder and under the other arm, after
the manner of a baldric; and into an eyelet of which is
attached a lanyard made of one inch and a half tarred
stuff, furnished with a wooden button at the free end.
Aided by these, the men seized the drag-rope, and, surrounding the boat to keep it upright, began hauling it
through the deep, soggy snow, which at times reached to
our waists. Whooping and singing, we at last carried
and dragged it as far as the depot of supplies that I had
deposited the day before; but here, very much to our
surprise, Mr. Dunbar announced that the farthest flag,
to which we were ordered to advance, was still half a
mile beyond. Orders are orders, particularly in a fix
such as we were in, which allowed of no discretion whatever, so forward we went. The first pull when we were
fresh and vigorous had not been especially distressing,
but before we had accomplished this second and unexpected march we were all utterly fagged out, two of the
men being unable to stand; so they were both left seated
in the snow, the one drawn up with cramps in his legs,
and the other with a similar attack in his stomach.

We found the camp in a violent state of commotion.

Immediately after we had left on our march, the floe whereon the camp was pitched began to break up and run into ridges. When we arrived, De Long, having seen the sick moved forward to the depot of supplies, was with half a dozen of men strenuously trying to get the boats and sleds across the gaping leads in the ice. The state of affairs was very dismal indeed; our beginning was discouraging, and it really looked as though, metaphorically, we would never get to Texas; many even said they did n't care. However, there was need of prompt action; the boats containing the provisions must be bounced across the leads at once; so all hands were placed on one boat or sled at a time, and when the passing floes came together we hurried it over; many of us with a firm grip on the drag-rope dashing into the slush and water "neck and heels," to be hauled out by our companions ahead. Thus, amid roars of laughter and good-humored banter, we succeeded late in the afternoon in again bringing all our baggage together. But the sleds had been so badly damaged that it was necessary to unload and lash them again, besides lightening the freight of the smallest ones. This caused another day's delay. Meanwhile the first cutter was fully half a mile in front of us; but as she lay in the centre of a large, solid floe-piece we were but little alarmed for her welfare. We had now learned several valuable lessons; namely, the importance of keeping ourselves and goods well together, of not permitting too great a distance to intervene between our depots, and of not transporting any of our baggage across a fissure or lead in the ice until we had first brought all of it up to the ferry.

But imagine our chagrin at failing to be able to haul together two of the lightest sleds, and being compelled to advance them singly. By this arrangement, in order to forward our eight pieces of baggage, we must pass over the course thirteen times, or, to make one mile good in

a straight line, we must march thirteen. Thus, because
of the devious nature of our course, the floe being broken
and hummocky, we would toil hard from seven P. M. to
six, seven, and often nine A. M., traveling from twenty-
five to thirty-two miles to be gladdened by a direct progress of only two or two and a half miles.

Profiting from the experience of the first day, we traveled more easily on the second. In the matter of lashings for the sleds we found hemp to be much better than
the raw (walrus) hide, upon which we had relied so
much. Perhaps in cold weather walrus hide may make
a better lashing, but I doubt it, and am of the opinion
that the only advantage attaching to its use is that upon
a pinch it can be eaten. Indeed, fresh walrus hide
roasted with the hair on is toothsome at any time, and
many members of our company feasted on it after con-
suming their rations of pemmican. We also learned
that the mere stupid exertion of strength, upon which,
backed by a little "luck," sailors are too prone to depend
for the overcoming of their difficulties, was not the proper
way for us to accomplish a good day's work. Nursing a
weak sled; bridging at certain times; going round a
hummock to avoid cutting out a road, — all these expe-
dients served us in good stead.

Our daily toil had little of variety in it. When all
hands had been called, the cook of each tent drew three
quarters of a pint of alcohol from the doctor, which used
in our stoves would in about fifteen minutes bring to a
boiling point thirteen pints of water, melted from the
moist snow that we found on the high hummocks. The
issue of provisions was made by the carpenter, each cook
drawing from "Jack-o'-the-dust" his amount of bread,
pemmican, sugar, and coffee, and the officer of the tent
seeing that the food was equally divided among the men.
We also had a half ounce per man per day of Liebig's
extract, rations of which were served out to each tent,

generally at midnight, for soup, or according as the officer saw fit to dispose of the hot water, the limit of which was governed by the supply of alcohol issued. To secure an impartial distribution of food in tent number four, I detailed Adolf Górtz, seaman, to divide the bread and pemmican into six equal parts, putting each part in a small tin basin or pan. These were then placed in the centre of the tent and each man ordered to take a pan, which most did with astonishing alacrity, Górtz and I appropriating the remaining two.

We usually took up our march at seven o'clock, sharp, continued it until midnight, allowed one hour for dinner and rest, and then endeavored to bring all the boats and sleds together by six A. M., for supper and sleep; but in this we were not always successful, our labors often extending to nine A. M. Then the camp must be made. The ground, generally selected by De Long and Dunbar, must be level, and the ice beneath the snow free of water and cracks. Frequently it was impossible to find such a situation; so a scramble would ensue for the best places upon which to pitch the tents, and this brought about so much contention that every one was at length forbidden to choose any particular spot until all the boats and sleds were in and arranged for the night. Then the word was passed, and several men from each party shouldered the tents, poles and all, and set them up on the best available spots in the near vicinity of the baggage. Camp made, the kettles were put on, each man, officers excepted, serving a week as cook; and, supper over, the sleeping-bags and knapsacks were gotten out. But before turning in we repaired our clothes and moccasins for the next day's march, hanging out such articles as were wet to dry. A watch of one hour for each man was set, beginning with tent one, and continuing on to and through number six, the officers and sick being alone excused from duty. If any of the sleds required lashing, it was done before turning in, unless the work was trivial, when the watch attended to it, — our aim being to permit nothing to check our progress except the necessary halts for rest and repairs.

Next to the labor of hauling the boats and sleds, our greatest hardship consisted in the almost constant wetting we received. True, we carried several extra suits of clothing for general use, but among so many they could be of little advantage, and we soon came to pay no attention to our frequent soakings. Our course was laid out with two rows of flags, between which it was my duty to take the straightest line practicable, and since this rendered it impossible for us to keep dry all day, we argued "as well early as late," and so pushed boldly through the ponds of slush and water which lay knee deep in our path, making detours only with the precious bread sled. As far as our moccasins were concerned, there was not a man in the working force at the end of the first three weeks who wore a tight pair on his feet. Traveling in summer-time through the water and wet snow, the raw hide softens to the consistency of fresh tripe, and then — what with the hauls on the drag-rope and the slipping of feet on the pointed ice — the moccasins are soon gone. Many, many times after a day's march have I seen no less than six of my men standing with their bare feet on the ice, having worn off the very soles of their stockings. Nor would it have been possible to avoid this, since we could not have carried enough "oog-joog" skin, of which moccasin soles are made, to have kept alone our boats in repair.

Many were the devices to which we resorted in order to keep our feet from off the ice. At first we made soles by sewing patch upon patch of "oog-joog." Then we tried the leather of the oar-looms, but it was too slippery, as was also the sheet rubber, which some of the men had thrown away. We used canvas; sewed our knapsack-straps into little patches for our heels and the balls of our feet; platted rope-yarns, hemp, and manilla into a similar protection, with soles of wood; and platted whole mats the shape of our feet. A large number marched with their toes protruding through their moccasins; some with the "uppers" full of holes, out of which the water and slush spurted at every step. Yet no one murmured so long as his feet were clear of the ice, and I have here to say that no ship's company ever endured such severe toil with such little complaint. Another crew, perhaps, may be found to do as well; but *better* — never!

CHAPTER IV.

RETREATING OVER THE PACK.

ON the first Sunday of each month, as had been our custom since leaving San Francisco, the Act for the Better Government of the Navy was read, with prayers; and saving this mild diversion, our daily routine continued without variation,— an occasional accident to the sleds, or an unusual amount of bridging or ferrying, alone delaying our forced march. When we had been on the retreat several weeks, Captain De Long secured a good observation of the sun, and learned therefrom, very much to his astonishment and chagrin, that we had drifted about twenty-four miles into the northwest. There was no doubt about it; his conclusions were confirmed by a "Sumner," and our situation now seemed absolutely hopeless. After daily marching from twenty-five to thirty miles for two weeks, to find that we had retroceded twenty-four miles!

In order to cross the streams of running ice at right angles, De Long now changed the course from south, true, to south southwest. We all knew that we must eventually come upon open water by marching due south, however much we drifted to the north and west. It was merely a question of time; yet we had but sixty days' provisions, a journey of five hundred miles before us, and we might not be able to take our boats to water, and then only to be frozen in.

Wrapped in my sleeping-bag, it was amusing to lie and hear the men prate of their past joys at the table. After enumerating every toothsome thing they had ever eaten, all would finally agree that the best dinners on board the Jeannette were those of Wednesday, — the " bean day," when " duff," the sailor's delight, was also served. And a wail would go up over the remembrance that, having these two delicacies in one day, it was impossible to do justice to both, so either duff or beans must be neglected. Then would follow confessions of what had been done with the surplus: the generous fellow telling how he had given what duff he could n't eat to the Chinese cook; the funny fellow, how he had presented his to Iniguin, just to see him swallow molasses, or had eaten it all and cried for more; and the mean fellow acknowledging that he would not even throw his to the dogs, but had kept both beans and duff, and consumed them cold.

De Long craved a few fried oysters, while Ambler and myself were wont, in fancy, to chuckle over a whole canvas-back duck, or turkey, or young wild goose; however, — " A *whole* one, you know, old fellow." Not, perhaps, that either of us would eat all of it, but then the luxury of carving and feasting on just such parts as we chose, each to his own taste and from his own goose — ah! —

" Ha, ha! " — from Chipp, — " you dainty little ones. A broiled partridge on toast, eh? A ten-cent plate of hash is what you 'll get instead — maybe hog's jowl and greens." This or such another sally generally awakened us from our day-dreaming, the bright fancies of which, alas! so few of us lived to realize.

Perhaps a week subsequent to the discovery that we had been drifting into the northwest, Captain De Long learned, from another observation, that we had at last made good about twenty-one miles. The men had become despondent and suspicious, rightly guessing the reason why the results of the first observation had been kept secret. So now, when all but one piece of baggage had been advanced, I announced the good news in a loud voice: "Boys, the captain says we have made twenty-one miles good during the past week, and that we now have a current in our favor."

A cheer arose from one end of the line to the other, and the last sled was rushed to the front with renewed vigor.

And now a bright vision arose before our eyes, cheering all. We had been marching toward the New Siberian Islands, and for several days a dark cloud hanging in the south-southwestern sky had been anxiously watched. Finally, at noon of the 11th or 12th of July, the sun shone clearly in the southward, and the land stood boldly revealed; its blue mountain peaks rising grandly aloft, the ice and water showing plainly below, while a white, dazzling cloud floated dreamily above — in all the most perfect scene of isolated or insular land ever viewed at a distance in the Arctic Ocean. It inspired us with new hope and life, and we toiled forward as to a second Land of Promise. Approaching nearer, the ice became looser, the leads more frequent, and game more plentiful. On two or three occasions we had seal for supper; and at length, just before effecting a landing, Görtz shot and killed a bear, whose carcass, as we were now detained on the ice for a day or two in the fog and sleet, very miserable in our wet tents, wet clothes, and on our soft, wet snow beds, was a most welcome addition to our meagre diet. With the empty pemmican cans for stoves, we fried his steaks, broiled his chops, roasted his paws, and made stews of his flank pieces, using his blubber for fuel.

Meanwhile the ice surged back and forth with the ebb and flow of the tide, tending steadily to the eastward. Should we drift past the island it would be utterly impossible to recover our lost ground; so on the morning of the second day we prepared to make a dash for the land, on which, though not in sight, we could hear the constant grinding of the ice and the calling of the sea-birds among the cliffs. Suddenly, as we approached, the sun, as though by an extraordinary effort, rent the cloud veil in twain, and lo! before us, so close that it seemed we might step on shore, uprose and towered to a height of 3,000 feet the almost perpendicular masses of black basaltic rock, stained here and there with patches of red lichens, and begrimed with the decayed vegetable matter of unknown ages, the bold projections fissured and seamed, and the giant rocks split and powdered by the hand of time. The sight was glorious. Involuntary exclamations escaped from all. It infused new life and vigor into us; and each man straightway became a Hercules. Now or never, thought we, and so seized the boats and sleds, rushing them upon a tongue of the ice-foot which our main floe grazed in passing. At last! The ice-foot rested on the beach, and now many of our company set foot on terra firma the first time in two years. A sorry looking set we were, too, gathering together our weather-beaten traps; sunburned, lean, ragged, and hungry. We had appeared quite bad enough while on the ice; but now, after our late terrific toils, camping under these great mountains, the tents looked not unlike ant-hills; while we, a group of vagabond insects, tugged away at a heap of rags, bags, and old battered boats as spoils. Supper over, we formed a procession, and with colors flying marched to the island, which Captain De Long took possession of in the name of God and the United States, naming it Bennett Island; and Lieutenant Chipp was directed to give the crew as much liberty as was possible on American soil. Very little, indeed, this was, and Jack growled at the " dry christening;" and even though he was just come on shore, with two years' pay, how could he spend it?

Camped under the frowning cliffs of the island, on a little strip of ice that swayed uneasily with the action of the tide, we watched the majestic procession of floe-pieces rolling and grinding by. On came the endless column, crowding and crushing, with rare and beautiful gaps between, revealing the deep blue of the sea, and we who had lived amid the wonders of the ice-world for two long years now stood with mouths agape and marveled as the grand parade of stately bergs sailed past; and when night had closed in around us we at last lay wearily down to rest.

There was a narrow channel of water about twenty yards wide between our ice-foot and the island, which we had crossed by means of certain stepping stones, by wading, and by floating ourselves over on an ice-cake or raft; and we scarcely had time to crawl into our sleeping-bags when the ground on which the ice-foot rested rocked and trembled with a noise like the roar of distant thunder, or the bursting of some huge berg. The next instant we were out in full view of a sight that it is permitted but few mortals to witness and live. A land-slide had started down the rocky declivity, and was now making its awful way toward us with irresistible speed. The spectacle was grand and terrific, but had the ice-foot extended to the shore without the intervention of the channel, we would either have been buried by the rushing mass or swept into the sea.

Our stay at Bennett Island was determined by the time required to repair the boats, allowing us a brief respite from our distressing labors. Two parties made extended explorations around the coast: the one under Lieutenant Chipp, in the second cutter, sailing along the southern face; while Mr. Dunbar with Alexia and the dog-teams sledded around the northern face, from the point called Cape Emma. Neither party discovered anything of importance, and, having each built and left a

TAKING POSSESSION OF BENNETT ISLAND.

record in a cairn, returned to camp laden with firewood, which they had found in considerable quantities on certain portions of the beach. With this we cooked savory stews of loon, gorney, gull, murre, and other sea-fowl, which had been killed in such numbers for a day or two as to do away for the time being with our issue of pemmican.

The men brought into camp all the peculiar or interesting articles that they found in their rambles. Among these were a bleached and decayed reindeer horn, chanced upon on the highlands, and a part of a head and horn resembling that of the musk-ox, but so very much time-worn that none of us could classify it otherwise than as a fossil, along with the shells which were seen in abundance. A seam of bituminous coal was discovered on the face of the cliff about one hundred and fifty feet above the level of the sea. It varied in thickness from six to twenty-four inches, and ran along in a horizontal plane for a distance of a mile or more. Samples of it were brought into camp and a fire started. It was soft and friable from long exposure, giving out considerable smoke, but it burned to a white ash, leaving little "clinker" or stone, and the refuse was nearly one half in weight of the original coal.

Our explorations and observations finished, it now became necessary to change the order of things, since we were about to take to our boats. These were duly repaired, and the loads lightened by casting away a lot of worthless clothing and other small gear. Then it was impossible to carry all our dogs with us, even if we could longer feed them, for each one ate nearly a pound of pemmican per day. So we retained the best seven as a light team for sledding, and the rest were taken behind a hummock, shot, and their bodies thrown into the sea.

The six tents and their occupants were doubled up in the three boats as follows : First cutter, Captain De Long

with part of the crews of tents four and five; Second
cutter, Lieutenant Chipp with the crews of tent two and
part of number five; Whale-boat, myself with the remainder of the crews of tents three, four, and five. There
was open water for a mile or so between the island and
the nearest floe, and into this we at last pushed off in
our deeply laden boats, — it requiring two trips to transport all the provisions, sleds, camp equipage, dogs, and
men.

We stepped our masts to sail or help the oars as much
as possible, and placed the sleds across the boats forward
or abaft the masts, at times towing them and the spare
oars. When a floe was to be crossed the boats were run
in alongside of the ice, eased of their freight, hauled over
on the sleds, launched, reladen, and we reëmbarked and
were off again as merrily as though on a summer sea.
In this way our first day's progress was very encouraging, and except some little but disagreeable sledding
we found, as we had anticipated, our new manner of pro-
gression a delightful improvement over the old one, as
well in the matter of labor expended as in distance accomplished.

Before leaving the island (August 6th), winter had
really set in. When we landed, the water was rushing
in torrents from the glacier, ice-cap, and snowy peaks,
and its noise could be heard in the silence of the night
for miles. But during our short sojourn there, how
marked the changes. At first, we could fill our tea-kettles
at any of a hundred purling streams which ran down the
mountain side; then it rained; but, before we left, the
streams were dried up, young ice was making, and the
bright red or green spots which had looked so cheerful
to our eyes were fast being clad in their winter garb.
A day or two of travel, and our beautiful island, only
seen at intervals between the snow-squalls, was, like
everything else around us, shrouded in white. And the

last we saw of it was a mere shadowy contour, curved like a whale's back, and lifted into the heavens as though to mingle its snowy purity with the silver glory of the clouds.

Before the young ice began to make and so unite the hummocks and bits together, it was an easy matter to put a floe of an acre or more in motion, but now this was fit work for a Titan ; and, if the freezing and cohering process went on much longer, it looked very much as though we should have to halt and wait until the ice became strong enough to bear us. But fortunately the winds kept the ice in constant motion, and so preserved comparatively open water. Sitting in our cramped quarters in the boats we now became very tired, cold, and wet, with little or no covering for our hands and feet, having rejected, in view of our long journey across the ice, all but absolutely necessary clothing, which was now worn into rags ; and to add to the discomfort of those of us in the first cutter and whale-boat, which were leaking badly, we were compelled to bail continually. The second cutter was tight, because being light and short she had rested easily on her sled without rocking.

So long as the wind which put the ice in motion made open water for us, our progress to the south was rapid, but not so when it crowded the pack together and made it appear as though we would never get out of the wilderness. At length, after a good day's run with a freshening breeze, we were finally forced by the gale, and the crowding of the ice, and the approaching darkness, to haul out on a floe-piece.

Pitching our tents near the edge, we ate supper and crawled, wretchedly wet and cold, into our sleeping-bags; but about midnight we were all summoned to shift the boats and tents, as our floe was breaking up ; and it was no less amusing than painful to see each other in various states of *deshabille,* — some barefooted, many barelegged, darting about in a howling snow-storm, securing our traps and carrying them to places of safety. We were up with dawn, and breaking our fast were at it again; and after a fairly good day's work brought up against an old rotten pack, full of holes and water spaces, around or across which it seemed impossible for us to make our way; so we hauled out on it, the snow starting to fall again, and the water to freeze. Pitching our tents we waited that day while the storm blew around us.

Having vainly wished and watched for a shift of the ice, we set out on the second morning across the skeleton pack, which was joined together by young ice half an inch or so in thickness. There was no picking a road; so we made a straight line for the nearest open water, across pools, ponds, holes, fissures, and hummocks, sinking to all depths from our knees to our necks.

Taking every lead that opened to the southward, or had most southing in it, we worked our serpentine way in the direction of the New Siberian Islands, the sea expanding, or, rather, the leads becoming larger and more frequent as we progressed. The next morning, after breakfast, the order was issued to heap snow in all the boats for water purposes, — De Long desiring that we should make our tea on board, and not haul out for dinner as had been our custom. So away we sailed with a fine following breeze and plenty of open water, — too much, indeed, at times, for our heavily laden boats. When word was passed to make tea and serve dinner, De Long was booming along in the lead, the whale-boat next, and Chipp in our wake. The ice was all in motion, and, where the lanes widened into great bays, covered with the first white caps we had seen for many months, our boats danced, capered, and scampered like circus horses. We were now dodging in and out the floe openings as best we might, acting quickly with tiller and sail to avoid coming in contact with the sharp edges of the ice. Now and then, when the boats were in line and not more than a hundred yards apart, the first cutter would shoot through a passage, followed by the whale-boat; but, before the second cutter could come up, the ice would perhaps shift and shut her out.

Finally, when we opened out into a bay where the slowly increasing breeze from the north was raising considerable sea, our boats, weighted deeply and their sailing as well as safety rendered almost impossible by the heavy oak sleds, began shipping water; and it became apparent to all that if the sea continued we must of a necessity lighten our boats. The ice, too, was crowding in upon us again, and we were working to the south and west in a narrow lane of water; all three boats being hauled on the wind to try and weather a point of ice, at the same time keeping clear of the edge of the pack under our lee, over which the sea was breaking fearfully.

When De Long and I hauled out, Chipp had dropped behind again, although all three boats were carrying every inch of sail they could stagger under; and when he eventually rounded to, Chipp, for the first time, complained about his boat. Until then she had been the favorite, and even yet, indeed, was considered sound and efficient, only she was overloaded by the heavy sled, which article of freight, indeed, came nigh to burying every boat.

In the face of the day's experience it was quite evident that if such weather continued, we could not carry the sleds with us across the open water between the islands and the coast of Siberia. So De Long very wisely directed us to cut them up into fire-wood, and when we started, as we expected to in the morning, to stow the pieces in the boats. But when day broke we found ourselves shut solidly in with not a speck of water anywhere visible, the whole of the northern pack having been driven down by the gale full upon the islands, which were now in plain sight. Although it was cold enough, the

4

constant motion of the ice prevented the floes from ce-
menting; so we could do naught but await a favorable
change in the aspect of affairs, and this we proceeded to
do, having accustomed ourselves to make the best of
every misfortune. Nevertheless, we had then no idea
that we would be kept prisoners for ten days; albeit, if
we had, there was nothing we could have done to liber-
ate ourselves.

In truth, our situation now looked worse than ever;
the provisions were rapidly disappearing; winter advancing; and the islands ahead of us were uninhabited; so
De Long sent for and consulted with Chipp and myself.
Talking the matter over, we agreed upon the impossibility of transporting the boats to land, at least with
their bottoms in. We then discussed the course we
should pursue in the event of our drifting through the
channel between the islands of New Siberia and Thadeouiski; unanimously deciding that we proceed from
point to point along the south side of the islands until
we reached the southwest point of the island Kotolnoi;
thence to Stolboi, to Wasilli, to Simonoski, and finally
to Cape Barkin, at the Lena Delta, where we felt assured we would find the native huts as marked upon our
charts.

Our existence had now become a mere question of provisions. Had there been a depot of eight or ten thousand pounds of pemmican on the New Siberian Islands
we could have wintered there with comfort; and when
I read all the plans for our succor suggested, while we
were absent, by people who assumed to know that we
were coming out by the way we did, I cannot help wondering why it was that some one did not propose such a
depot with a guard to watch it. Yet, as in other things,
our aftersight informed us of much that our foresight
had overlooked.

About noon of our tenth day in camp the ice seemed

looser than usual, and we found ourselves closer to land.
So we hauled the boats a shoit distance, and launched
them in a swirling mass not unlike the rapids to some
great cataract. Moving rapidly in all directions, now
closing and now opening, the ice, at times, would form a
solid bairier in front of us; and, while we considered
whether to journey east or west, the wall would suddenly part and open a passage-way, shutting peihaps as
soon as we had fairly entered. And in this bewildering
manner we continued on our course until night closed
in around us, when we were obliged to haul out again
and camp, after a hard, though good, day's work. Next
morning we were up bright and early. Launching our
boats, we caught a brief view of the land, when the sun
shone through the fog bank, and concluded that we were
now well down between the islands of New Siberia and
Thadeouiski. The ice was running through this opening
to the southward like water in a mill-race, and the fog
gathered densely about us; but on we ran in mid-channel, now and then catching glimpses of the eastern headlands of Thadeouiski. Ere night we had come out at the
southward of the islands, and before us, as far as the eye
could reach, rolled the blue free sea, although the ice-
blink showed away to the south. Following the coast
to the westward, we at length, after great difficulty, effected a landing, and for the first time in two years and
a half enjoyed a good sleep on *terra firma*, realizing Dr.
Ambler's oft-repeated wish that he might once again
"renew the electrical conditions between his body and
the earth," or, as Dunbar briefly expressed it, might
"sand his hoofs;" and this we all did upon the mossy
tundra on the high ground of Thadeouiski.

During the evening we all, officers and crew, scattered
over the island in quest of game or any objects of interest relative to our position as a shipwrecked party. We
found several decayed and tumble-down huts of the ivory

hunters, and one of the sailors said he saw moccasin
marks in the muddy beach of a river, but this was before
he learned to what an extent the imprint of a reindeer
hoof in the mire will spread when washed by a receding
river. A few black ducks, caught late in the season with
broods to raise, still remained paddling shyly about in the
open water. Snow was settling on the hills, and young
ice was making along the shore and in the ditches. The
reindeer had taken to the valleys among the distant hills,
there to remain until the return of spring-time and sunshine, for in a very few days the silence of an Arctic
winter would rest upon the island.

Keeping a bright lookout we pushed along the coast as
rapidly as the shoal water would permit, now and then
grounding. Long windrows of driftwood were thrown
up on the beach and crowded far back from the water-
mark by the ice. At the time of our sojourn along the
coast the interior of the island looked high and mountainous. Hills and valleys were covered with snow and
ice, and the rivers had all dried up, the sun having ceased
to give sufficient heat, even at midday, to melt the snow.
The low, irregular coast-line resembled a series of huge,
peaked or cone-shaped furnaces, which, however, upon
closer inspection looked not unlike villages of conoidal
tents or huts ; but the earthy portion of these islands, we
soon learned from examination, is rapidly being washed
away into the sea. In early summer the turbulent
streams coursing down into the valleys cut great ravines
in the mountain sides, and, later on, the snow melting
along the ridges of the hills eats out transverse and
smaller ravines ; and so through ages the general erosion
has proceeded until nearly all the soil has been washed
into the Arctic Ocean. And now that there is not sufficient surface on the peaks of the cones for the snow to
lodge and run off in little rivulets, the erosion goes on
through the slower process of freezing and melting, thus

expanding and contracting the masses; and evidences of the gradual leveling are plainly to be seen in the rounded earth mounds of all sizes at the base of the cones.

Although greatly retarded by the shoals, we made a good two days' run in our boats past the islands, hoping the third day to be able to camp on the eastern end of Kotolnoi. But the wind headed us off and a shoal stood in our way. We had tried hard all day to round the shoal, which we found making twenty or twenty-five miles to the southward of its charted position; the wind was increasing, rendering the navigation of the boats a cold and wet task; darkness was approaching, and so to avoid a night in the boats we ran in under the shoal. Considerable sea was now rolling in, and it would be decidedly unsafe to attempt the hauling of the boats ashore, since it would be impossible to launch them through the surf should it blow a gale for any length of time. After several ineffectual efforts to land without running back to the point whence we had started in the morning, it finally became apparent that we must pass the night in our boats. And a memorable night it was.

The southerly wind drove the ice in upon us, and at the same time forced us toward the shoal, over which the sea was breaking with great fury. We were without anchors, and so with reefed sails we did our utmost to obey the order to keep together. The night was dark as pitch, and our only guides were the roar of the surf under our lee, and the glare of the ice on the other side when the sea surged over it. About midnight we had driven perilously far in towards the beach, and the order was passed to anchor the boats as best we could; but to no purpose, — the sea was too strong, and we would have been forced into the breakers and doubtless drowned, had we not succeeded in getting under way in time, and standing back to the northeast on our track of the day before.

Considering the violence of the wind and sea during

the night, our anxiety was great when we found the second cutter nowhere visible. Running alongside of a grounded hummocky "bit," we pitched our tents and stretched our legs, and while we yet breakfasted the second cutter hove in sight. Chipp reported a bad night of it, and his crew looked much more worn and battered than the rest; but we had no time to spare in sympathy, for the tide began to rise, and the waves breaking over our hummocks soon washed us out; so tumbling our effects into the boats, we made off again before a moderately increasing breeze and rounded the sand-spit off the east end of Kotolnoi. As the wind blew stronger, the sea ran high, and in a little while it was all we could do to stagger along under single-reefed sails, keeping ahead of the waves, which washed over us constantly. It was wonderful how we avoided cutting or staving our boat to pieces on the sharp-edged bits; but fortunately while we thus ran briskly the ice was much broken, and there were no floe-pieces calling upon us for halts and hauls. Within several hours the second cutter was again out of sight, and De Long concluded to continue running until we could find a large solid floe-piece, and there await Chipp, who must needs take care of himself. We had scarcely secured our boats in a shallow cove washed out in the ice, where they lay like ships in a dock, when the water suddenly disappeared as if by magic, and we found ourselves in the midst of a wild mass of broken ice, apparently as hopeless of navigation as the pack which balked us and brought about our unfortunate "Ten day Camp."

Pitching our tents as night came on, we supped and crawled into our sleeping-bags, well worn out and most thankful for rest. Next morning the gale yet blew with vigor, showering snow. Still no sign of Chipp. We hoisted a black flag at the mast-head of the first cutter, and hoped it might bring him to us. Toward evening, while Iniguin was watching on a high hummock, the second cutter was at last seen skimming briskly along in the open water. She sighted our signal, and drew up within a mile of us, and soon we observed Chipp and Kühne making their way across the ice in our direction. It is needless to say how rejoiced we were to see them; but after supper they visited around in the four tents and recounted to us their experiences.

The morning smiled on us, and the sea tried to show its blue face through the dense fog that had closed in and shut the land from our sight. Soon we were all well under way, and Chipp, scudding along the canal for a mile or two, finally came to a halt, the ice ahead having packed into an impenetrable mass. As we ran on we had observed another canal inside of the one we were navigating, and likewise a passage connecting the two; so now we turned back to it with De Long in the lead, and sailed merrily along until dusk, when we again found ourselves in a *cul de sac*, the land showing fair and bright to the northward of us, and the mountains raising their snowy peaks far inland. We all three rounded the point together and hauled out on the inside of a long sand-spit making eastward from Kotolnoi toward Thadeouiski.

We were now camped on the eastern end of Kotolnoi. Driftwood was abundant, so we gathered great heaps of it and built a rousing fire, before which we warmed our fronts, froze our backs, and burned or shriveled up considerable of our saturated garments, in our anxious endeavors to profit by the first really good camp-fire we had enjoyed since leaving the United States. A night of grateful release from our prolonged fatigue, and the next day opened gloriously. With a view to further rest and the stretching of our cramped limbs, but more especially to a good reindeer stew, we remained on the island ; those who chose going on the hunt, with their bodies full of vigor and pockets full of cartridges. To-

ward evening they straggled back; some had gone incredible distances, all had found plenty of tracks, but
none had seen a live deer.

Next morning we were out bright and early, launching
the boats in high glee at the expectation of making a
good day's journey along the inshore water. Rounding
the point of the sand-spit, we stood to the westward,
many of the men walking on the beach for exercise.
This beach was strewn with various kinds of driftwood.
Some of the lumber showed marks of the friendly axe,
and how eloquently such silent signs of civilization.spoke
to our hearts, recalling distant scenes and friends. Dinner over, we were forced to make a portage of about half
a mile, hauling the boats on their keel runners along á
little ridge of snow above the high water-mark. They
were soon launched again, and away we gayly went, a
number of the men continuing to run along the beach
for exercise, keeping pace with the boats, which picked
them up when their progress was checked by the creeks
making out from the land. Advancing thus, we at
length, toward night, hauled out and camped on the high
ground a short distance back from the beach. During
the afternoon we had left the low sand and mud shore,
and had arrived at a long line of perpendicular cliffs of
shale and slate; from which we inferred that we were
now on the southwestern coast of Kotolnoi Island, whence
we might start at once across the open ocean for the
Lena Delta, via the island of Stolboi. Accordingly we
made a close stow of our goods, filling all the spare vessels with snow for a water supply, and on the morning
of September 7th set sail before a fresh breeze from the
east northeast. Standing to the southward, we shortly
came up with a large floe alive with small running hummocks and stream ice. It was blowing stiffly, the sea
was lumpy, and our boats careering at a lively rate.
Pumping and bailing to keep afloat, we suddenly came

unawares upon the weather side of a great floe-piece, over which the sea was breaking so terribly that for us to come in contact with it meant certain destruction. It was floating from four to six feet above water, its sides either perpendicular or undershot by the action of the waves, which dashed madly over it, the surf flying in the air to a height of twenty feet; and, where the sea had honeycombed it and eaten holes upward through its thickness, a thousand waterspouts cast forth spray like a school of whales. Round about, down sail, and away we pulled for our lives. De Long being fifty or a hundred yards in advance of me, and so much nearer danger, hailed me to take him in tow, which I did, and together we barely managed to hold our precarious position. The second cutter was away behind again, but upon coming up seized the whale-boat's painter; and so we struggled in line, and at last succeeded in clearing the weather edge of the floe. It was a long pull and a hard pull. The sea roared and thundered against the cold bleak mass of ice, flying away from it like snow-flakes and freezing as it flew; the sailors, blinded by the wind and spray, pulled manfully at the oars, their bare hands frozen and bleeding; and the boats tossed capriciously about with the wild waves and the unequal strain of the tow-line. Drenched to the skin by the cruel icy seas which poured in and nigh filled the boats, the over-taxed men, as they faced the dreadful, death-dealing sea and murderous ice-edge, found new life and strength and performed wonders.

This, indeed, proved a day of trial and tribulation to us; the restless condition of the ice requiring nice navigation, and the low state of our provisions calling for prompt movement and the avoidance of disastrous delays.

While we were under the influence of the land the wind had been even and steady from the south and east, but as night closed in it became fitful, blowing in heavy

squalls, and the sea ran high. Our boats were well
bunched together, and although it was now pitch dark
we could yet for a while discern each other looming up
out of the black water like spectres and plunging over
the crests of waves. Presently the second cutter faded
away, but as mine was the fastest boat of the three I
experienced no difficulty in following De Long. Indeed,
in my anxiety to obey the order, "Keep within hail," I
at times barely escaped running the first cutter down.
Now that we were sailing night and day it became necessary to relieve the helmsman. Not that any one could
sleep; but then a rest was needful, from the increasing
vigilance required in guiding the boat clear of the multiform dangers that arose constantly before us. Yet the
least error of the helmsman, when his ear caught the roar
of the sea breaking over the edge of a floe or ragged
pack, and the waves which came tumbling after, mountain
high, would certainly engulf us. Saving a few minutes
when I was otherwise engaged, the main sheet was never
controlled by any one but myself; and cold work it was,
too, with a pair of mittens on my hands made of cotton
sheeting which I had originally used as coverings for my
fur mittens.

Toward midnight we approached the weather edge of
the pack, the roar of the surf reaching our ears long
before we could see the ice. I involuntarily hauled the
whale-boat closer on the wind, and by so doing lost sight
of the first cutter, but the terrible noise and confusion of
the sea warned me beyond doubt of the death that lay
under our lee. Presently out of the darkness there appeared the horrid white wall of ice and foam. Not a
second too soon. "Ready about, and out with the two
lee oars if she misses stays." This, of course, from the
heavy sea, she did; and quick as thought my orders were
obeyed. As we turned slowly round a wave swept
across our starboard quarter filling the boat to the seats.

Ye gods! what a cold bath! And now we were in the
midst of small streaming ice, broken and triturated into
posh by the sea and grinding floes, and this was hurled
back upon us by the reflex water and eddying current in
the rear of the pack, which was rapidly moving before
the wind. With bailers, buckets, and pumps doing their
utmost, the two lee oars brought us around in good time,
and we filled away on the other tack, the waves still
leaping playfully in as though to keep us busy and spice
our misery with the zest of danger. Finally we ran into
a field of streaming ice, which, calming the fury of the
sea, afforded us some shelter.

When day broke neither of our companion boats was
in sight. The wind had moderated greatly, and we were
now in quiet water among the loose pack, — perhaps the
most miserable looking collection of mortals that ever
crowded shivering together in a heap. We looked, indeed, so utterly forlorn and wretched that just to revive
and thaw, as it were, my drowned and frozen wits, I
burst forth into frenzied song. Of a truth, as we sat
shaking there, our situation was nigh desperate; we were
down to an allowance of a pint of water to each man per
day, now that De Long was separated from us; but upon
the suggestion of some one in the boat I set up the fire-
pot and made hot tea. We were thus breakfasting when
the first cutter hove in view. I at once joined company,
and shortly after the second cutter made her appearance,
and we were again together. The sea soon calmed, *les
misérables* thawed out, the morning became as pleasant
as the memorable May mornings at home, and we again
were bright and alive with hope.

Soon the sun shone brightly and warmed us into jollity, and when we halted for dinner De Long secured a
sight which placed our position to the westward of Stolboi Island, now plainly visible.

At dusk, having accomplished a good day's journey,

and been well heated by the sun, we pulled out on a floe-piece, cleaned the boats, wrung our clothing, and after the evening meal, crept, still soaked, into our sleeping-bags under cover of the tents, where we slept the sleep of the just. The next morning we were up and out, enjoying the early sunshine of a beautiful day. The ice in our vicinity was not heavy now, being apparently of but one year's growth, and unbroken by collision — ice that seemed to have floated about in a dead sea devoid of currents or islands, which to the northward break and mass it into hummock and floe-berg. We toiled strenuously all day at oar, tow-rope, and sail, until ten P. M., when the water began shoaling rapidly and we heard the roar of the surf; so mooring our boats to a grounded floe-piece, we ate our supper by candle-light in the tents, and again, well satisfied with our day's journey, lay down to rest.

When morning dawned it revealed to us the mud cliffs of Simonoski, not more than five hundred yards distant.

Landing for dinner we found the tracks of deer, and those of a bear or a wolf. The water procured here, although fresh, was discolored and unpleasant to the taste, savoring of the bog from which it was taken, and being filled with animalculæ and red grubs. We had proceeded a couple of miles when a fine deer, attended by her fawn, was observed running along the edge of the cliff as if in alarm. Our hunters soon killed and dragged the carcass of the doe to the edge of the cliff and dropped it upon the beach. It dressed about one hundred and twenty pounds, and we had each a clear pound of sweet venison, washed down by a quart of tea, — a royal gorge, indeed. And when, after supper, the hunters having fruitlessly scoured the island to its end for a sign of the fawn, we at last turned in, the wet sleeping-bags troubled us but little, for now, the first time in many months, we enjoyed the delightful and al-

most forgotten sensation of being replete and distended
with palatable food, a delicious frame of body and mind
enhanced by the pleasing prospect of a jolly good soup
on the morrow.

That day (Sunday) we passed upon the island, some
of the men employing their time in another unsuccessful
hunt for the fawn, while I set busily to work altering
the cover of the whale-boat ; and a record was deposited
on a high point of land.

Since Saturday, the 10th of September, the weather
had been dark and gloomy, with occasional showers of
rain and some snow. We were all wet and miserably
cold, the moss was soaked, and our camp-fires, by reason
of the scarcity of wood, afforded us little comfort beyond
cooking our food. The wind had been almost constantly
blowing in fitful gusts, approximating to a gale during
Sunday night, and on Monday morning it was yet fresh,
while the sea was covered with white caps. Still, though
the weather looked ominous, there was no certain indication of a coming gale, and a delay, further than was necessary for the rest and refreshment that we had already
obtained, would be extremely dangerous, since a day's loss,
now, might count a week in the near future.

About eight o'clock in the morning we sailed away,
and with a good breeze under the lee of the island, ran
briskly to the southward in the direction of Wasilli.
Passing the channel between the two islands, our boats
careened and we then felt the full force of the wind.
Wasilli was soon in the distance, and just before noon
we hauled in alongside of the floe for our meal of tea and
pemmican, — and it was the last dinner we ever ate together. We had now arrived at the edge of the ice, with
Cape Barkin, our point of destination, only ninety miles
or less distant. Dinner over, we filled all our vessels
with snow for drinking water; every one jolly in the hope
that with our present breeze, should it not grow too

heavy, we might be able to reach Cape Barkin and the land after one night at sea. Chipp and I conferred together a long time, pacing up and down the floe. My boat being the fastest of the three, I anticipated no trouble; and Chipp, though his boat was lighter than before, seemed to think that he could keep abreast if the first cutter did not carry full sail.

De Long verbally directed both of us to keep, if possible, within hail, and reiterated his orders in case of separation.

"Make the best of your way," said he, "to Cape Barkin, which is eighty or ninety miles off, southwest true. Don't wait for me, but get a pilot from the natives, and proceed up the river to a place of safety as quick as you can; and be sure that you and your parties are all right before you trouble yourselves about any one else. If you reach Cape Barkin you will be safe, for there are plenty of natives there winter and summer." Then addressing me particularly, he continued: —

"Melville, you will have no trouble in keeping up with me, but if anything should happen to separate us, you can find your way in without any difficulty by the trend of the coast-line; and you know as much about the natives and their settlements as anyone else." This was our last conversation in a body.

As soon as we had embarked De Long led off under full sail, laying the course southwest. We sped forward at a good rate, but the sea had risen considerably, and began to bother us when we had cleared the ice. Owing to the superior speed of the whale-boat, I encountered some difficulty in preserving my position astern of the first cutter. I had taken in one reef, and for that reason Chipp was keeping well up with us. But the sea grew steadily heavier, the boats jumped and jarred until it seemed they would lose their spars or mast-steps, and it soon became necessary that we reef our sails. My boat,

and I think De Long's, too, was closely reefed at three
o'clock. The first cutter, any way a dull sailor, was
loaded very deeply, having on board, in addition to its
share of weights and provisions for thirteen men, all the
records, books, papers, specimens, etc., etc., beside a
large oak sled for transportation purposes.

The second cutter was now performing very badly in
the heavy sea, and at times kept barely in sight. The
first cutter, an excellent sea-boat, stood up splendidly
to her work, but freighted as she was, even her weather
cloths could not prevent the waves from breaking continually over her. The sea, to be sure, was moving much
faster than the boat; so it combed and broke across her
stern, or, running nigh her whole length, would dash
against the weather cloths and tumble in, soaking the
men and at times almost swamping the boat. Towards
seven o'clock it was blowing a living gale, and it seemed
impossible that we could struggle longer in such a sea.
The danger to the whale-boat was imminent, since in
trying to slacken speed so as to keep, as ordered, in the
wake of the first cutter, the swifter waves forged ahead
and breached clear over the stern, threatening to overwhelm us. Fine manœuvring with helm and sail was
out of the question, though at the suggestion of some
of the men it was attempted once or twice with almost fatal results; and it was while I was endeavoring
to deaden our speed by hauling the boat closer to the
wind that we ran far past the first cutter and well up
on her weather bow. It now looked as though we could
not possibly regain our position without heaving to;
when at this juncture De Long signaled me to approach,
probably within hail. Should I run down towards him,
I would certainly shoot far ahead, so there was but one
thing to do, namely, to lower the sail, and reduce it below
a close reef, — several men accomplishing this by gathering it in at the foot, and holding it firmly with their

hands against the fury of the wind and the thrashing of
the boat.　It was a trick I had practiced in bad weather
ever since we reached open water, and one that had answered admirably.　But now in slackening our speed the
waves came tumbling in and filled the boat.　This, for
the time being, naturally alarmed the men, who clinging
tenaciously to the foot of the sail suddenly found themselves steeped to their hips in icy water.　They would
as promptly release their grip, and then such a tumult
of flapping sail, pounding boat, and demoralized baggage
would ensue as might startle old Tom Bowline himself.
After several of these mishaps, succeeded by vehement
bailing with buckets and pans, and no little growling on
the part of the crew, I perceived that we had drifted almost to within hail of De Long, who was gesticulating
and shouting something to me altogether inaudible above
the roar of the elements.　Just then a monstrous sea
came combing onward and deluged both of us, but chiefly
the whale-boat, which nearly filled.　It started me to my
feet, and I shouted down the wind to De Long that I
must run or swamp.　He appeared to realize the peril of
our situation at once; for the next instant, as the sea swept
over and around us, he waved his arm in an energetic
manner motioning me onward or from him, and at the
same time hallooed some message which was lost in the
noise of the gale.　However, I felt that we understood
each other; that if I would save my boat and crew I
must run for it; that to lay alongside of De Long meant
quick destruction; and that if either of the open and
overladen boats should swamp or roll over, the other
could not possibly rescue the unfortunate crew.

So when De Long waved me permission to leave him,
I hoisted sail, shook out one reef, and as we gathered way
the boat shot forward like an arrow, and the spray flew
about us like feathers.　Heretofore we had been running
dead before the wind on our southwest course for the

land, but the heavy sea and lively motion of the boat caused the sail to jibe and fill on the other tack, whereupon we would broach to and ship water. For this reason I hauled up the boat several points or closer to the wind, and our condition at once improved. Now that we were separated I resolved to concern myself directly with the safety of my own boat; so that when one of the men said that De Long was signaling us, I told him he must be wrong, and further directed that no one should see any signals now that we were cast upon our own resources.

The whale-boat was leaping forward at a spanking rate and fast distancing the first cutter, when, hearing another of the crew exclaim that De Long was signaling Chipp, I turned around and looked back over my left shoulder towards where I expected the second cutter would be. For an instant she was not to be seen, but presently I saw her far off in the dim twilight rise full before the wind on the crest of a wave, and then sink briefly out of sight. Once more she appeared; an immense sea enveloped her; she broached to; I could discern a man striving to free the sail where it had jammed against the mast; she plunged again from view; and though wave after wave arose and fell, I saw nothing but the foam and seething white caps of the cold dark sea. When last seen, the second cutter was about one thousand yards astern of us, the first cutter probably midway between, and there is no doubt in my mind that she then foundered. A conversation with the only two surviving members of the first cutter (Nindemann and Noros) has confirmed me in this belief; for they witnessed the scene as I have described it, and state that it was the general opinion of De Long's crew that I had shared the same fate simultaneously with Chipp.

5

CHAPTER V.

VOYAGE OF THE WHALE-BOAT.

Weathering the Gale. — Our Sea-Anchor. — Siberia in Sight.

LONG before our separation, and while the sea and wind were growing in force and fury, the crew debated whether or not our boat could outlive the gale; and many were the wishes that we might sight the stream of running ice in which to take shelter. To me it seemed another case of "Night or Blücher"—*ice* or *heave to*. The latter alternative could only be accomplished by means of a drag, or sea-anchor, such as the "double-enders" used on our coast during the late war. It was only a question of size and proper weighting, which we thoroughly discussed, educing many suggestions that were not more novel than absurd. I finally ordered a drag to be made of three tent-poles, lashed together and covered with canvas so as to form a triangular parachute. The small watch-tackle supplied us with rope; and the iron straps, block-hooks, and brass tips from the tent-poles, gave sufficient weight to submerge and hold the anchor in position.

And here, to the reader unacquainted with the use or philosophy of a sea-anchor, or drag, a brief description of the same may not be out of place. During a gale the sea generally runs with the wind even against tide or current, though, indeed, these may do battle, and so produce a state of neutral agitation. Still the rule is as I have stated it; and consequently it becomes necessary for

every vessel, large or small, when running before the
wind in a heavy sea, to keep constantly ahead of the
waves, — in other words, go faster, — else they will roll
over the stern and swamp the boat. In great storms,
too, the vessel may be so strained and racked that the
timbers and fastenings will be wrenched apart, and she
will founder, literally thrashed to pieces. Or again there
is a more common danger, though equally fatal, where
the sea, striking the vessel on either quarter, or rushing
past, hurls her from her course. In this case the rudder is powerless to save, being momentarily out of the
water; the sail jibes or fills on the other side; and the
sudden alteration of course and reduction of speed is instantly taken advantage of by the next wave, which
boards or hurls itself against the vessel, in most instances
destroying her, however large, and, if an open boat, certainly overwhelming her. It is to avoid these perils that
the sea-anchor is prepared. So long as a ship remains
tight and light she will rise to meet the waves, her bows
cleaving them sharply in two, if properly shaped, with
a graceful buoyant swell towards the rail; and even
though part of a wave be carried over her bows, the top-
gallant forecastle is there to shed it off. Now the main
object is to hold the vessel's head to the sea. With sailing vessels this is done by using particular sails, and woe
betide the one whose helmsman is careless of keeping
her head just so; for should she broach to or fall off into
the trough of the sea, and a wave board her, all movable
as well as many stationary articles will be swept from
her decks. It is an easy matter to keep a steamer's head
to the sea, and by slowly turning the engine gain sufficient headway to hold her in command; or, assisted by a
little sail to prevent rolling, she may be laid to in safety.
Yet, if too much speed be given the steamer, the waves
will dash over her forecastle and sweep her decks, as

But to keep the head of our little craft to the sea we must resort to the drag, or sea-anchor. If a parachute be made and so loaded as to sink properly below the surface of the water, and a rope be attached which passes through or is fastened to the bow of the boat, it can plainly be seen how the parachute, if of sufficient area, will, as it drags slowly through the water, keep the head of the boat toward the sea. The boat, catching the wind, is hurried along, and would soon be turned sideways and rolled over like a log were it not for the anchor, which gently but firmly resists the furious tugging of the would-be suicide. Should the drag from any cause float to the surface, or the rope break, its efficacy is at once lost, and doubtless also the boat; or, if it hangs perpendicularly under the boat when a sea rushes upon it, the drag will become an additional weight for the bow to lift, failing in which the boat's head will be buried; perhaps the stern, too. It should be borne in mind that a sea-anchor is drawn through the water after the manner of an aerial parachute dropped weighted from a balloon.

So now, not only the propriety of making one, but its size and weight, became momentous questions with us, since in our party of eleven there were none who had lain with a drag in an open boat; and those I had seen on steamers during the war could give me no adequate idea of the kind now required in our emergency. Hence I had nothing to rely upon but my good judgment, though I listened attentively to the many authorities around me, even if I did not seek their opinion. Jack Cole, my main prop as a seaman, declared that the drag would come home on us if not weighted more than it was. Cole and Manson made it, I having selected the canvas. Danenhower, seated beside me in the stern sheets, held the end of the boat-fall, and so singled the three anchor ropes, preserving an equality of strain on the three corners of the drag. Leach (seaman) was steering, and I, as before and after, steadily attended the sheet. By this time my hands were swollen, blistered, and split open by the cold and stagnation of my blood. When poor Jack Cole protested that the drag was not heavy enough and would surely come back on us, I had nothing else to weight it with other than our cook-kettles or fire-pots, so I determined to launch it as it was. The first cutter had been lost to sight for more than an hour; it had taken us two hours to prepare the drag; and it was now nine o'clock (of the night of September 12th).

Finally, with the fire-pot ready to slip on the rope in case Jack Cole's fears should be realized, we were prepared to test the efficacy of our anchor, when, at this juncture, Mr. Danenhower asked me if I would permit him to put the boat about. I hesitated a moment, and then replied no, that I would do it myself. But the next instant, concluding that, if there was any especial point in the seamanship of putting the boat about in a gale of wind of which I might be ignorant, and he as a professional sailor aware, it was my duty as commander to avail myself of it or any other chance that might insure the safety of my men; so I granted his request, standing by, however, with a view to any emergency. The oars were gotten out, the helmsman directed to watch the sea, and the drag was carried forward to the bow and placed in charge of fireman Bartlett, the rope being coiled under his feet, and he was ordered to see everything cleared when word was given to throw out the drag. It is a well-known fact among sailors that the waves of the ocean follow each other in succession of threes; that is, after a large wave the two which follow will surely be as heavy in appearance, at least. When such a succession had occurred it was but the work of a moment to "starboard the helm, lower away the halyards, and gather in the sail." Then, — "Give away starboard, and back port" was the order; but before we could veer around, an awkward, lazy wave came tumbling in torrents upon us. Pumping and bailing with might and main, we kept the boat before the sea and awaited a calm; until, after a hurried consultation, we put her about again, and as she came around head to the sea, Bartlett cast clear the drag. As he did so, the boat's head dived down into the hollow of a wave, and I saw Bartlett lose his balance and pitch headlong forward, apparently into the sea. But with his right hand he had grasped the halyards, which were flying clear of the mast, and the next instant, as the boat arose on the crest of a wave, he was hurled sprawling back against the mast, where he clung for his life. When I saw his frantic struggles, my heart leaped into my throat lest he would plunge overboard, and had such been the case, no matter how great his vigor and power of endurance, his fate would certainly have been sealed. And now, for a short time, we were forced to use the oars in a most clumsy manner, in order to keep our bow toward the sea. The drag was too light, and began to rise to the surface and drift rapidly to leeward. The boat, held in place by the oars, yawed about, shipping considerable water, and Jack Cole gave vent to his expected "I tould yees so." For this I was prepared, however, and soon had the copper fire-pot running out along the rope towards the drag, which it promptly sank and caused to rest much better. So the oars were brought into the boat, and the weathercloths again raised over the tops of the stanchions and backs of the men.

And now, how long we must thus lie anchored would altogether depend upon the duration of the gale. Miserable we were, indeed, with the sea dashing constantly over us, and our strength severely taxed in bailing out the flood with pump, buckets, and pans. Danenhower and myself had vacated our two seats in the stern sheets to permit a freer use of the steering oar; and I now divided the men into two watches; detailing Cole and Bartlett to look after the forward part of the boat, and Leach, Wilson, and Manson to take regular turns of two hours each at the steering oar. The others I directed to lie down on the thwarts or seats, and get what rest they could. Sleep was impossible, but it was a great relief to stretch our limbs after their long cramping, and feel the warm blood flowing again towards our almost frozen feet. The weather-cloths were frail bulwarks, and the waves broke over them on either side. Then "Quick, quick, before another one comes!" and we, weary wretches that we were, faint from hunger and thirst (for we had no water left), would renew our exertions, and bail clear our little craft ere the next wave could tumble in and wreck us. And the intervals — they were lucid with the cruel spray which dashed and froze upon us. Thus we passed the night, an incubus of horrors, and at daybreak there was no abatement of the storm. We had among our medical stores one quart of brandy and one of whiskey, and a request for a drink being made by one person and eagerly seconded by others, I drew forth the latter bottle and told Bartlett to divide it as equally as possible among the ten men; not caring for it myself and feeling rather chagrined that any one else should ask for it. In the division, Bartlett dealt with a too liberal hand, and alas! for himself, went dramless along with me.

Daylight did nothing but enhance our misery, since it enabled us to witness each other's wretchedness. For breakfast I issued a quarter of a pound of pemmican to each man, the same allowance I had made for supper the night before. I could not and did not eat my ration, small as it was, and so returned it to the common fund. I had now placed the men on one half the ration they had been receiving before the separation, or three quarters of a pound of pemmican instead of one and a half pounds, hoping by this economy to make our four or five

days' supply last twice as long, and in this I succeeded.
The vessels we had filled with snow were all deluged
with salt-water; and though each of the other boats was
possessed of five and ten gallon water breakers, yet the
one belonging to the whale-boat, it will be remembered,
was abandoned on the ice with the story of the Jeannette
between its oaken ribs. The others complained of thirst,
but I never felt the need of a drink until we came upon
the sweet water of the river. I constantly chewed a
piece of wood, which induced a flow of saliva, and my
body, as indeed the bodies of my companions, exuded but
little. Again, throughout the whole of my Arctic experience, I accustomed myself to do without water for
drinking purposes; and I am confident that outside of
our usual allowance of tea and coffee at morning and
evening meals, and a small cup of black coffee at noon, I
did not drink three pints of water during the time we
were on board ship and on the march. This saved me
much of the suffering that the others endured; for while
on the ice I have seen many of the men, when they
thought they were unobserved, drink from the pools and
eat of the soft salty snow, which De Long had forbidden
them so to touch.

As the day wore on, the wind, which had shifted,
seemed to subside, and the sea became more broken and
turbulent. The heavily-laden boat now sunk deeper
than ever, and, almost water-logged, shipped water, as
she rose and fell, over forecastle and stern. Still we
stuck bravely to our work, and poor Iniguin, who had
been little at sea, sat squarely down in the bottom of the
boat and bailed for dear life. He and Charley Tong
Sing were, in fact, our most effective bailers. About
three or four o'clock in the afternoon the clouds began
to roll away, we caught glimpses of the blue sky, and
there was every evidence of the gale having spent its
fury. At this season of the year the sun was exactly

twelve hours above the horizon, and when it shone we encountered no difficulty in shaping our course by my old watch. The sun arose at six A. M., was due south at noon, and set at six P. M.; so I had two checks on the watch, besides the one at noon by compass, which, being a prismatic instrument, did not swing as easily as a mariner's compass. It was fitted too neatly to its case, and the water gaining entrance glued the circle and clogged its movement; so that while afloat we found it better to steer by sun and moon when they were visible.

At five o'clock the sea and wind had calmed sufficiently to permit of my getting under way, which I did at once, laying the course about southwest for Cape Barkin. The wind had shifted more to the eastward, and we ran rapidly along, with the moon and stars now shining brightly. At six o'clock on the morning of the 14th we were sailing through the young ice, and keeping a bright lookout for the land which we expected to raise every moment, when the boat suddenly brought up on a shoal. Long and anxiously we gazed in the hope of sighting the low beach or mountains back of the Delta; but to no purpose. Bartlett thought he saw land; but as no one else confirmed his guess, I concluded that the shoal made off shore beyond the range of our vision; so we pushed off and stood to the eastward. My orders were to proceed to Cape Barkin, where I would be sure to find natives, with whom I could arrange to be piloted into the river to a native or Cossack village.

I felt assured of our general position, knowing, from the northeast gale which had steadily blown while we lay to, that we were now to the west of Cape Barkin, eastward of which I also knew was the Bay of Borkhia. So I kept the boat to the eastward all day, at the same time progressing, whenever an opportunity offered, toward the southward. The weather was genial and warm, the sun glowing clear in the heavens, and to all appearances there never blew a gale over so placid a sea. Morning, noon, and night we ate our quarter of a pound of pemmican without water or drink of any kind. Six o'clock — and as the sun sank in the west a cloud arose in the east. The wind began to blow in fitful puffs, and every man foretold another storm. I knew the season was well advanced and the time at hand when fall gales must be expected, but could not believe that we were about to suffer another siege similar to the one we had barely weathered but a day or two before. It seemed too inclement, too merciless; yet if the worst came we had only to face and fight it fairly, however brutal and unequal the contest between our feebleness and the cowardly rage of the storm king.

And, in the first place, it would never do for us to be caught napping on the shoals. I had lost all faith in our ability to reach Cape Barkin or the shore except at the mouth of one of the rivers. Bartlett had been sounding all day with a tent-pole; but our efforts to work toward the southward or beach had been regularly unavailing; so, as darkness approached, I decided to stand off or along shore to the eastward, having discovered that the water deepened in that direction. Accordingly I reefed sail and ran on until six o'clock of the following morning, when we found ourselves in nine fathoms of water, — additional proof that we were now in the Bay of Borkhia, and had struck the coast-line about twenty miles to the east of Cape Barkin. I now proposed to attempt an entrance into one of the many branches of the Lena, and so put the boat about on a southwest course. The wind was light and the weather fine; indeed, when I had run from the coast twelve hours and trusted to making it again the next twelve I had calculated without my host, for shortly the wind died away and we were fain to use the oars, still very grateful for our good fortune in escaping the prospective storm-struggle. But the current of

the river set out so strongly to the eastward, bearing us
on its bosom, that whereas I had at first hoped to make
the land on this day, I now saw, with regret, that all
our rowing would scarcely do more than hold the boat in
its course. Day died into dusk, and all through the
night we toiled, by watches, at the oars and bailers,
quickly casting out the sea before it could solidly freeze
in the bottom of the boat, for it changed to slush the
moment it tumbled in. At dawn of the 16th we were
again in shoal water, indicating our approach to the east
coast. The men were now painfully athirst, and as we
progressed kept tasting the water, all the while begging
permission to drink of it; and it was surprising how anxious they all were to just sip it and see if it was fresh
enough; but noticing that the tasting process was only a
ruse for drinking, I prohibited it altogether, promising
to make tea as soon as the water was fit.

At this period high lands, apparently a mountain-
range, were seen to the southward, but nothing greeted
our sight to the westward, where the low lands of the
Delta should be. I was counseled and urged to steer in
toward the mountains; but my main object was now to
join company with the natives, who, according to the
charts and our best information, roamed all over the
Delta from Cape Barkin to Bukoffski Moose (Bull
Point). Our provisions were about exhausted, and the
meagre ration of three quarters of a pound was telling
on the spirits and energy of the crew. So, having still
a desire to obey my orders and make, if possible, Cape
Barkin, and having, as well, full confidence in the truth
of my chart, which distinctly declared "winter huts of
natives" to be there, I held my course to the westward,
and presently raised two low spits of land with a gap of
four or five miles between, evidently the mouth of a
river. Towards this I made directly, and by following
up the sweet water soon ran into a swift current. Mean-

while the men were confidently assuring me that the water was fresh; so I ordered a pot of tea to be brewed, which, proving too salty, I threw away. A mess of canned mutton broth and ham, that I had heated and watered well into a pleasant stew, then atoned, however, for the loss; and we had breakfasted.

CHAPTER VI.

ON THE LENA DELTA.

Sailing up Stream. — The Hut. — Our Frozen Limbs. — Meeting
with the Natives. — Learning their Language.

BY this time we were well within the mouth of the
river, making W. N. W. Considerable driftwood lay on
the southern beach, and I tried to effect a landing in
order to afford all hands a most needed opportunity of
stretching our limbs, and coaxing our blood into circulation, for our feet, legs, and hands were now entirely
bereft of feeling : all save those of one or two persons,
who being without any particular occupation had availed
themselves of their idleness to remove their foot-gear
and rub with a towel, about twice an hour, their limbs
as high as their knees, — expressing the while great surprise that others did not idle away their time in a similar manner. I could see the water shoal and ripple
along the shore, and as there was but very little swell
directed the boat to be beached ; but she took the bottom
one hundred yards off shore, and, in spite of all our pre-
cautions and exertions, was carried by the swell almost
broadside on the beach, in great danger of capsizing or
rolling over. This was not due to any lubberly handling
or guiding of the boat, but the men were too numb to
act promptly, too weak to lift or work their oars ; and it
was very noticeable and discouraging how the cold had
robbed us of our vitality, and produced a dullness of
mind, movement, and speech among the whole party.

But we were soon afloat again, and I now consulted my chart, a pencil copy of De Long's, which, in turn, was a copy of a small chart published in the " Geographische Mittheilungen," by Petermann, the eminent German geographer, in connection with a scientific paper on the coast of Siberia and the Lena Delta in particular. From it I learned that all the branches of the Lena on the east side of the Delta, with one exception, discharged their waters to the north of east, and that the branch in which I had entered emptied to the south of east; or, the course up the river as it lay before me was about W. N. W., and because of its great width at the mouth I concluded that we must be in the main branch, which flowed by Bukoffski Moose. The presence of a large island corresponding to one laid down on the chart confirmed me in this opinion, but I was surprised to find so little water in so broad a river, it being necessary for us to sound continually with a tent-pole in order to avoid getting ashore on either bank. Our common dangers and miseries had bred a closer fellowship among us than the relations between officers and men usually admit of. We conversed together at all times of our many past escapes, and of the uncertainty of our future. Of one thing we were glad and sure. This was the main branch of the river; we had assuaged our suffering from thirst, and had now a bountiful supply of sweet water. The hot tea seemed to have thawed out our tongues, and we speculated freely upon the fate of our companions; the general opinion being that ours was the only boat which outlived the gale. And so, " Hurrah for the whale-boat," said we.

But a matter which concerned and troubled me most was the possibility that De Long and Chipp had succeeded in reaching Cape Barkin; and hence should I not obey orders and proceed there too? My verbal instructions, indeed, had been to first seek a place of safety for my own crew, and, if the other boats in arriving at Cape

Barkin had met with no natives, my joining them there would be no relief, but only a senseless addition to our general misfortune. Mr. Danenhower sat beside me in the stern sheets, and we naturally discussed the situation together. He strongly advised me to go to sea again, and work up to Cape Barkin, where "we would be sure to find natives;" though it was only some forty miles to the north of our present position. The river grew more tortuous and narrow, assuming the appearance of a blind stream issuing from a swamp, so I finally told Danenhower that if it did not show better by noon, I would put the boat about and try for Cape Barkin and the natives. It was but a day's journey, and I had no hesitancy in attempting it, until I glanced at my men, — weak, hungry, and hollow-eyed, — and then remembered the shallow coast, the stormy weather, and our recent hardships; while ahead of us I had now not the faintest doubt of finding succor. Still, — orders, alleged relief at Barkin, and the possibility of the other boats being there and in danger. At twelve o'clock we struck a shoal, and agreeably to my expressed intention I ordered the boat about and announced that we would return toward Cape Barkin. It was quite evident from the look of grave surprise on the faces of many, that the crew did not wish to risk the chance of another gale at sea. There was considerable murmuring, but the boat was turned around, when suddenly fireman Bartlett said, —

"Mr. Melville, I don't believe this river is as small as you imagine; there is plenty of water if we can only find it, and if you will but think a minute you will see that the river even here is as large as the Mississippi at New Orleans."

Our fellow-suffering, I repeat, had been a bond which bound us all together; I had listened attentively to every suggestion the men saw fit to offer me, cheerfully adopting any that, to my mind, might conduce to our comfort

or safety, and equally careful of turning aside, without
hurting the feelings of my counselor, any advice not pertinent to, or out of the question. So, giving heed to
what Bartlett said, and well recalling the appearance of
the Mississippi at New Orleans, I ordered the boat about
again, and proceeded up the river. Thus did his timely
admonition, like the proverbial straw which alters currents, luckily deter me from cruising to Cape Barkin,
where, the following spring, I found nothing but the
remains of old huts which had not been occupied for
years.

We now labored forward with renewed vigor, born of
hope, — the hope that we would soon meet with a village,
a hut, a canoe, a man, — saviors that our previous reading
had led us to believe were abundant in the district round
about us. I did not regain, however, my lost faith in the
game with which the Arctic regions are reported to
abound. A few black ducks, geese, and swans, late in
rearing their young, paddled about, awaiting the maturity of their broods ere they would follow the other feath-
ered tribes in their winter flight. Several seals exhibited
themselves at odd times; or perhaps it was but one and
the same seal, for they are very curious, and will pop up
and gaze at a boat or a man until a bullet ends their observations. Doubtless the reader will wonder why I did
not tarry to secure even this solitary seal. But I knew
too well the uncertainty of shooting from a boat at an
object one hundred yards distant, not much larger than
a double fist, and which in all probability would duck
under at the flash of the gun. I have seen as many as
forty shots, requiring two hours for their discharge, fired
at this elusive game, without other result than the loss
of powder and patience.

Toward dusk the wind grew cold and blustering. We
were surrounded by sand-spits and low islands, while in
front of us there seemed to be two main channels sepa-

rated by a large island which towered upward like a great
fortification. After many trials we at length found a
way through the shoals, and during the delay of more
than a hour had sighted a black uncertain object, near
to a collection of sticks apparently forming a series of
traps. The prospect of passing another benumbed and
sleepless night in the boat was horrible; so, when we
discovered the dim object on shore to be a hut, our joy
was almost as great as though we had suddenly chanced
upon a modern metropolis. After making several ineffectual efforts to land abreast of the hut, we finally
moored the boat in a friendly little cove, and then, dis-
embarking for the first time in five days, attempted to
stretch our limbs. I say attempted, for most of us were
powerless to control them. As for feeling in feet and
legs, we had none; and my fingers could not perceive
the difference in size between a rope and a needle.

We took possession of the vacant hut. It was old and
dilapidated, having been rudely constructed of small
round timber and split poles. Its general dimensions
were, at the base about eight feet square, the sides four
feet high, sloping or with a batir of about thirty degrees; the whole being covered with mud, *tundra*, and
lichens. There was a fire-place some three feet square in
the centre of the hut, over which, in the roof, was an
opening about two feet square, formed by a rude frame,
notched in, and supported by transverse rafters, which,
as well as the sides of the primitive dwelling, were covered with earth and sods. It was a hunting hut, used
temporarily by the natives during the summer season,
when they hunt the deer or geese; and had been vacated
but a few days before, since the fresh offal of birds and
fish lay in, and around it. The hunters had evidently
been accompanied by their children, for we found a number of little toys or playthings; among others a rough
cutting in wood representing a man seated on the back

6

of a reindeer. There was nothing, however, to indicate
where the recent tenants had gone, or whether they
intended to return ; though, to be sure, it was natural
to suppose that, with the hunting season over, they had
journeyed back to their village for the winter. The
traps were old and unset, the majority of them used for
catching the fox, while a few that were larger we then
thought to be intended for the wolf.

After unloading and securing our boat, we collected a
pile of wood, and soon had a camp-fire roaring and crack-
ling merrily. Our bag of tea, knocking about in the
bottom of the boat, had been thoroughly saturated with
salt-water; still, with the stew and soup made of the
birds shot at Simonoski Island and reserved till now, we
drank it, and supped with relish. Then carrying up our
few wet clothes, we all, barely able to lie down, crowded
around the fire, sitting huddled together in a heap. Up
to the present time, beyond the torpor referred to, there
had been no sign of our limbs having frozen, and I believed the frost to have been sufficiently driven out, by
our splashing around in the icy river, to permit of our
approaching the fire with impunity. So, we enjoyed
the genial warmth, and by comparison were in perfect
harmony with our new environment, even though it was
a tumble-down hut through whose wide chinks the cold
blast filtered perhaps a little less freely than it might
through a rail fence. Long we talked, and uninterrupt-
edly, reviewing the past and guessing the future, always
mourning the loss of those dreary " ten days," for had
we been so much earlier, — ah, what might not have
been, indeed !

Presently most of us were tortured with agonizing
pains in our hands, feet, and legs. Sleep was utterly out
of the question; and many were forced to leave the fire,
and even desert the hut entirely, in order to avoid the ir-
ritating heat. Our legs, upon examination, presented a

terribly swollen appearance, being frozen from the knees
down; and those places where they had previously been
so frozen and puffed as to burst such moccasins as were
not already in tatters, or force the seams into gaps corresponding to the cracks in our bleeding hands and feet,
were now in a frightful condition. The blisters and
sores had run together, and our flesh became as sodden
and spongy to the touch as though we were afflicted with
the scurvy. To move caused us the most excruciating
agony, and it seemed as if we were about to be worsted
in the end by what we regarded as our best friend, — fire.
Packed closely together in the hut, crippled, and nearly
blinded by the smoke, it was no wonder that in staggering about we trod unintentionally upon each other's
feet. I had removed my moccasins, and one of the men,
in reëntering, planted his whole weight upon my left
foot; the skin gave way from the ankle down, and shot
my friend (or enemy for the time being) off to one side,
like a ship slipped from its greased launching way.

We hailed the morning with delight, and, feebly breaking our fast with the scant quarter of a pound of pemmican, reëmbarked, full, at least, of good cheer, at the
outlook of a fine day's journey, which by my calculation
would carry us to the first village marked on our chart.
Following the main branch of the river to the westward,
we eventually ran into a small archipelago, threaded in
all directions by shallow little streams, from which there
seemed no escape. It was about noontime, so we repeated our dose of pemmican, and while some stayed to
brew a pot of tea others set out to reconnoitre the lay of
the land, and explore the various leads in the labyrinth
of streams. At the same time, Newcomb seized his gun
and tried to steal upon a flock of wild geese which he saw
on the *tundra.* But alas! either the wind was in their
favor, or the men made too much noise; at any rate the
geese flatly refused to tarry and contribute to our suste-

nance. On the reconnaissance another deserted hut was seen to the southeast of our position, but not visited; and, as the water seemed to lead nowhere in particular, I determined to head the boat up stream, reasoning that like every other thing it, too, must have an end, which we, should the river remain open and our provisions last, must some time reach. And thus it was. In an hour or two we were again in a navigable stream, a mile and a half wide, and flowing swiftly. With a fresh breeze we sped along against the current at a rapid rate, passing a long high island tallying to one on the chart, and continuing on without difficulty until we came to a fork in the river. Here, I first essayed the northwest branch, but discovering it to be a mere pocket or mud-hole, turned back and followed the west or west by south branch; when the wind died out, and we were obliged to resort to the oars. The current was strong and swift, our rowers weak and dispirited; but several miles ahead there was a tall promontory where the river turned abruptly to the southward; and towards this I steered, keeping close in shore out of the force of the current, and encouraging the men with the assurance that twelve or fifteen miles beyond the promontory, where we would camp over night, there was a village according to the chart, and that next day we would surely arrive there. The pull was a long and laborious one, and it was well-nigh night when we rounded the point and landed on the miry, shelving beach, which we designated the "Mud Camp."

Setting up the two tents, we gathered enough soggy wood to build a smoky fire, and then turned in. Danenhower, Newcomb, Cole, and Bartlett climbed the hill, some sixty or more feet above us, and slept in a small hut erected there by the natives as a deer lookout. The night was cold and blustering; there was a light fall of snow, and at dawn the weather looked lowering. The wind blew stiffly from the westward, and the broad ex-

panse of river, which here ran north and south, was rolling like a little sea, its bosom covered with white caps, and
the young ice making along its shores. We were mostly
so disabled as to move about with difficulty, Cole, Leach,
and Lauterbach complaining particularly of their frozen
feet; and as I wished to keep every one as dry as possible, I winded the boat about on shore, and stowing all
the cripples, myself included, tried to push off; but to no
purpose. Some of the men, at length, leaped over the
bows and set us free, but before they could jump back
again the boat was whirled amid stream, and it was only
after great exertion that we could pick them up again.
I then sailed under one reef up the river toward the settlements. Danenhower now asked permission to act as
coxswain for the day. His hands and feet were in good
condition, and, although one eye was bandaged up, he
seemed to see well enough with the other; so I gladly
consented, pleased to have a relief for my helmsman,
albeit there was none for me, since I would not intrust
the keeping of the sheet to other than my own hands, now
cracked and swollen out of shape. The wind came rushing down the hills and through the cuts in "willa-was,"
careening our boat until the water seethed along her lee
rail and soaked us in spray. Soon a herd of four or five
deer was seen scampering across the crown of a distant
hill, and we hopelessly sniffed the meal from afar. For
our general condition now caused me considerable anxiety; with such fierce gusts and in our maimed and
weakened state, with a half blind steersman, should the
boat strike a snag or shoal and capsize, we could scarcely
escape drowning in the fast running, freezing water.

We had been under way little more than an hour when
two large, well-preserved huts hove in sight on the west
bank of the river. Our hopes at once rose high — were
there natives here? A closer approach decided in the
negative. Signs there were of recent occupancy; but no

smoke — that infallible emblem of human habitation and
cheer — issued from the chimneys.   What a beacon and
a joy, indeed, to the weary wanderer is the smoke which
floats from tenements in every clime!   The blue, peaceful
cloud is a welcome anywhere, whether seen, as I have
seen it, softly curling above the quiet cot in the jungle of
Africa, the tangled forest of South America, the *tundra*
plain of Siberia, or, stained with the lurid light of the
fire, shoot seemingly upward from a snow-bank, beneath
which is buried the busy, bustling hut of the Tunguse or
Yakut, whose hospitable inmates, warned by the howl
of the native wolf-dog (for the Esquimaux dog does not
flourish in northeastern Siberia), are making ready for
the coming guest.

However, I ran the boat into a little cove at the mouth
of a creek, where the huts, apparently a fishing and
hunting station, were located.   A large quantity of drift-
wood and a number of native utensils were distributed
about, proving that the huts had not been permanently
deserted.   Further delay was dangerous; but we had
rested so little during the past two days, that I decided
to avail myself of these elegant quarters and the chance
arrival of natives until the following morning.   There
were several upright posts in front of the huts, and to
one I lashed fast a tall staff, flying from its head a black
flag, with a view to attracting the notice of natives or
our companion boats, should either advance along what
I took to be the main branch of the river.   The boat
hauled out and cleaned, I divided the men into two parties, assigning a hut to each; and soon before blazing fires
we had dried our clothing and lain down to rest.   Day
dawned fair and clear, finding us all in excellent spirits
after a sound and much needed sleep, and it mattered
little to me how badly my limbs were frozen; my blood
flowed freely, and I never lost my head.   And now, indeed, I had every reason to count upon our meeting

with the natives or their villages at any time. Previous
to leaving, I lashed a paper around the flag-staff at about
the height of a man's nose, stating that I had landed at
that place the day before with my company all well, and
had proceeded to the southward in search of a settlement.
Then, pushing off, we followed the west bank of the
river, until, greatly to our surprise, we ran into a wide
bay. It mystified me, since I believed that we had been
journeying up the main eastern branch, and had turned
south into the main river itself when we rounded the
promontory two days before. Here, now, we were in a
great bay from fifteen to twenty miles wide, with the
land only visible in spots around the horizon, and the
water so shallow that our boat was almost constantly
aground. It was an easy matter to lay a southern course
across the bay with watch and compass, for the weather
was good and the sun bright; but the difficulty was to
find water enough in this vast area to float us. We were
in a confusion of sand-bars, shoals, and channels; the
currents running in every direction, and yet all tending
to the eastward and northward. As we gradually approached a high headland to the southward, it grew
loftier, looming up like a mountain, and on it several
huts were presently discernible. Long and anxiously
we gazed for some sign of life, and I told the men that
we would eat our dinner there if we could land. But
the currents and shallows crowded us to the eastward;
no smoke issued from the huts to encourage us to pro-
long our efforts, although we still kept a southerly course,
passing within a mile of them; and, at length, entering
a channel which carried us to the eastward, I determined it folly to further wear out the crew in vain attempts to visit these doubtless abandoned huts, and so, it
now being long past our dinner hour, succeeded in reaching the bank below them.

There was no game in sight, so we swallowed our usual

tea and pemmican. An examination of some shelter
huts on the banks gave additional proof of the recent
presence of natives. As I had done at every landing,
the compass was set up and the general features of the
country observed, as well as the direction of the currents.
The men explored around and about as far as their inclination carried them, until I summoned all hands to
reëmbark, intending to follow the stream against the current which ran out of our course to the northwest; or,
rather, since the water flowed from the southeast, to keep
against it we would be forced into the northwest. I was
debating this point with Danenhower, and at the same
time gazing up the river, when lo! and hurrah! I suddenly saw approaching us three canoes with a native in
each. Not knowing whether they were friendly, but remembering that the Tchuchees were very hostile at times,
as were also the natives at Cape Prince of Wales and Cape
Barrow at Behring Strait, I hastily got afloat in order to
meet the strangers on equal terms, and ordered our fire-
arms to be kept in readiness, although under cover. As
we pulled towards them, I beckoned and signaled, but
they fought shy of us in evident fear or suspicion. I
then addressed them in English and German, and we all
smiled and laughed at my successive fruitless attempts to
open up a conversation in every modern tongue of which
I had the slightest smattering. Finally, two of the
canoes shot past us, but the youngest and seemingly most
fearless of the natives drew alongside of our boat to
receive a piece of pemmican which I directed one of my
men to first taste and then tender him.

His name, as we afterwards learned, was Tomat. At
least this was a section of it; for the natives are known
to each other by many names, baptismal and parental,
according to peculiarities of nature or stature, age or
avocation. Thus, "Vasilli Kool Gar" (William-of-the-
Cut-Ear); "Starry Nicolai" (Old Nicholas); " Georgi

Nicolai" (George Nicholas); "Bulchoi and Malinki Nicolai" (Big and Little Nicholas).

At any rate, as Malinki Tomat approached, I told one of the men to pay no attention to him until he was close upon us, and then to seize his canoe. This was done, much to Malinki's discomfiture and fright. He urged us to free him, intimating the danger of capsizing; but we had now drifted down stream abreast of our late dinner camp, and I motioned to him our intention of landing again. By this time he appeared to realize that we meant him no harm, and only required his friendly assistance, for when we were moored he hailed his two companions, who had hauled out on the beach a mile or so below, and they quickly joined us. Then our friendly intercourse began, and one member of my party, more gushing than the rest, kissed each native on both cheeks; which salutation they received with stolid amazement. I ordered a pot of tea to be prepared, and while it was brewing exhibited our guns, hatchets and other equipments, all of which interested and pleased them. Meanwhile, the sailors, in a more business-like way, had overhauled the canoes, and found some fish, a goose, and a piece of venison. The natives in presenting these articles to us said, " Cushat, cushat," accompanying the utterance with a gesture indicating the act of putting food into the mouth. This was the first word of our new vocabulary — *Cushat,* to eat. And we at once instituted a general search for more; eliciting the information that their caps were *shapkas,* and the red and yellow handkerchiefs they wore around their necks, *platocks;* which they also made us understand had come from Belunga. This was exactly what I wanted. The village of Belun was marked on my chart, and to it, or their Belunga, it was now my endeavor to have them conduct us. We displayed our knives; they showed us theirs, saying, " Knoshocks," and the next instant, " Belunga," pointing at the same time up the river to the westward, with a sweep to the southward. They likewise spoke of *coperts*, meaning merchants; and then Malinki To- mat mimicked the motions of a blacksmith forging a knife. I could have shouted for joy at the thought of finding a follower of Tubal Cain here in the wilds of Siberia; and although the natives had exhibited little brass medals hung about their necks as proof of their Christianity, and crossed themselves and gone through many of the signs and genuflections of the Greek Church, still, to me the most welcome and assuring evidence of our approach to civilization was Malinki's pantomime of the village smith at Belun. I knew that religion in some form or other is found everywhere that man can eke out an existence, or the Christian agent penetrate; but then there is so much religion in remote parts of the earth, and so little civilization, that I reasoned : Since the arts and sciences go hand in hand, even if they do not all centre in Belun, yet the iron-worker, one of the most essential contributors to industrial prosperity, is there; and all will be well.

We had arranged logs into seats around three sides of a fire, and sat there drinking our tea, which the natives called *chi* and seemed fully acquainted with. I poured a spoonful of alcohol into each of their pint pots of tea. Tomat did not relish the mixture, but the other two apparently knew whereof they drank, and Feodor, who proved to be a criminal Yakut exile, begged for some pure alcohol. I gave it to him, and observed that it yielded the same mellow fruit in the savage that it does when planted in the Christian. His spirits soared and he wanted more, but I stowed away the demijohn in the boat, and caused him to know that it must not be tam- pered with. After our first course of tea we put the kettle on, and, cooking the goose, fish, venison, and a patch of raw hide from the rear of one of the company's panta-

loons all into one glorious stew, divided it equally around
and supped. I then proceeded to inform our guests of
my desire to reach Belun.

To accomplish this I resorted to object teaching, —
drawing a picture of a village with a large dome in the
centre, surrounded by a number of lesser size. Tomat
said "Belunga" at once, but took exception to the choice
variety of domes or spires. This being altered to his
liking, I then pictured the whale-boat, with sail, mast,
and spars in place, and all hands on board, which both
promptly pronounced to be "Flotska." Next I designed
a man in a canoe in the act of paddling, with two companion canoes in the middle distance, and signified to
Tomat that his was the first canoe, which would be followed by the whale-boat, and then his countrymen. I
lastly indicated the direction of Belun, and motioning
him on, uttered the word "Belunga." He comprehended
the situation immediately, but cried, —

"Soak, soak" (no, no); "boos, boos" (ice, ice);
"pomree" (die)! and casting himself upon the ground,
he closed his eyes, and otherwise simulated the appearance of a corpse.

We were all at first very much puzzled by *shitty*, an
exclamation to which they gave frequent vent, and supposed it to be a corruption of the word city, which they
may have acquired from the Russian traders, since the
latter tongue has borrowed heavily from the German,
French, and English. But upon close attention we
shortly learned that it was an expression of surprise or
wonder. Tomat was bright, and taught us the names of
all his garments and accoutrements. He owned a little
gun — *fintofki*, and was delighted with our breech-loading
Remington, though somewhat awed at the size of the
bullet and the loud report when Bartlett pulled the trig-
ger, and more particularly so at the hole bored by the
bullet in the stump of a tree. They all gazed with admi-

ration on our axe and hatchets, but were very proud of
their own, made of un-steeled iron. The blade of the axe
is carried in a leather or hide case to protect the edge,
and is whetted with the steel which accompanies their
flint and punk bag.

Still I could not induce them to pilot us to Belun.
They made many excuses, chiefly the great danger and
distance; and there was the ice forming on the river,
and the scarcity of food and clothing; at which they displayed their tattered moccasins and other raiment. All
this while, however, they labored to convince us that
they would very much like to go to Belun, if possible,
since there they could get all manner of good things,
inter alia, plenty of *vodki* (rye whiskey). And after
showing us how they would drink a great deal of it, by
putting down one imaginary article of clothing after another in trade or pawn, until at length they could get or
drink no more, they then, to the amusement of all, lay
down and indicated how very sick they would be, or, as
the sailors have it, "Hog oh!" Meanwhile we had
gleaned that *balogan* was house; *spee*, sleep; *olane*, reindeer; and *ballook*, fish; and by means of these few words
I shortly acquainted them with our wish to proceed at
once to a place where we could eat and sleep. They
willingly assented, and to show me how thoroughly they
apprehended my meaning laid their heads in their hands,
closed their eyes and snored; then with cheeks distended
they, in fancy, spurted water into their hands, and went
through the act of washing their faces, after their custom, saying, "Cushat, cushat, olane, ballook." This being altogether satisfactory to both parties, we pushed off,
and in less than a half hour, under the guidance of our
new friends, arrived safely beneath the huts on the promontory, which we had tried so strenuously to reach that
day. Hauling the boat and canoes well up on the beach,
we conveyed most of our gear to the huts; I being par-

ticularly careful, before I abandoned the demijohn, to
empty all of its contents into a small alcohol breaker, of
which the natives knew nothing, and which I carried upwards with me. We found the place to be a deserted
village, formerly known as Little Borkhia. It had badly
gone to decay, only several huts remaining habitable,
while a graveyard in close proximity was well filled with
Yakuts or Tunguse. This, I afterwards learned, is the
state of affairs all over the Delta; the cities of the dead
are much more populous and prevalent than those of the
living. For, as the guides told me later on, when a death
occurs in a *balogan* or hut, it is instantly abandoned;
and chancing upon a thickly settled cemetery where the
graves were marked by rudely hewn crosses, my guides
would say, —

"Yakut pomree manorga; Yakut crass manorga" —
(many dead Yakuts; many Yakut crosses).

We appropriated two huts, built rousing fires, made
tea, and the natives, setting their nets, caught some fish
for our supper. In the mean time our pantomime communications progressed with more or less understanding.
The names of our three friends, we found without ex-
changing cards, were Tomat, Karranie, and Feodor, the
last being a miserable dependent upon the other two.
Tomat informed us that Karranie would go in search of
an old man, Vasilli Kool Gar, whom he called "Ta Ta"
(father), but who, I afterwards learned, was his father-
in-law, and would also bring *olane cushat*, deer to eat.
Our meal ended, after some additional indulgence in the
delights of conversation we all turned in. This was the
first *balogan* or hut we had seen which affected a fire-
place, a wooden floor, and berths arranged around the
walls. It is needless to say how wondrously dirty it was,
or how pungent the odor of ancient fish and bones; yet
we were very glad to be so comfortably housed, for outside a wild snow-storm raged in the night.

CHAPTER VII.

UP THE LENA.

My Unsuccessful Effort "to go it alone." — Vasilh Kool Gar. — My Duck Diplomacy. — Fears of Scurvy. — Arii, the Deserted Village. — Spiridon, the Ugly Starosti. — Siberian Ice-Cellars. — Jamaveloch.

BEFORE retiring I noticed the natives paid their devotions to a small brass icon, in the left hand corner of the hut, farthest from the door. The bed or berth in this corner, no matter how the dwelling faces, is invariably the guest chamber, and over it the icon or icons (for there are often a dozen or more of them) are placed on a little shelf, together with a number of small wax tapers the thickness of a lead-pencil and about three inches long, which are lighted on special occasions, or, in the case of wealthy natives, kept burning throughout their devotions. We passed a restful night, and awoke to a good breakfast of fish caught over night in the nets which the natives had reset. Karranie had not yet returned, and I began to doubt the fidelity of the natives, discerning in their manner, however friendly, a certain fear of us, and suspecting from their actions that they contemplated stealing away and leaving us in the lurch. I tried to induce Tomat to embark with us, or pilot us in his own canoe, but all to no purpose; and when I directed one of the men to seize and force him into the boat his piteous demonstrations of dread determined us to go on without him. The condition of my men, the low state of our provisions, and my anxiety to reach Belun, where I could communicate with the Russian authorities and arrange for a search for De Long and Chipp, would brook of no delay. So I took what fish the natives had, and Mr. Newcomb traded a knife and neck-comforter for a net, of which we hoped to make excellent use in case of an emergency. Then with much reluctance I shoved off, and left our good Tomat standing tearfully on the beach.

He had persisted in assuring us that we could not get to Belun, but had pointed out the way to the villages to the southeast. Here I was in a quandary. The west northwest current was running strongly. Danenhower cautioned me, "Melville, you should go to the south; not that way."

But I nevertheless did desire to keep the boat's head against the stream ; still the arguments of my companion and the existence of the villages prevailed, and I endeavored to work along what I conceived to be the southwest coast of the main river. But the shoals intervened, forcing us off to the eastward, and we made very little southing. At length, getting the boat far out into the bay where we had been the day before, I sighted two tall headlands to the southward, and believing the river to debouch between, endeavored all day to reach them. The weather was raw and windy, the water rough, and it seemed impossible for us to proceed a mile in any direction without lodging on a sand-bank. At two o'clock in the afternoon I decided to return to Borkhia. The wind meanwhile had moderated and was fair, but there was no channel, and the water breaking over the spits and bars drenched us to the skin, and froze in the boat. The men were exhausted with constant rowing, bailing, pumping, and sailing. Bartlett had called out the soundings from the time we started, and, wet from head to foot, his clothing had frozen stiff; the tent-pole he used as a sounding-rod was a mass of ice ; and his hands were swollen and cracked in a horrible manner. The thawing out and comparative luxury of the night previous had doubtless rendered us more tender and susceptible to cold and pain, for certainly the complaints were now greater than ever before. My legs from the knees down were covered with sores and blisters, causing me the most intense agony. Leach, Manson, and Wilson, being younger than Cole, appeared to work the boat quicker than he at my word of command, when Bartlett shouted the soundings from the bow, but they had been on duty all day and were now utterly fagged out. Though I had continually held the sheet, I did so mechanically, my hands having been robbed of all feeling. Towards night Danenhower volunteered to steer, but the wind and snow blinded him; and after several attempts to lay a course according to the direction of the wind as it blew against his cheek (I having rolled up the ear-guards of his cap for that purpose), with the almost disastrous result of jibing the boat, I ordered Leach to relieve him at the tiller, — concluding to anchor for the night under the lee of the first shoal we encountered, and like St. Paul await the coming of the day. After trying once or twice to round a point of land on which we could see the white waves dashing, — our boat close on the wind and the lee oars in motion to keep her clear of the shoal, — we at last succeeded, and found ourselves in deep quiet water behind the bar.

But how to anchor without an anchor or even an excuse for one! Our only resource was to fasten the boat to a stake, and even this article being denied us I directed Bartlett to drive three of the brass-tipped tent-poles far down into the ooze; and then, that they might not be loosened or pulled out by excessive strain at the tops, I caused a loop to be made at the end of the painter, and this being sunk drew all the strain to the bottom of the poles. And thus the boat was held all night against the wind, which at times blew fiercely. However, to prevent the loss of the poles should they in any way be set free, I had a slack, light line securely fixed to their tops. Then

with paddles driven down on either side to keep her head
to the wind, and one man to watch that she did not drift
from her moorings, the rest of us stowed ourselves away
for the night as best we could. Sleep, to be sure, refused
to slide into our souls; yet covering ourselves with the
weather-cloths, some pieces of canvas, and the mackintosh
that we had used to spread inside the tent while on the
march, we composed ourselves for rest. But it was bitterly cold, and what with the rain, sleet, and snow, our
sufferings were intense. Those of us not already frozen
soon became frost-bitten, and we who were frozen before
were now rendered almost helpless by the fresh freezing
of our limbs, which swelled, and stretched the moccasins
to bursting.

At dawn the storm had abated and the river was tran-
quil; but the scene that presented itself to our gaze was
one of depressing desolation. Every shivering soul in
the company showed plainly the anguish he had endured
the night before; the ropes were frozen and covered with
rime, and the snow-fall of several inches was spread over
the boat and every object in it. The hills and mud flats,
which a few hours earlier were green and black on the
horizon, now glistened in their wintry coat of whiteness;
and so altered was the appearance of the surrounding
country that we could barely distinguish our landmarks
of the previous day. It seemed, however, a simple matter to return as we had come   I had worked by compass
to the southeast, and should find no trouble in working
back to the northwest. But there was such a bewildering variety of opinion that I permitted every one to venti-
late himself on the matter until nearly noon, when, the
bank looking very familiar, I concluded to land and
make our meal of tea and fish. Presently we had a fire
blazing warmly, and while some prepared dinner, others
who had argued most vehemently regarding our location
explored along the bank, and no sooner rounded the point

7

of land under which we had camped than the huts of
Borkhia arose before them in plain sight. Everything
deserving the name of game had long since left the regions round about us, and with the coming of the snow
and freezing of the ponds, the few dilatory ducks and
geese that had abided the maturing of their young were
now winging their southerly flight in straggling pairs or
skeleton flocks. Only the gull and other carrion birds
hovered in sight, or, floating aloft, looked down and
gloated on our misery. Our meal eaten we pushed on,
and towards evening, as we approached the bluff, discerned
with pleasure four natives running to meet us. They
aided us in hauling the boat upon the beach, clear of the
young ice which was fringing the shore, and I noticed
our first three acquaintances were now joined by an old
man whom they pushed forward as their chief, calling
him " *Starosti*," "*Commanda*," " *Ta Ta*," etc., etc., and
who stood with cap in hand saying "*Drastie, drastie.*"

My own, and the limbs of Leach, Manson, and Lauterbach, were so badly frozen that we were forced to crawl
on our hands and knees; Bartlett, Cole, Newcomb, and
others, though severely attacked, were not disabled ;
while Danenhower and Iniguin were the least affected
of all. The natives assisted us to the hut, where they had
a cheerful fire, and a supply of fish and venison; and observing in the hands of the men several gulls which Newcomb had shot, intimated that such were not fit to eat, and
gave us fish instead. Yet I cannot understand why they
object to eating gulls when they have so often to resort
to food infinitely more disgusting. I remember that one
of the Jeannette cabin mess asserted that young gulls
were sold in the markets of a great seaport town of the
United States, and esteemed a great delicacy by the *élite*
of that place; and though I am willing to give my personal assurance that many things are more unpalatable,
as we all had cause to know before leaving the Delta,

still I do not propose to eat a gull or any other carrion
bird when better food is at hand.

I supplemented our supper with an extra kettle of tea,
of which the natives were very fond, and then set about
informing the old man of our utter poverty and our great
desire to be shown the way to Belun, or some other settlement. He grasped the situation, and signified that after
a sleep we would all proceed to a village. I tried to
persuade him to conduct us to Belun, but he joined the
younger men in assuring me that it was impossible, owing
to our want of food and clothing, and the rapid formation
of ice in the river. I was determined, now that I had laid
hands on the natives a second time, not to loosen my grip
under any condition ; and, indeed, had so announced my
intention to the company after our ineffectual attempt to
find a passage up the river; saying then that I would return and make prisoners of the natives, seizing their canoes and equipment and compelling them to pilot us forward. Happily it was unnecessary to enforce obedience,
for after breakfast, the following morning, we set forth
on our journey. Before shoving off it was important
that the old man, Vasilli Kool Gar, be instructed to
avoid the shallows, since our boat, being larger, drew
much more water than the canoes. Bartlett explained
the matter by catching him by the shoulders and pointing out to him the water-line of the boat and then that
of the canoes. Vasilli seemed to understand, and, as a
proof that he did, he cut a mark with a knife on his dou-
ble-bladed paddle, after measuring with it the distance
from the ground to the indicated water-line on the whale-
boat. This seemed a triumph of intelligence, and we
thenceforth had full faith in our new pilot. I now advo-
cated again that we go to Belun, but met with the same
positive refusal. Ice, cold, starvation, and death, the natives said, would surely overtake us; and they drew in
the snow diagrams of the course of the river, with the lo-

cation of the villages at which we would have a halt, winding up their argument by feigning a tragic death scene. So with the promise on the part of Vasilli that we would eventually reach Belun, we at last started.

For a while Vasilli ran along the coast in the direction I had traversed the day before, continuing to the southeast as long as the whale-boat would float. Two of the young natives were stationed out ahead on either bow of the old man's canoe, finding the channel, and thus they piloted us over the same flats we had crossed before. But the water in time becoming more shallow and the shoals more numerous, Vasilli abandoned the course he was pursuing and steered off to the eastward, occasionally as high as the northeast. Towards evening he sent the other canoes ahead, and remained behind encouraging us to advance; growling, muttering, and entreating, yet always good-humoredly, while he laughed with us at our vain attempts to understand the constant chattering and gibbering kept up between us. We had been pulling hard against a strong current for several hours, and it really looked as though we would never round a long sand-spit which stood in our way, when suddenly a bright light shot upward from the beach, a mile or so in front of us. The other natives had forged ahead and built the first guiding fire that had greeted our eyes since we left Unalaska. It infused fresh vigor into the crew, and soon we were hauling out our boat on a bleak, snowy beach, back of which arose a lofty *tundra* bank full of fissures and cuts.

I erected the two tents for our accommodation, giving the natives the boat sail to set up as a shelter, and directly the evening was over we all lay down to rest. The snow-fall of the past few days had made a soft bed for us, save where the driftwood that littered the beach pushed through and rendered it lumpy. Our sleeping-bags, too, from frequent soakings, were denuded of their hair, full of holes, frozen hard, and so shriveled up as to be almost useless. Still was our thankfulness in itself an all-sufficient comfort; thankfulness for the friendly care and guidance of the natives. I had hidden away our small residue (about twenty pounds) of pemmican, and persuaded the natives that we were entirely out of provisions, in order to urge them forward; and, at the same time, I induced them to set their nets and catch some fish, while I secretly issued a small ration of pemmican in the tents. Old Vasilli overhauled our pots and kettles, but finding no food, supplied us from his canoe-box with a few little fish, of which we made a watery stew, long drawn out.

It was very cold, and the wind that arose when the sun went down blew half a gale. We kept a great fire going most of the night in safe proximity to the tents, but shivered, nevertheless, and shook and froze, and morning found us more unfitted than ever to cope with our undiminished difficulties. A pint of hot fish soup and a quarter of a pound of pemmican to each man (the natives being included in this meal) composed, with the tea, our breakfast; and the tents, covered with ice and snow and frozen like boards, having been rolled up after a manner and stowed in the boat, we once more pushed off. Clear of the shoal water of the large bay, we were now conducted through the serpentine windings of the river out to sea, around an island to the north of Bukoff-ski Cape and into the river again, accomplishing a good day's journey, and, as darkness fell upon us, we arrived at two deserted huts on the north side of the eastern discharge of the main branch of the Lena; which I subsequently visited twice during my second search for De Long. One of the huts was in a much better state of preservation than the other, but both together could not shelter us all; so, from preference, some of the men and natives set up a tent for their accommodation. The natives caught two or three small fish, to which Vasilli gingerly added several from the box in his boat. Newcomb during the day had shot a pair of ducks, and these I magnanimously presented to Vasilli, telling him that though we had nothing else to eat, we yet felt called upon to tender our little all to him, on whom devolved the responsibility of feeding and speedily bringing us to a place of safety. My magnanimity was not without its effect. He peered into our kettle and box, and finding them empty wished to return the ducks, and proffered us the last two fish in his canoe, assuring me that his larder was now as vacant as ours.

Several of the party were so enfeebled by the additional freezing of their feet and legs that the boat was not hauled entirely out, as heretofore, but the gear was heaped upon the beach, beyond the reach of the water. The huts, according to custom, were built on the bluff above the river, and in our frozen condition it was no mean task for us to struggle up this height. I had altogether lost control of my legs, and so in ascending the steep bank leaned on the shoulders of old Vasilli and Karranie, who afterwards assisted Leach and Lauterbach in the same manner. Our supper of fish soup was prepared and served in the usual way; Bartlett dividing the food as equally as possible into the pans, which were placed on the ground, each man seizing his pan, and Bartlett and I appropriating the last two.

Meanwhile Vasilli had cut up his ducks, using a great deal of water in cooking them, and after we had supped he generously gave a portion of them, together with some soup, to my party. Many, many times afterward have I seen Vasilli relate to others, and go through the pantomime of my presenting him with the ducks, when we were almost dead from hunger, cold, and exhaustion. The artifice worked well, and I felt the good effects of it all through my first terrible journey in search of De

Long; and even to the time when, leaving the Lena Delta, I distributed my few remaining fish and small stores among Vasilli and his neighbors.

Crawling into the huts at night, we invariably removed our foot gear in order to ease our swollen, blistered, and bleeding legs and feet. On this particular night, after I had withdrawn my moccasins, all the natives, one after another, examined my legs, and, pressing their fingers into the tumid and spongy flesh, watched anxiously to see if the indentations would disappear; and, when they had remained a considerable while, the natives, shaking their heads, consulted together with all the gravity of medical doctors, and apparently reached the conclusion, that, though I was in a very bad condition indeed, yet under the circumstances they could do nothing for me. What worried me most, however, was the fear that scurvy was about to break out among us. From my knowledge of this frightful disease, it seemed too anomalous that we, having endured by far the greatest hardships of any Arctic expedition on record, should be exempt from its fatal clutches. We had passed unscathed (for Alexia's case was a mere suspicion) the crucial test of a long march in wet clothing, under a discipline of severest toil and lowest diet, such as the whalemen, hunting from the floe-edges at a distance from their ships, sink under and perish. All this and more our gallant fellows had undergone; but now, though frost-bite could account for the sores, blisters, and curling up of toe and finger nails, yet I could attribute the dead swelling of the limbs to nothing else than scurvy; which seemed also responsible for the soreness of gums that Danenhower and Newcomb complained of. And yet time has satisfied me that my fears were groundless, notwithstanding the long interval through which we had lived without antiscorbutics of any kind.

Next day we were early out and on our way again,

now rowing and now sailing, or both, when practicable,
to lessen the labor of the crew. Our pilots would occasionally lead us into difficulty by forgetting that their
canoes drew only three inches of water and the whale-
boat twenty-six. Still they were mindful of our weakness and misery, making as many short cuts as possible ;
and towards noon we found ourselves on the bosom of a
broad, deep stream, speeding merrily along under oars
and sail. Vasilli had dispatched a couple of his young
men ahead, and was holding on to the side of our boat,
signing to me that his arms were so tired he could paddle no farther. Yet he appeared desirous of detaining
us pending the reconnaissance and return of the canoes,
insisting that we rest the oars, and, though the wind
was fair, lower the sail. Presently there appeared in the
distance a village of considerable size ; but no smoke
arose from the chimneys, and as we drew near the bank
our hearts sickened to find no welcome from man or
beast. It was a mystery to us at first ; our stealthy
approach to the place, and its oppressive solitude and
silence. I even suspected that the natives had gone on
in advance and hurried the people away ; but, upon
closer inspection, I learned that the island, for such it
was, had been deserted for months. Then it dawned on
me that this was a winter village, whose inhabitants had
not yet returned, and that Vasilli had forwarded the two
young men to establish this fact, intending to tarry there
if the natives had arrived, and if they had not, to journey on to the southward to another village which he
knew to be inhabited. But as we had run by the turn
in the river, Vasilli concluded to stop at Arii, the abandoned settlement ; so we landed and occupied one of the
huts. It was in good order, and the window places were
closed against the weather by slabs of wood. We rum-
maged among the huts and store-houses, but found absolutely nothing to eat ; Newcomb was more successful

with his gun, shooting several ptarmigan, of which we
made soup. Vasilli, having sent one of the natives to
the neighboring village, bared his arm and showed where
it had been pierced by a bullet or spear near the biceps.
It was shriveled and nigh powerless, and he indicated
that he could go no farther, but had sent for a native
who would pilot us on.

We built a fire and made hot tea, and during the afternoon discerned a canoe and a native pull-away boat
approaching us. The latter resembled a whale-boat in
shape, being sharp at both ends, but having a much flat-
ter bottom. The planks, split about one and a quarter
inches thick, ten inches wide, and long enough to reach
from end to end, were clinker-built, and fastened through
with wooden pegs three-eighths of an inch in diameter.
The frames, of birch or spruce, were about three feet
apart, and the stem and stern-pieces connecting with the
keel, which was chiefly inside, were huge sticks of timber, with rabbets cut in so as to make a flush finish at
both ends. The work was roughly, though strongly,
done; and a boat of this kind, from six to eight feet
beam and from twenty-five to thirty feet long, will probably weigh about three times as much as a whale-boat of
the same general dimensions, even though it is without
an iron or copper fastening. The seams are caulked on
the outside with reindeer moss and the fine rootlets of
the peat moss, dug out of the *tundra* beds and washed
clear of earth.

Our friend Karranie was in the canoe, and the row-
boat contained two men and two women, three of them
rowing while the elder of the men steered. This, Vasilli
made me understand, was the head man of the village,
and a worse-looking old pirate I never saw. He was
short and thick-set, and his eyes, arched over with beet-
ling brows, glittered far back in his head like two little
balls of fire. His hair was cut close, and a like fate, it

appeared, had befallen his ears, from the bottom of one
to the other of which nearly extended his long, firmly-set
lips, which opened above a great square jaw. The body
of a giant resting on the legs of a dwarf, — and this was
Spiridon. The two women who accompanied him — one
of whom was Mrs. Spiridon, and the other his sister —
had each lost her right eye ; and, though meeker in behavior than their husband and brother, were quite as rascally in looks. The young man was a boisterous, devil-
may-care youth, all rags and tatters, a savage counterpart of the numerous ne'er-do-well young man of our
large cities, who is content to laugh and live at his
friends' expense.

Spiridon, with the women, retired at once to his house,
while Capiocan, the youth, was shortly fraternizing with
the sailors. Vasilli came in to tell me that the *starosti*
had arrived; so, in company with him and Mr. Danenhower, I called upon the great little man. He was
very stolid and stupid, and would not exert himself to
talk or be at all agreeable. A large pot of tea, which I
ordered to be made in our hut, was carried over, and this
we drank from earthenware cups furnished by the women.
Spiridon then informed me that Capiocan, who was his
protégé and an excellent pilot, would conduct us to the
next inhabited village. Vasilli here explained to the
starosti that we had nothing to eat ; so before we started
he handed me a boned goose, inside of which were stuffed
four other geese. Soon we had gathered our few effects
together and· were journeying on rejoicing ; our good
friend Vasilli, hood in hand, bowing and wishing us farewell from the beach. Capiocan and Karranie embarked
with us, and Feodor paddled along in his canoe, at times
in tow of the whale-boat.

At first there was no difference of opinion between the
natives in regard to the course we should pursue ; but
presently we came to a place where each one indicated a

different course, and as Capiocan was our authorized
pilot, I went in the direction pointed out by him, and
soon brought up in water too shallow to float the boat.
This was by no means a strange event in our experience,
for it seemed as though the natives would never understand that the whale-boat drew two feet of water more
than a canoe. We had run into this cut with a fair
wind, and were forced to struggle out of it against wind
and tide. The natives had assured me that we would
reach the inhabited village that night, but we were delayed so often in navigating the crooked and shoaly
stream that it became necessary to camp again, which we
did in a couple of old huts on the river bank.

I boiled our boned geese, and they were far too
"gamy" to titillate the palate of the most advanced
gourmand beyond the purlieus of the Arctic regions;
where putrid meat is in general demand, not so much,
perhaps, from choice as from necessity; for, though there
is a perennial abundance of ice, yet during the summer
months the sun is scorching hot, and, unless the natives
build ice-houses, the game they kill in July and August
will spoil as readily at the Lena mouth as at New York.
The store-houses and huts are built on the high banks of
the river to escape as much as possible the floods which
at times inundate the whole of the Delta, so that the customary Siberian ice-cellar is here impracticable. Then,
again, it is an undertaking of considerable magnitude for
these people to excavate a cellar with the tools they have
at command, which consist alone of a wooden spade
tipped with iron, — the tip of an inch and a quarter being split so as to fit on both sides, and held in place by
iron hooks, which are driven through the spade and
turned up on the back. The iron fitting is sold by the
traders, and the spade is soon fashioned out of a straight
piece of tough spruce. It is a tool in use by all the natives, forming a part of their winter kit to clear away

the snow from their fox-traps.   But the ground is solidly
frozen to an average depth of forty-seven feet in the district of Yakutsk; and when a cellar is to be dug a fire
is first built, thawing out a few inches of earth, which
is then removed, and the excavation so continues on
through a succession of fires until a proper depth has
been reached, when the sides are retained with small
round timber, which the next winter freezes in as firm
as a stone wall, and so gives the finishing touches to a
cellar of perpetual gelidity.

And this is a long digression, considering that I merely
wished to say that our ancient and odoriferous geese had
not been kept in an ice-cellar; but as it was a long time
since we had obliterated a good, honest meal, and as it
might yet be longer before another opportunity presented itself, we absorbed them and turned in. Next
morning it was astonishing how well every one felt after
the night's rest.   To be sure, those of us who were still
frozen remained so, and hobbled painfully along; but
when sitting in the boat we were all vigorous and strong
in mind, and, above the hips, in body.   The suffering in
feet and legs was borne without a murmur until the expiration of every second hour, when it became necessary
to shift the relief at the oars.   Then, indeed, the injured
ones anathematized with a vengeance, nor were the re-
torts always couched in conciliatory or loving terms.   Yet
on the whole, every one was considerate of the others'
comfort, and there was very little ill-feeling displayed beyond these momentary and pardonable outbursts of temper; and, if the crowded condition of the boat be remembered, — two men sitting on each thwart, and the limbs
of nearly all as sensitively painful as though scalded from
knees to toes, — it need cause no wonder that at every
movement of the boat a shriek of agony should burst from
some member of the company.   Towards noon we turned
a long sand-spit in the river, and came into view of a low

island, on which was pitched a village of probably a dozen *balogans*, *palatkas*, and store-houses, and one spireless church. Feodor, in haste to herald our coming, shot ahead, and under the guidance of our two pilots we ran quickly after, scanning the village with eager eyes and hearts. Soon we saw the smoke curling above the huts, and then the crew found voice, —

"I see a man!" "I see two men!" "Look at the dogs!" "Hurrah! There's a woman!" "No, *women!*" "Look at the young ones!" etc., until finally, as we drew near to the island and the water shoaled, two or three canoes put off from the shore, in one of which was a regular red-headed Russian. We all noticed him at once, and sang out together, "There's a Russian!" He was evidently pleased, and, bobbing his head, cried "Ruskie, Ruskie!" Then we plied him with a hundred questions in English, French, Spanish, German, Swedish, and every crooked tongue of which we had a smattering; even descending to Iniguin's dialect, whom I told to address the young man in such Russian as he could command; but it was a flat failure, since Iniguin doubtless addressed him in Asiniboine or Chinnook.

CHAPTER VIII.

AT JAMAVELOCH

Nicolai Chagra. — An Impressive Pantomime. — The " Red Fiend."
— Over-ripe but Green Geese. — Devotions. — A Balogan described.

THE *starosti* of the village, Nicolai Chagra, pointed
out where the deep water lay, and soon our boat was
moored to the shore. With clear heads but weak limbs
we all clambered out as best we could, chiefly on hands
and knees. The whole village, to be sure, had come to
greet us; men, women, children, dogs, and all. On the
beach was a number of boats, canoes, sleds, and their
equipment; hunting and fishing gear lay around; and
there were places for the drying and repairing of nets,
together with flats, on the tops of which some fish were
drying. When most of our gear had been discharged and
the boat secured, some of the women and children took
hold of the sled on which I was seated, watching the operation of unloading, and dragged me off a distance of seventy-five yards to the house of Nicolai Chagra. Leach
and Lauterbach, who were entirely disabled, followed me
on another sled, or hobbled after on sticks. Nicolai having ushered us in with considerable ceremony, we made
mutual attempts at conversation, and I endeavored to post
him on the state of our affairs. Meanwhile the rest of the
crew came marching into the house in a body, armed with
pots, kettles, and sleeping-bags, very much to the conster-
nation of Nicolai, who, flinging his arms around his wife,
whirled her about and into the corner of the room. He

had stationed me in the post of honor, or guest corner, under the icon, and, seeing his dread, I told the men to retire for a little while until I could explain to him who and what we were, and what we wanted. But presently they reassembled in the hut by degrees; the natives, too, crowded around us, and soon we were all on terms of good fellowship and in the best of humor. The boiler was immediately put on and tea made. It was salty, but we relished it, the natives particularly, to whom it was a luxury in the summer-time when traders were scarce, and all food other than goose and deer meat more so. Mrs. Nicolai Chagra had set a kettle of fish to cooking, and soon we had a panful of it, boiled, indeed, without salt or seasoning of any kind, and yet, to us, the most delicious mess we had ever eaten. While the fish was getting ready, our host had passed around a small quantity of tried deer-fat. There was altogether not more than a couple of ounces of it, and this he broke into little bits and served to us like sticks of molasses candy. Some members of the party, carried away by their imagination, pronounced it the sweetest morsel they had ever tasted. Had there been enough of it to satisfy one man, he would perhaps have deemed it a jolly good feast; but, hungry as I was, it seemed to me but a modicum of rancid deer tallow fried in a dirty pan filled with deer hair.

I ate a piece about the size of a copper cent, nor cared for more; but some of the men, I noticed, were anxious to have a second or third chance at the "pan o' fat." Throughout the entire expedition I never lost my taste for good things when they were available. On board of the Jeannette I ate mechanically — as a duty; ate to keep up my vigor; for, although ours was the best provisioned ship that ever crossed the Arctic Circle, yet so regular was the diet that many of us came at last to loathe the sight and smell of the canned food, which at the beginning of the cruise had been most palatable. It

palled on us as the partridge did on the man who had
engaged to eat one every day for a month, and I question
if one fourth of all the food issued was consumed.

While the dinner was preparing I undertook to tell
Nicolai Chagra the story of our shipwreck. Yapheme
Copaloff, the exiled Russian, was all mouth and ears,
and seemed more intelligent than the natives. With a
red and blue crayon pencil I pictured the American flag
on a slip of paper, whereupon Yapheme exclaimed,
"Hurrah Americanski!" and then explained that he
had been a soldier at the fortifications of Vladivostock,
and had seen many American ships. But to make the
Yakuts understand, I drew a vessel which Yapheme
called a *shipka,* and to the natives said, *bulchoi flotska*
(a large boat). Then having learned the Yakut word
for ice, namely, *boos,* I explained that it had crushed
the ship and she had sunk. Yapheme comprehended
quite readily, but the natives were not so bright, and
after a long confabulation between the two, I made use
of a large piece of wood, calling it the *shipka.* On this
I placed four smaller sticks, *viatkas* or *malinki flotskas*
(little boats), and thirty-three chips as the crew. I next
put the table in motion to illustrate the rolling sea, which
they called the *byral* (sea), and showed how the *boos
byral* (sea ice) had come in on the ship. Then with a
great agitation of the table I spilled the boats and men
from the ship, and threw the latter, along with a
malinki flotska, under the table, to represent how it
had gone under the ice. All appeared to be enlightened
by my pantomime, and the "ohs," "ahs," and sighs of
men and women, expressed their sorrow and pity. I
then told off eleven sticks as my crew, and put them on
board one of the three remaining boats; the other two
being assigned thirteen and nineteen sticks respectively.
Advancing all three for many days and sleeps, I next
blew a gale of wind, and amid its roar, or *poorga,* and

the surging of the *byral*, rolled over two of the boats and
dashed them under the table. But one *malinki flotska*
remained, and once more telling off myself and men as
the sticks, I finally moored our boat at Jamaveloch, the
name of the village.

The women seemed very much affected by the story,
and examined our frozen limbs, shaking their heads in
compassion, and even weeping over our miseries. After
dinner, Nicolai gave us each a leaf of tobacco, quite a
windfall to those who used it. I did not; but I heard
some of the men remark that it was the vilest stuff they
had ever smoked, not excepting the tea leaves and coffee
grounds they had used on the march. So now we continued the custom of drying our tea leaves for those who
desired to smoke them, much to the astonishment of the
natives, who cut up about equal quantities of wood or bark
with their tobacco. Our large pipes likewise occasioned
them surprise, theirs being very small, and similar in
shape to the Japanese tobacco-pipe, the bowl holding a
ball of tobacco about the size of a large green pea. The
smoke ended, we all lay down to a good sleep, the house
being darkened for that purpose by placing boards against
the inside of the ice-glazed windows. Some of us slept on
the berth-places, others stretched themselves on the floor
in or on their sleeping-bags, and were soon snoring peacefully. We, however, whose limbs were frozen so badly,
found no "balmy," for each beat of the heart forced the
blood in vigorous and painful circulation through our
throbbing flesh. At dusk we all either awoke, or were
awakened by the natives preparing our supper. The in-
variable beverage of tea was handed around, and Mrs.
Chagra, assisted by some of her female friends, put on a
large kettle of ancient but hardy geese, which had long
and honorably served the natives in raising numerous
progeny of their kind. But they had been slaughtered
during the summer, when in pin-feather, and hung in

8

pairs, with their bills interlocked, across a pole out of the
reach of dogs and foxes, and as they had neither been
plucked nor dressed, the juices of their poor bodies naturally gathered at the extremities ; hence, ere freezing
weather set in, the dead geese had generated another and
more prolific family within themselves. So when such
are heated for the purpose of cleaning, the natives are
usually saved the trouble of opening them, for the whole
after-part of the fowl drops out of its own accord, —
anything but a pleasing sight to contemplate, particularly if the agony, or inside, be long drawn out. Still we
ate of the boiled geese, and heartily.

Before retiring for the night Nicolai Chagra prepared
a number of small wax tapers and arranged them in front
of the icons. I speak in the plural, for he had a row
of them facing the south, placed on a little shelf in the
northwest corner of the house. They were brass images,
ranging in size from one to four and a half inches square ;
some being merely heads of particular saints, others,
groups of three or more figures, crucifixes, medals, plain
crosses, etc., such as the Greek priests sell or trade to the
Yakuts. The tapers lighted, their ends having been
melted and so stuck on the shelf, all the natives, old
and young, with the women in the background, fell into
line and went through their devotions, apparently with
some extras in honor of our safety. The service consisted in multiform genuflections, bowings and crossings,
with long pauses between, during which they fixed their
eyes on the ground as though in rapt meditation, and at
times they prostrated themselves, kissing the floor, and
touching their foreheads; but they uttered no audible
prayer.

When it was over, they all fell back as though to give
us a chance, and Chagra, bowing towards me, waved
with his hand an invitation to myself and people to follow them in the service. I thought he looked a little

disappointed at our not accepting, and so, to avoid giving offense, requested the crew to go through the motions. Jack Cole, whose honest good spirits were always redundant, vociferated at the top of his voice, much as though he were calling a watch to duty, —

"Come, fellows, all o' ye, and say yer prayers ! "

Whereupon, followed by many of the men, he took the lead in the performance of an entirely original ceremony. Nicolai then extinguished the tapers, and we all turned in ; some of us, as before, on berths arranged around the room, and the rest using the floor as a common bed with the natives, inclusive of our pilots and the Russian exile, Yapheme Copaloff, who had apparently attached himself, unasked, to my party, as guide, counselor, and friend. It was evident that he regarded himself as quite the superior of the natives, although dependent upon them, at times, for food, shelter, and clothing ; yet, as the white man, American or European, usually does, he had assumed a lofty air, and the natives were forced to bow to it.

A description of Chagra's hut, the best in the village of Jamaveloch, will be a fair example of the finest of that class of permanent dwellings, commonly known as *balogans* or *yaurtas*, on the Lena Delta, and throughout the districts of Yakutsk and Verkeransk.

The main or inhabited portion of the building is rectangular in shape, and built of hewn timber, the base dimensions being about twenty-four by sixteen feet. The timber is placed on end in the earth without sills, all four sides leaning inward about ten degrees out of the perpendicular ; or when the height of the hut inside is, say, eight feet in the clear, a plumb line dropped from the top would fall about two feet from the side of the hut near the floor. The timber is neatly hewn and squared down to seven inches in thickness, the width varying from seven to seventeen inches, and placed together as tight as possible, the joints appearing remarkably close indeed, when the rude implements of the natives are considered. These consist alone of a "pod-bit;" a very rank-angled chopping axe with a handle not more than twenty inches long; and a two-handled drawing-knife curved to a radius of about three inches, — the saw being a tool unknown to them.

Horizontal sills or plates are laid on top of the inclined sides of the hut, and on these in turn a girder, seven inches thick and twelve inches wide, is placed transversely of the structure, and midway between the front and rear. A stanchion resting on a block in the centre of the floor runs up through the hut and supports the girder, and for the matter of that, the roof too, which is formed of hewn timber such as the sides are built of. These rest on the front and rear plates, or stringers, abutting together on the central girder, and so giving a slight pitch to the roof; and the seams in the whole structure are caulked with reindeer moss. A low door, three feet high by two feet wide, is cut in one end, and the sill raised about two feet above the ground, to keep out the cold air. On each side are cut two windows eighteen inches square, and occasionally one window of similar size is cut in the rear. The chimney and fire-place are located in a direct line midway between the centres of the roof-ridge and door, facing the rear. The fire-place is made of a box sixteen or twenty inches high and four feet square, back of which is the chimney, woven of wattles and small round timber, its projection being supported by two knee-pieces like those used in boat building. These also answer as supports for a small mantel-shelf and the upper end of a rude wooden crane, by means of which the natives swing their large kettles over the fire. The entire chimney and fire-place are coated with a fine sedimentary clay, dug from the pond bottoms, and gradually baked until thoroughly hardened. The box, which is filled

with earth, may either be pinned together, or, as is often
the case, the four corners may be held in place by eight
stout stakes driven well into the ground.

All well-built huts are floored with slabs, split with
wooden wedges. The interior arrangements are as follows : A low "transom" or bench, raised about eighteen
inches above the floor, runs around three sides of the
room, but not around that in which the door is cut. It
is about two and a half feet wide, used as a bench during the day and converted into sleeping berths at night ;
for it is divided off by partitions, usually three or four
feet high, though now and then reaching to the ceiling.
At the end opposite the door are two berths, and along
either side there are three, making eight berths in all,
and when they are double or intended for the accommodation of two sleepers, a small ledge-piece, which swings
out on hinges made of thongs and is supported by several
small sticks, can be readily adjusted at night and let
down during the day. The location of the huts is not
determined by any rule of compass or sun, but they are
here and there situated with their backs to the prevailing
wind ; although this caution is by no means generally
observed, and I noticed no regularity in the facing of huts
in any village, those set across the wind appearing as
stable as the others. The disposition of berths, however,
I found the same in all parts of the Delta and northeastern Siberia. As viewed from the door, the rear right
hand corner is invariably occupied by the host and his
wife ; the one opposite in the left hand corner is always
set apart as the guest chamber, and over it is the shelf
of gods or icons. The three berths ranging along the
right hand side of the hut are devoted to the use of
the immediate family, the married sons and their wives
being next or near to their parents according to age or
other condition. On the left, the near relatives start in
at the berth below the guest chamber, commencing with

the senior aunt or uncle, and terminating at the door
with the stranger or dependent. In all huts there are
little spaces of four or five feet between the last berths
on either side and the front end. On the right this
space is where the pots, kettles, and other culinary utensils are kept; on the left a small supply of dry wood is
stored for kindling and bad weather. Extending from
side to side of the hut, and suspended by thongs in
front of the fire-place, is a light trestle-work, made of
poles and slabs, on which the frozen food is placed to
thaw, and likewise the fish intended for the dogs. For
the latter is a practice always observed when possible;
the dogs, in cold or bad weather, being furnished with
hot food. Narrow shelves fitted over the berths to support small ornaments; a ditty box for the reception of
valuables, such as needles and thread; a tea-cup or other
fancy article, and several small, rudely constructed tables,
compose the furniture of the hut. Feather pillows are
quite common among the natives, and their bed-clothing
is of skin, the mattress being made of two, three, or as
many reindeer skins as the prosperity of the house will
permit. In almost every inhabited hut I saw one or
more old person, male or female, who occupied a corner
near the door; a sort of "granny," generally blind, always miserable, poor, ragged, and dirty — living on the
few scraps of food to be found in the refuse of the household. I never could learn whether this personage was
a parent of my host or hostess, who would only observe
that he or she was one of the old and poor. At any rate
these aged pensioners of either sex are kept constantly
employed, whether blind or not, manufacturing and repairing the horse-hair nets. Blindness, I may say, is a
disease peculiar to the people of this region. Dr. Capello, surgeon-general of the district under the command
of General Tschernaieff, informed me that forty per cent.
of all the natives north of Yakutsk are totally blind, and

sixty per cent. are partially so, or have lost one eye, and I cannot remember having visited any hut north of Yakutsk wherein one or more of the occupants was not afflicted with some affection of the eye. Syphilis prevails to an awful extent among them; and by their mode of washing, which consists in filling their mouths with water, spurting it into the hollow of their hands, and then rubbing their faces, they manage to rub into their eyes the virus from their diseased mouths. The glare of the snow, too, their filthy habits, and the smoky atmosphere of the huts, all give rise to or aggravate this terrible affliction.

When the wooden structure is finished, a row of upright timber, three or four feet high, is planted around, and about two feet, from it. The intermediate space is filled in with earth during summer and trampled down with the feet; and finally a layer of soil and *tundra* sods, a foot thick, is raised up to and on top of the hut, being well packed and tramped. The *balogan* is now complete; in shape, a frustrum of a pyramid of rectangular base, and, in external appearance, an earth mound or butte, from which, indeed, it can only be distinguished at a great distance by the presence of smoke. An outer apartment is next added to the front of the main building, generally of the same width, but not so high, and only about half as long or as substantially built. At right angles to it is a still smaller and frailer structure, fully strong enough, however, to withstand the gales, and support the snow, which during the winter months assail and envelop it. A peculiarity of this attachment is, that it is invariably located on the right hand side of the outer building as one enters it.

These three apartments are permanent and constitute the habitation proper, but, as winter approaches, a light, temporary building of poles is erected, and covered with snow. It is taken down when spring opens, and stowed

away for the next season. I have seen another similar
structure intended as a winter lodge for the dogs, or
bitches with pups, and in which they were fed from a
log trough. These two are only built when the ground
is covered with snow, which is heaped upon them with
wooden shovels, and, later on, the natural snow-fall fills
in every crevice. The first outer apartment is used as
a general store-room. In it all the fur clothing, fishing
gear, dog harness and sledge equipment are kept; and
here the visitor deposits his outer garments. It also acts
as a weather porch to the main dwelling, and, at times,
the fish or salted fish-bellies, which are traded to the
Russians, are stored therein. The smaller attachment to
the right, entered through a swinging door, is used as a
provision store-room, in which the winter food supply of
venison, fish, or geese, and all furs intended for trade, are
kept. No light is admitted into these outer buildings,
unless, perhaps, it be through a sky-light of crystal-like
ice in the roof, the object being to keep them always
cool. The contrary is, of course, the case in the main or
living apartment, where the windows are glazed with
translucent ice, abundantly collected in the fall and reserved for winter use. Considerable light shines through,
though, to be sure, outside objects are not visible; but
the heat on the inside during the day gradually destroys
the windows, which are scraped clean every morning
with a little iron instrument made for the purpose. For
at night, when the fire is permitted to die out, a coating of
rime collects on the inner surface of the ice-pane, produced by the exhalations of the sleepers,—a board always
being placed on top of the chimney to prevent the escape
of heat; and I have seen forty persons *sleep* in a hut
whose dimensions were sixteen by twenty-four feet, and
seven feet in height. The glazing operation is readily
accomplished. When the fresh-water ponds have frozen
to a depth of six inches, blocks of ice are cut out and car-

ried to the house-tops, out of the way of the dogs. So
that, now a window requires glazing, the old melted pane
is knocked out from the inside, and a fresh cake of ice,
chopped to the correct size, instantly inserted; the chinks
being "puttied" in, as it were, on the outside with wet
snow, taken from a ready kettle or boiler, and, as this
freezes immediately, a "pane of ice" eighteen inches
square and six inches thick has thus been set in a few
minutes. Before going to bed, boards, fitted for the purpose, are stood in the window places to protect the ice
against the heat of the hut; and it is interesting to watch
the gradual destruction of the panes by the currents of
warm air, governed by the location of the fire-place, the
depth of window recess, and the partition of berths.
Slowly they melt outward, until, finally, the blessed day
appears, and with it the necessity for more ice.

Such is a general description of the residence of Nicolai Chagia, and, with slight modifications, of the hut in
which myself and party spent the thirty days following
our sudden return to Jamaveloch.

And now, to revert. We awoke the next morning
(September 28th) greatly refreshed by our night's rest,
and, after an enlivening application of cold water to our
heads and faces, breakfasted on boiled fish and the ever-
present tea. The day was stormy, but I made Nicolai
understand that we must proceed at once to Belun. He
expostulated vehemently, saying that the *poorga, car,*
and *boos* (wind, snow, and ice) would surely cause us to
perish. Hobbling outside, I took a look at the weather.
It was blowing briskly, and the heavy, driving clouds
portended a gale of snow. So there was nothing to be
done but await a lull, and it came sooner than I expected.
At ten o'clock the sun shone through the dark masses of
vapor, the wind sank to a gentle breeze, and presently I
had hustled Nicolai Chagra and our two pilots into their
canoes, — Yapheme, "The Red Fiend," accompanying us

in the whale-boat. Nicolai had furnished me with sixty
fish as provisions, besides stowing a small piece of venison
in his own canoe, but as he told me it would require fifteen sleeps to make the journey to Belun, I replied that
the supply of fish was not sufficient. He pointed to the
nets in the three canoes, signifying that fish would be
caught as we progressed, and I was satisfied, remembering old Vasilli's faithful catches. Before starting I
lightened the whale-boat as much as possible; presenting to Nicolai one of the tents and poles; an empty alcohol keg (having poured that precious fluid into the
India-rubber bottles that we had originally filled with
water and lime-juice) ; our large axe, and some other articles. At last, when the natives had kissed each other
good-by after an unusual amount of "hoodooing" before
their fetiches, and our cripples were limping to the river
and into the boat, Leach, whose feet were shockingly
frozen, begged to be left behind, saying that he would
rather stay at Jamaveloch than risk the chances ahead.
He and Lauterbach had lost all spirit; but it was at
Leach's rapid and sudden change from cheerfulness to
despondency that I was more astonished. Of course, I
would not listen to his entreaties ; so, very reluctantly,
he embarked with the rest.

CHAPTER IX.

SIBERIAN LIFE.

Balked again. — The "Balogan Americanski." — A Row with the
Starost. — Catching Fish — Deer and Geese Hunting.

FOLLOWED by the hearty well-wishes of the villagers
and the tearful remonstrances of Mrs. Nicolai Chagra, we
set out upon the river under oars or sail, as best we
could, with the canoes leading the way. Shortly after
starting we came up with young ice running in compact
streams; and the wind increased until the boat was almost unmanageable. Our pilots had changed their course
in order to round a point of land, and so gave us a dead
pull to windward, which, in our lamed condition, was no
child's play. The boat was loaded down to its utmost
capacity, and grounded repeatedly on the shoals, making
several narrow escapes from capsizing, while the seas
rolled over us. Presently the natives became frightened
at the surrounding ice, and seeing how fast we were bailing and the utter impossibility of our gaining any headway, motioned me to turn back, they having already
cleared the shoals into deep water and started on their
journey home. However anxious I was to reach Belun,
it was now imperative that I be cautious in my actions
and not risk the lives of those intrusted to my care, for,
should the river freeze us in between Jamaveloch and
Belun, the probabilities were that a majority would perish from cold and starvation, since but two or three in
the party could walk, and even their powers of endurance were very much reduced. Besides this, and the al-

most insurmountable difficulties of navigation, how could
I proceed forward without the natives ?

So, back we turned, and in less than an hour were
again at the village. The inhabitants were out to greet
us, and, when the lame had scrambled up the steep bank
and the boat had been discharged, the natives urged me
to haul it out on land. This I, at first, refused to do,
still hoping to make the passage to Belun in the boat,
and fearing that by careless handling they might further
damage it, for it was already in a sorry condition, leaky
and rickety. Still, gathering from their vigorous talk
and pantomime that they feared the wind and ice would
carry it out to sea, I finally consented, and the sequel
proved the soundness of their advice. So I witnessed
the operation, sitting with Leach and Lauterbach, because of our infirmity, on dog-sleds, and when it was
over we were all conducted to the hut of one Gabrillo
Passhin, a deer-hunter, pending the preparation of our
new quarters,— a vacant hut which was put in order for
our reception, and into which we shortly moved. And
it was while storing our boat gear and equipment of all
kinds in the outer apartment that, very much to my chagrin, I found that the bag containing the sixty fish for
our journey had been captured by Nicolai Chagra.

And now for the present we were dependent for food
upon the generosity of the natives. Our hut, though
airy and the chimney smoky, was still, withal, in a fair
condition ; and I portioned off the space as equally as
possible, making such regulations for the government of
the men as seemed to me indispensable to their health
and comfort. Of the seven double berths in the hut, five
were each occupied by two men ; Danenhower and myself sleeping singly in the remaining pair, for by the addition of the " Red Fiend " my party now numbered
twelve. I divided the men into two daily reliefs for carrying wood and water, or ice, of which there was an

abundance on the island or along its shores. It was out
of the question for Leach to perform any duty, and I
also excused Messrs. Danenhower and Newcomb, since
there was little or nothing to be done beyond the taking
of judicious exercise. The office of cook was filled by
the men best able to walk about and carry wood; Charley Tong Sing, the steward, doing duty the first week,
and Manson, Wilson, and others succeeding him. I assembled the crew and reminded them of the circumstances under which we were placed. How after a long
march we were now nearly naked, and entirely subject
to the bounty of the natives for what little food we might
obtain, and, as we would doubtless be obliged to stay
there for some time to come, the wisdom and necessity
of conducting ourselves in a peaceable manner must be
patent to all. Then, too, after our hardships, there was
danger of scurvy or other sickness breaking out among
us, and the only way to contend against this was to live
as we had before, in good fellowship, keeping as bright
and cheerful, and as dry and warm, as possible, constantly exercising without fatiguing ourselves; and when
the river had frozen over an effort would instantly be
made to communicate with Belun.

Nicolai Chagra daily furnished us with four fish,
weighing in all about sixteen pounds, and of these we
made a *long* soup, that is, a soup economically lengthened
out with water. I still adhered to my old plan of equally
distributing the contents of the kettle in pans, and so in-
sured a fair division of our food; although it was amusing at times to see two persons seize the same pan and
struggle over it, until, through feebleness or complaisance,
either surrendered; or to watch those who, their hunger
overcoming their manliness, would, with a watery mouth
and rapt eye, gaze on the pans in process of filling, edging their way the while towards the one they accounted
the largest, and at the word "Go!" grab it triumphantly.

Yapheme took his soup and fish along with the rest of us, and we led a comparatively happy life in our " Balogan Americanski " (American House). Beyond the petty tiffs growing out of arguments, in which, as is generally true, there was more talk than logic, little or no quarreling occurred among the crew. And their discussions usually ended in a loud guffaw at some happy hit made by a party to the argument.

Yapheme taught us the Russian and Yakut languages, and acted as our interpreter. The men constructed checker-boards and chessmen and repaired their clothing. The first night in our new hut I drew up a letter to the officer commanding the district, outlining the circumstances which had brought us to Jamaveloch and requesting him to forward a copy of my letter to the American minister at St. Petersburg. Copies were written in French, German, and Swedish; and a few old letters or envelopes belonging to some members of the party being added as proof of our identity, all were done up in a package, and then securely sewed within an oilskin bag, cut from an old piece of clothing. Mr. Danenhower and myself walked over to the starosti's hut, and impressed him with the importance of dispatching the package at once to the Commandant at Belun. He understood, and promised to send it as soon as possible; and, in order to urge him to action, the package was sewed up in his presence by his wife. He then told us that it would be ten or fifteen days before the bay could be crossed in safety. A light fall of snow had occurred during the night, and the river, as far as the eye could reach, was covered with ice, save in mid-channel and a few spots. I now knew why the natives were so anxious for us to haul the whale-boat out on the bank above the river. I had supposed they dreaded a storm; but it was for fear the boat would be frozen in, and then a storm coming on would break up the ice and carry it out to sea.

So now there was nothing else to be done. Mrs. Chagra gave us some fish to eat, and we trudged back to our hut to await the solid freezing of the bay.

At this time our situation seemed very uncertain to me. We had not yet become fully acquainted with the temper of the natives, about whom very little was written or known. On board ship there had been a record of a Russian officer, who, accompanied by his wife and thirty or more Cossacks, had attempted to winter at the mouth of the Lena; but who, although bountifully supplied with food, and in full communication with the natives, had, with his whole party, perished of scurvy. What then was the outlook for us, who had already accomplished a wonderful retreat, and who, utterly worn out, lame and half-famished, were now living on decayed geese and a very limited supply of fish? We certainly could not exist through the winter at Jamaveloch; for, should scurvy, perchance, favor us with its absence, I felt confident that either typhoid fever or poisoning, as a result of the food we were eating, would break out among us.

We possessed but very few articles to barter away to the natives, and they, indeed, could spare as few as we. Our clothing was worn out, and we repaired it by sewing patch on patch. Soaking our swollen limbs in warm water we soon found to be a pleasant, temporary relief from pain. The frost-bite and sores healed rapidly, the swelling subsided, and to my great delight we all gained in strength and happiness — all save Leach, from whose toes the flesh had wasted away, exposing the bones. Gangrene had apparently set in, and if they were neglected for a day the odor was unbearable. Bartlett was his constant attendant; daily preparing a kettle of hot water in which he bathed and cleaned the sores, and, with a jack-knife in hand, pared away the flesh in a masterly manner. But at this time Leach seemed ill all over.

He had lost heart, and said he did not care whether school kept in or not. Although a brilliant fire was continually burning, and he sat so close to it that his jacket blazed, he yet complained of the cold, and when told that he was on fire said that it made no difference to him, he would get another jacket. Indeed, to all appearances it was quite an effort for him to tolerate the kind attentions of "Dr." Bartlett.

Thus our daily routine ran on. The geese, which comprised one of our two meals per day, had been killed during the summer while nesting, and as a consequence were inordinately poor and proportionately tough. Yet if this had been all, our food had been agreeable; but, as I have mentioned elsewhere, the geese, which tide the natives over the hard times they experience between the going of the deer and the coming of the fish season, though dead, are often found to be alive again, and anything but pleasant to the taste. It was Bartlett's daily duty to obtain our supply of four geese and four fish from Nicolai Chagra, and we had been living on good terms with our dusky neighbors, until, one morning going as usual on his honest errand, Bartlett was surprised to have Chagra hand him three instead of four fish for our breakfast. Of course, he remonstrated, when Chagra, after considerable talk and gesticulation threw, in great anger, a partially rotten goose at Bartlett; who, thereupon, in good American style, sprang at the repentant *starosti* and chased him through the village. We now received information that fish, and, indeed, food of any kind, was very scarce in the village, and there was danger of our supplies being entirely cut off. Yapheme cautioned us against bartering away with too liberal a hand our small store of articles; but what I most feared was that the natives, being somewhat wandering in their habits, might, unknown to us, fold their tents like the Arabs, and as silently steal away in the night, and so

leave us in the lurch. For after all, whether by bartering or not, we must depend upon them for our food, and
when none was left, they doubtless would travel from
place to place in search of some; eventually, perhaps,
quartering themselves on their more fortunate neighbors,
the men, women, and children lightening the burdens of
their friends by assisting them to fish and make nets.

Meanwhile the "Red Fiend" continued to instruct us
in the mysteries of the Russian and Yakut tongues, with
which we all became more or less familiar. Iniguin, our
North American Indian from Norton Sound, was quite a
curiosity to the villagers, and at once sprang in great
favor, when it was made known that he, too, had been
nomadic like themselves. They were Tunguses, they
said, and we informed them that Iniguin was an American
Tunguse; and soon he was visiting around among his
copper-colored brethren and sisters, who began straightway to make and repair his moccasins and clothing;
until, finally, it was noised around that Iniguin had
found a sweetheart in the village, which he blushingly
acknowledged, and, in praising her good qualities, said,
"Him plenty good little old woman."

In daily visiting the edge of the bay to test the
strength of the ice; in spinning yarns; in speculating
on the fate of the first and second cutters, and the length
of our detention at Jamaveloch — thus we passed the
tardy hours. Some of the men would go down to the
shore and watch the natives hauling in their nets; and,
as the fish became more plentiful, would augment our
scanty supply — no matter how.

And, while on the subject of fishing, I will enlighten,
in a very few words, those of poor De Long's critics who,
I have noticed, wonder why "the fool did not catch fish,
in which the rivers abound?" The fact is, that he did
try to catch fish with the only means he had at hand —
hook and lines, which I recovered and brought back to

9

the United States. But fish are by no means procurable
in the Lena River at all seasons of the year, and when
they are (for a couple of brief months) the natives take
them entirely by nets, not knowing, indeed, what a hook
or line is. It would have been manifestly out of the
question for us to have burdened ourselves on the march
with a net or nets, and hence it should be plain to the
reflecting reader why De Long and his party did not sub-
sist on fish.

It is not until October, when the ice covers the bay,
that the fishing season sets fully in. Then the finny
tribes run as elsewhere, ascending the rivers to spawn
and descending later on. The nets are made entirely of
white horse-hair, white manes and tails forming a laige
proportion of the stock of the few traders, — the natives
preferring the white since it is not so readily seen by the
fish. Their net-making is one of the most tedious pro-
cesses imaginable ; and at it the women, old and young,
and the blind stranger or other pensioner who sits wearily behind the chimney-place, are almost constantly employed. The first operation consists in pulling out five,
seven, or nine hairs, and "fairing" the ends at the roots,
which the blind accomplish by pressing them against the
tongue and so tying the knots. After which, during a
second handling, they are twisted into a strand, and then
knotted by short pieces into the net, a small stick serving
for the size of the mesh, but no needle being used to pass
the strand, as the custom is with our fishermen. I noticed that the tongue and teeth play a prominent part in
the knotting or weaving of their nets ; which, when finished have from a two-and-a-half to a three-and-a-half
inch mesh; are from a fathom to a fathom and a half
deep; and are from fifteen to twenty fathoms long. The
top lines are kept afloat by means of light wooden buoys;
and the bottoms are held down by a series of weighted
hoops, six inches in diameter, made of wood, split in two,

after the manner of cane-seating for chairs ; and to which, when twisted into shape, and neatly sewed with flexible rushes, stones are fastened. The idea is to keep the nets free of the mud while the floats preserve them in a vertical position.

When a net is to be set after strong ice has formed, a row of holes is cut across the proper channel, and a long pole is pushed under the ice from hole to hole, carrying with it a horse-hair rope a little longer than the net, the length of which is consequently less, by a couple of fathoms, than the distance apart of the extreme holes. Then by hauling through the rope, and paying out the net, until it is all under the ice, the ends being fastened to stakes, the net is set and ready to capture the fish in its meshes. The middle and intermediate holes are now permitted to freeze over; but it becomes a necessary and most arduous duty to keep open those at either end, in order to haul out the net, which is done every morning and night, or, if the fish are running in large quantities, as often as every four hours, no matter how low the temperature or how high the wind. To break and keep open the holes, an ice-pick made of iron, fitted on the end of a short, stout pole, is used, — the broken ice being cast out and thrown up into hills by means of an oval-shaped wooden sieve attached to a pole.

The natives here, and, indeed, all along the coast of Siberia, live upon the game peculiar to each season. In the spring-time they lie in their canoes, ambushed under the high river bank, and await the coming of the reindeer, which have favorite crossing-places on their annual migrations to the north. The herd marches across the *tundra* until the water edge is reached, when the leader strikes boldly out for the opposite shore. They wade and swim unmolested until the whole herd is well out in the stream, and then the hunters dash forth in numbers from under cover of the bank, each armed with a long

spear or lance, which rests in the crotch of a forked deer-
horn placed in the bow of the canoe, in order to keep
the spear in readiness and protect it, as well as the oc-
cupant, from harm. As the hunters dash whooping and
yelling into the midst of the herd, the deer are panic-
stricken, and, losing the guidance of their leader, strike
out in all directions. Although excellent swimmers, the
poor animals, which can fly like the wind over the smooth
heath or *tundra*, are now at a disadvantage ; for the natives are in their most congenial element, and, nimbly
plying their paddles, dart and flash about from one victim to another, working quick and sad havoc among the
stately drove with the deadly thrusts of their lances.
The action is continued while a living deer is in the
water, and, when there is none left, the floating carcasses
are towed to the shore, where the women and children,
if at hand, assist in cleaning and preparing the meat.
Meanwhile those of the herd which escaped injury have
scampered away in safety until the next crossing is
reached ; while the fugitive wounded are followed up by
the young hunters on the opposite shore, or sometimes
are tracked by the dogs.

In autumn, when the herds are wending their way to
the south, the slaughter is repeated, and thus are two
seasons of the hunting year filled in, during which the
natives are comparatively well fed ; while through the
summer and winter they rely for food upon the fish and
geese. These latter are sometimes killed with bow and
arrow. Another means of securing them consists in running a line of horse-hair nooses across a point of land or
convenient place frequented by the geese. These nooses
are fastened to short, flexible rods after the manner of
fishing-poles, which are then stuck into the ground, and
the snares are arranged so close together that it is impossible for the game to thread its way through the line un-
caught. The geese settle on the point of land to feed ;

whereupon the native boy or woman approaches them, and they gradually retreat full into the real danger, and the nooses tighten around their necks, until the whole flock has been driven through the line of poles or frightened away by the fluttering wings of the captives, which the natives soon dispatch with heavy sticks. During the nesting season the eggs are also gathered in large quantities and buried in the earth until winter; their state of incubation, howsoever far advanced, mattering but little to the accommodating taste of the native, who, in fact, makes use of all kinds of eggs and finds no fault with the fresh ones. And though when eating them raw the mere presence of a young bird in the shell does not seem to perturb him, yet I have noticed that everywhere he is particular, when frying his eggs, to pick out the yellow feathers from the pan. Yapheme supplied us at different times with these eggs, which we fried in the orthodox American style — *sans* feathers. There was, at first, some little discussion in the hut as to the propriety of using the over-ripe eggs, but I finally concluded to cook them all together; and thus the identity of the poor little geese was lost in the " scramble."

CHAPTER X.

KUSMA TO OUR RESCUE.*

Incidents in our *Balogan.* — Kusma. — Faithless Spiridon.

MANY were our projects, at one time or another, to make forced marches from Jamaveloch to Belun. Aye; but we had neither food, clothing, sleds, nor guide, and the distance was two hundred and eighty versts — across the bay, over a mountain range, and along the ice-gorged, but still broken, Lena River. It pleased me to sit and listen to the numerous and diverse plans proposed for our relief by one and all; and I thought to glean a word of wisdom, a ray of hope. But there were too many "ifs" in all the schemes. "If we had " — "if we only had " —

However, with the exception of a little difficulty which grew out of the trading proclivities of one of my party, the days passed pleasantly enough with us in the village. We had repaired our clothing and health, our limbs were rapidly healing, and we were now on good visiting terms with our neighbors. In the difficulty mentioned I did not think our Yankee peddler nearly so shrewd as the Russian *copert;* and he desired to retreat after the bargain had been partially executed. The matter was referred to me, and, sleeping upon it, I decided that a bargain was a bargain, even though one of my people be the loser, and notwithstanding he alleged that in addition to the copert's advantage in the trade, he was also indebted to my countryman for medical services in having cured him of a bad cold by means of homeopathic doses from a private store of sugar-coated pellets. Saving this

and a well substantiated charge preferred against one of
the men of having stolen and eaten from our scant store
of venison, there were no disputes, no ill-feeling among
us — the familiar sailor growls counting for naught, of
course, since they ended, as usual, in smoke. But I
must note an exception in the case of Wilson, who,
acting as cook for the time being, did once hang up the
geese over his bunk in order that they might thaw out
during the night and so be ready in the morning to pluck
and clean for breakfast. They thawed, it is true, but
too freely, for the juices and intestines dropped out and
fell upon Leach, who was slumbering alongside of Wilson.
And yet even the commotion occasioned by this incident
was pleasantly smoothed over by some one saying that
Leach should not growl, since he was getting more than
his share of goose, and if he didn't like or want it, he
had only to put it back again.

One afternoon, while we were still waiting for the bay
to freeze, and wondering what the day might bring forth,
our "Red Fiend" entered the hut in a flurry and very
ceremoniously introduced his friend Kusma Germayeff,
a Russian *soldat*. He was a bright, intelligent looking
man, and I at once hoped far more from him than from
any one we had yet met; and so told him who we were,
as Yapheme had indeed done before. I complained to him
that the natives were dealing deceitfully with us; that
Nicolai Chagra was feeding us on putrid geese, which, I
feared, would sicken us unto death; that we were soldiers
of America, and that General Tschernaieff, Governor of
Yakutsk, would surely punish any of his people who
would permit us to suffer for want of anything in their
possession that would contribute to our health and comfort. And, finally, I said that if he, Kusma, would
journey to Belun with my letters, and bring me back
food, clothing, and reindeer teams, I would give him the
whale-boat and five hundred roubles; provided, however,
he start at once.

That was impossible, he said. The bay was still open in places, and although he had succeeded in picking his way across it from the main-land, it had been a very dangerous feat. He was a small trader, and had undertaken the risk in order to barter with the natives; and though he had but few provisions at his hut, he would send us all he could spare; and thereupon handing me what salt there was left in his salt-box, which he carried with him, he assured me that in four days, on Thursday (*Chick-verk*), he would come again. I could purchase a reindeer for food, he thought, from a friend in his village, who would accept my promise to pay; and as Mr. Danen-hower suggested that some one should accompany Kusma and secure it, as well as any other procurable provisions, I consented to his going, with orders to return as speedily as possible.

It was night when they started, and Danenhower came back the next morning, bringing with him some leaf tobacco; some sugar; five pounds of salt; about five pounds of rye-meal, and the dressed carcass of a young deer, weighing about ninety pounds. The venison was a great luxury to us, but greater than it by far was the salt, a taste of which had not been afforded us for weeks. Indeed, we had saved in all but four pounds of salt from the ship, and this amount quickly disappeared among thirty-three men, though we only used it in bear, seal, or walrus stews. When it had entirely given out, we found a substitute on the retreat in the shape of salt water; but if this be added to a stew too liberally, or in the early stage of the cooking, it renders the mess bitter and unfit to eat, owing to the presence of the bitter and purgative element in sea-water, which, in the manufacture of salt, is drawn off before the sodium chloride is deposited. Salt water, of course, cannot be procured at Jamaveloch, which is pitched along the fresh-water estuary between the mouth of the Lena and the outlying islands in the

Arctic Sea. Hence, at the Delta salt is worth a rouble per pound, and is used as sparingly by the natives as is cayenne pepper in our households; a thimbleful of salt amply supplying a family of ten persons for a day — if, indeed, they be so fortunate as to get any in a month.

With these fresh provisions and an increased supply of fish, our prospects began to brighten, and Kusma's expected coming to make final arrangements for his journey to Belun was now the topic for our hourly and almost constant comment. The question arose whether it would not be advisable to send one of the party along with Kusma to facilitate matters and stir up the Russian official, who, as a class, is the most notorious of procrastinators. Bartlett, in whom I placed full confidence in all things, asked permission to go, and I was inclined to grant his request, but as Mr. Danenhower remonstrated, saying it would be derogatory to him, I held my decision in reserve. I was in a quandary. Captain De Long had ordered me not to permit Mr. Danenhower to do any duty, and, although I was now independent of De Long, still I did not feel called upon to disobey his orders, for it was probable that upon my arrival at Belun I might there find him or Chipp; albeit the general opinion was that both had been lost in the gale.

At length the expected day arrived, and with it betimes came Kusma, true to his agreement. Again I went over the points in our contract, and urged him to make haste. He assured me that he could perform the journey to and from Belun in five days. Then, when I asked if he could take a courier from my party, his prompt reply dissipated my perplexity. No, he could not. Why? He had but seven dogs and would have to secure more, since one man with his equipment and dog-food would weigh four hundred pounds, and, consequently, the addition of a courier with outfit would increase the load to eight hundred pounds. Then, too, if he must go quickly, he could not be burdened with a man requiring care and attention; he must have, if any, some one who would be an aid to him, not a charge. If alone, he could go and return in five days, but otherwise not. And this settled the question.

I had prepared certain dispatches to the United States Minister at St. Petersburg, and to the Secretary of the Navy; but as I expected so soon to see or hear from the Russian authorities, I did not think it wise to send them; for since I had written the originals our situation had materially improved, and I had no desire to alarm the world with news until I was altogether sure of it; hence I refrained from telegraphing information of De Long's fate until I had first viewed his dead body. Yet, as the sequel will show, my delay in sending the dispatches made no difference in the final results, beyond postponing the arrival of the news in the United States of my landing at the Lena Delta.

On the 14th of October Kusma left Jamaveloch, promising to return in five days. Would he keep his word? I had learned by this time that lying is not considered a sin either by the natives or the Russian peasant; on the contrary, if cleverly done, it is rather regarded in the light of an accomplishment; and during the whole of my stay in Siberia I found it practiced everywhere, as well in the most trivial as in the most trying circumstances. Immediately after Kusma had departed for "Tamoose," his place of abode, I was suddenly reminded by Danenhower that I had forgotten to instruct him to spread the news, as he traveled along, of the loss of the other two parties, and to offer in my name a reward of one thousand roubles to any person who would bring me information of their whereabouts. To make good this neglect, I sent over Danenhower, upon his own request, to Tamoose. He returned the next day, and informed me that Nicolai Chagra, our *starosti,* would accompany Kusma to Belun. This, at first, was a very surprising and unwelcome piece of intelligence to me, for I could not understand it to be otherwise than some arrangement between the two calculated to defeat or interfere with my plans ; and it was not until some time afterward that I learned that Kusma, being a criminal exile, was prohibited under a penalty from visiting Belun unless accompanied by the staiosti. And they started together, I heard, on the 16th, leaving us to count with impatient anxiety the days of their absence.

On his visits to Tamoose, Mr. Danenhower had been told by Kusma and the natives that Cape Barkin was only forty versts to the northeast of us, — a wretched untruth, the distance being about one hundred and ten versts, in a bee-line. However, Mr. Danenhower was anxious to proceed there on a search, and I reluctantly gave him my permission, but cautioned him not to cross running streams or broken ice, or to jeopardize himself in any way so as to delay my party beyond the return of Kusma, four days thence. He had previously gone to Tamoose and back in company with Spiridon, the villainous-looking Tunguse whom we had met at Arii, the deserted village, and of whom we had all formed an adverse opinion. But now, to my great astonishment, Mr. Danenhower asserted that he had found him to be a most excellent dog-driver, and that he "could do with him just as he pleased."

So, everything being satisfactorily arranged, Danenhower and Spiridon started off for Tamoose to secure the few pounds of tea and tobacco which Spiridon exacted for the hire of himself and dogs. And then this wily guide, having his pay in advance, carried our astounded shipmate to his hut at Arii, where they supped and passed the night; and when Mr. Danenhower demanded of him that they begin the journey to Cape Barkin at once, the stonily stolid but astute native was nothing if

not a sphinx. Threats and cajolery were alike unable
to move him. The faithful one who would go anywhere,
or do anything at slightest beck or bidding, now that he
was paid, even refused to carry back his master to Jamaveloch and the " Balogan Americanski." But sure
enough, in due course of time, in came " Dan." with the
woful visage of the Knight of De la Mancha, and as he
gradually unfolded to our anxious ears the mournful tale
of Spiridon's duplicity, the loss of tea and tobacco, and
the consequent and ignoble defeat of the " first organized search " for our lost companions, — first a smile stole
round the hut, then a titter, and finally a loud guffaw,
when the " faithful one " was denounced as an " infamous pirate."

Still the good and honest Vasilli Kool Gar was ready
with his dog-team to retrieve the day for our doleful and
would-be hero. To be sure, it was necessary to obtain
another fee of tea and tobacco for old Vasilli ; so off they
went, sleeping that night at the hut of Kusma, and starting the next morning for Tarrahue. Mr. Danenhower's
report of the trip to me was that they ran along, much
to his perplexity, to the southeast instead of the north
or northeast, in which direction we all knew Cape Barkin to be. They journeyed so about forty versts, and
came in sight of a large island, which they were prevented from visiting by the insecurity of the ice ; then,
sleeping in a hut over night, they made several attempts
the next day to cross the black and treacherous ice to
the island, and, failing in which, returned instanter to
Jamaveloch and the bosoms of their friends.

Mrs. Kusma called twice upon us, bringing presents of
tea and tobacco. The latter she handed around by the
leaf, giving an equal number to all ; and then she took
tea with us, and a share of our fish. Later on she sent
us some rye-cakes, fried like flap-jacks in fish-oil ; but
meal being a very scarce article among the natives, she

could afford us but two or three apiece, weighing about
two ounces each. The few pounds of meal given us by
Kusma I had stored away, and only used in small quantities as thickening for our fish-soup. And speaking of
Kusma, we were now becoming impatient of his prolonged absence. Five days, the time for his return, had
come and gone, and still no Kusma. Among ourselves
we now discussed the situation over and over again,
gravely considering the possibilities of making the march
to Belun; for the Balogan Americanski was very much
of a Liberty Hall, where perfect freedom of speech prevailed, and I only interfered to check unhealthy famil-
iarity or prevent the progress of quarrels. We all indulged, to a greater or less extent, in song; some of the
men played games; and Bartlett once roasted a piece of
venison before the fire as a tidbit. It was toothsome,
but by no means as economical as our customary soups,
the hot liquor of which we sorely missed.

CHAPTER XI.

A STEP FORWARD.

Kusma's Coming. — Nindemann and Noros. — I start for Belun. — Siberian Dog-Sleds. — A Storm. — Ku Mark Surt.

MANY and long were our anxious looks from the hut-top, but all in vain, for a sight of Kusma. The natives now came oftener to see us, and, at times, brought fish or hauled us a load of wood with their dog-sleds. We were all in a fair condition save Leach, whose great toe had become black and was rapidly sloughing away, notwithstanding the constant care of "Dr." Bartlett, who did all to save it that mortal could with a surgical outfit composed of hot water, a jack-knife, and some ointment which Danenhower had carried in a tin box for the relief of his eyes.

Mrs. Kusma came over one day to tell us that an officer had died at Belun, which was probably the cause of her husband's detention. But from her manner it was plain she lied, although I subsequently learned that a petty official had indeed died, sometime near the expiration of the five days. Cold weather had now set rigorously in, and the driving winds and snow-squalls pierced through our ragged clothing, chilling us to the heart. But inaction was worse than death by the roadside; and I almost yielded to the tempting arguments of the men, some of whom, with Bartlett at their head, volunteered to haul Leach on a sled along with the provisions, if I would only give the order to start. And yet when I glanced at my half-naked party, still suffering from the frost wounds of a few weeks before, and listened to the howling pitiless blasts without, it seemed the height of desperation and of folly to venture forth upon such an undertaking. We now had plenty of fish, and with a proper guide might have made the march; and so, goaded on by my own harassing impatience at Kusma's intolerable delay, and our enforced idleness while perhaps our shipmates were dying for want of our assistance, I at length announced that we would load the native sleds with fish, haul Leach on another, and with a native guide proceed to Belun. But where was the guide? And when I made the announcement, I was at once opposed by Mr. Danenhower, who said that it would be madness to attempt the march, that half of my party would perish, and that he doubted if any one in our condition could outlive the journey. Bartlett, who never lost his wits, and seemed ready for anything, urged the trial; but looking around on the miserable objects about me, at the scant and tattered clothing, and the crippled feet and legs, I finally and resolutely determined the risk to be too great and too profitless. For, why incur such danger and court again our intense sufferings? Was not our messenger, Kusma, expected hourly? Had not the five days allotted for his trip elapsed and been almost doubled? So my proposal to march the party two hundred and eighty versts, and play a game of "mock heroics," luckily fell through and ended in talk.

On the afternoon of the 29th of October, the thirteenth day after Kusma's departure, a couple of sleds were seen approaching across the bay. Of course there was instant stop to our converse, and every man of us ran forth to greet our eagerly expected courier. Never was absent lover welcomed more joyfully than Kusma and his dusky companion, Nicolai Chagra. When the usual salutations and the unloading of the sleds were ended, and the tea-kettle had been put on for our friends' refreshment, Kusma was plied with questions as to the cause of his delay. In his endeavor to explain how the ice in the Lena River had broken up and run out, he interlarded his disjointed story with a vague account of his having met two deer-sleds in charge of some natives, who had with them two Americans almost dead from cold and exposure, and who, in turn, had spoken of the death of many of their comrades. All of this, and more too, Kusma related in a confused jargon of Russian, Yakut, and Tunguse, when, suddenly recalling his scattered wits, he reached inside of his clothing and drew forth two letters and a folded scrap of paper, which he handed to me, explaining that the first letter was from the Cossack commandant, and the other from the *Malinki Pope*, or young priest, of Belun. But the prize paper was the dirty, crumpled scrap, which, as I unfolded and deciphered it, opened our eyes in astonishment:—

"Arctic steamer Jeannette lost on the 11th June; landed on Siberia 25th September or thereabouts; want assistance to go for the Captain and Doctor and (9) other men.

"WILLIAM F. C. NINDEMANN,

"LOUIS P. NOROS,

"*Seamen U. S. N.*

"*Reply in haste: want food and clothing.*"

Questioning Kusma again, I learned that Nindemann and Noros were *en route* for Belun; that they had been found in a hut called "Bulcour," at the first bend of the river to the westward, twenty miles to the southward of "Tit Arii" (wood island); that they were very sick, having suffered greatly from hunger and cold; and that he, Kusma, had understood that many of their comrades had perished. But consulting the note, I saw that only one man was missing, since it read, "The Captain and Doctor and (9) other men;" the nine being emphasized by parenthesis. So the immediate query that arose in

DE LONG AND COMPANIONS WADING ASHORE

our minds was, Who could the unfortunate one be? No one guessed that it was Ericksen, a North Sea fisherman, and one of the finest men in the ship's company. A royal Dane, sure enough, who had worked himself out, and frozen his feet during the gale by his too constant application at steering the boat, when it was unsafe to shift the men at the tiller or steering oar.

While we were yet guessing who the missing man referred to in the note could be, I decided that the proper thing to do was to see Nindemann at once and learn the whereabouts of De Long and his party; so I told Kusma that he must instantly load the sleds again with the small supply of food, and carry me back to Belun. He protested that it was impossible. The dogs were lame; they had been running along for several days, had worn out their feet, and could not start on another journey until sufficiently fed and rested. But I would not brook a delay like this, and insisted upon his going or sending immediately to Arii, ten versts to the northward, for a fresh team of dogs, so that we might depart that same night or the following morning. A messenger was accordingly dispatched at once to Arii, and we renewed our cross-examination of Kusma.

He explained how he and the *starosti* had crossed over the mountain range to the east bank of the Lena, and found the river all broken up and the ice running out in huge masses. It so tossed, rolled, and jammed up into mountainous bergs that the stream was impassable, and as their road then lay along the river or on its icy bed, it seemed as though they would have to turn back for want of provisions. Still they held on at a *povarnia* (cook-house) until the ice again made over the Lena, when they managed to creep along the edge of the mountains which rise precipitously above the river.

As may readily be supposed, it is a peculiarity of these northern rivers that their waters are mainly de-

rived from the melting snows in the months of June and
July; when the Lena, for example, overflowing its banks,
spreads here and there to a width of sixty miles or more.
As the season advances, the waters decline in volume,
and during the month of August the river rapidly sub-
sides, until by the first of September it is flowing at
low ebb.   This is the period at which all melting ceases
and the young ice begins to form, and as the process goes
on, the water which bore the ice on its bosom falls away
for want of further supply from the south, and the great
sheet of ice, with nothing to uphold or sustain it, tumbles
in, and is carried away by the swift, unchanging current, for there is no tide.   Then the ice grinds, swirls,
and piles upward, while the river rises in its might and
drives the ice before it like so much brushwood rolled
before the wind; and these freezings and floods continue
until late in autumn, when the river, dwindling to little
or nothing, quietly seeks beneath the ice its muddy way
to the sea.   And such was the cause of Kusma's delay
in reaching Belun.

But upon his arrival there the Cossack commandant,
Bieshoff, would only permit him to tarry and rest one
night, hurrying him back to Jamaveloch, with a small
amount of provisions and the letter from himself and
the *Malinki Pope*, and also the verbal message that he,
Bieshoff, would be at Jamaveloch, *postle zoftria* (day
after to-morrow), bringing food, clothing, and reindeer
enough to convey the whole of my party to Belun.

Some uncharitable person has circulated the report
that Kusma's delay was due to his having become intoxi-
cated while *en route ;* but I am glad to say that such was
not the belief of Bieshoff, or the finding of an official in-
quiry instituted at my request by Epatchieff, the *espravnick* of the district.   It was then shown that there was
not a drop of *vodki* or spirits to be had between Kusma's
house and Belun; that Bieshoff permitted Kusma to re-

main but one night in Belun; and that, since he followed
him the day after by reindeer teams, and Kusma arrived
at Jamaveloch a day ahead, the report was consequently
a base slander, and had its origin in the meaner qualities
of man's nature, which, in some people, are supreme and
ungovernable — people whom we see belie, belittle, traduce, and abuse those of their fellows who earnestly try
to do their duty, to be a little better, or to do a little
better, than the soulless curs around them.

Poor Kusma *did* go and return as quickly as it was
possible at that season of the year; and remember, please,
there is no beaten track or road in these regions. The
face of the country changes its appearance every season,
and only those accustomed to traveling it can find their
way without compass, directed by the mountain peaks
and furrows of snow thrown up by the prevailing winds.
Kusma and all the natives in the vicinity interested in
our care or guidance were rigidly examined in my presence by Epatchieff, in obedience to an order from General Tschernaieff, and but one verdict could be reached,
which was, that they all did everything that lay in their
power for our health, comfort, and safety; that Kusma
carried the messages from Jamaveloch to Belun as early
and as speedily as possible (he, indeed, being the first
person to cross the country and river that season, at great
personal risk and sacrifice); and that his devotion and
suffering were certainly deserving of something better
than suspicion and slander. However, it is gratifying to
know that the most severe, as well as senseless, strictures
were evolved by critics 10,000 miles or more distant from
the scene of action, by persons who would doubtless think
it a terrible hardship if they were obliged to breakfast
before ten A. M.

I was unable to leave Jamaveloch on the night of Kusma's return, so everything was made ready for my departure on the morrow. It was October 30th, and be-

times in the morning, bright and early, came old Vasilli Kool Gar with a fine team of dogs. His sleigh was
old for so long a journey, but a new one was to be procured on our way. Before going, I gave Mr. Danenhower verbal orders, which afterwards, when paper became more plentiful, I put into writing, wherein I instructed him to immediately set out upon the arrival of
Bieshoff with the deer-sleds and clothing for Belun, and
there await my arrival. It was my intention, I informed
him, to intercept Bieshoff on the way and turn him back,
in order to have him accompany me on my search for the
missing party of the first cutter; but failing in this, I
would hurry on to Belun for the purpose of learning from
Nindemann the particulars and whereabouts of De Long
and party, and hence Bieshoff's arrival at Jamaveloch
would announce the fact that we had failed to meet.

I took with me the remains of what clothing I had
saved from the retreat, consisting of the shreds of an undershirt and pair of drawers which had done duty since
June; a pair of thin cassimere trousers which I had not
only used for months after leaving the ship, but had also
worn in China during my cruise previous to joining the
Jeannette, and the legs of which were now lopped off below the knees to furnish material for patching and quilt-
ing that portion of a man's main garment soonest inclined to decay; footless stockings, seal-skin moccasins, a
blue flannel shirt which I had worn for a year, and my old
seal-skin coat, shrunk, shriveled, full of holes, and devoid
of lining. These with a fur cap and a pair of canvas
mittens completed my costume; but I carried my faithful
old sleeping-bag with me, and hauled it up over my feet
and knees to keep them from freezing; and then with a
small supply of perhaps five pounds of bread, some tea, a
pound of pemmican which I had stowed away for just such
an emergency, and a lot of frozen fish, we at last started
on our journey to Belun, with the thermometer ranging
anywhere from ten to twenty degrees below zero (Fahr.).

It was but a short distance across the bay to Kusma's dwelling at Tamoose, where we were to procure the new sled that would stand the rough usage of travel across the mountains and over the broken ice of the river. Arrived at Tamoose, we at once busied ourselves in getting the sled in order, when lo! to my surprise, I learned that it would have to be built; that is, new runners and stanchions would have to be put under the sled we had in view. There was no use fretting. Our own conveyance was worn out and worthless; so the new one must be built, and at once, and I was at least pleased to watch it grow into shape under my eyes, which it did so smartly that before evening we were altogether prepared for our journey.

This was the close of October 30th, 1881. A memorable day, for about one hundred miles distant from Tamoose it sealed the sad fate of De Long and his comrades; and five months later, when I found their bodies, turning to the last written page of De Long's note-book, or "ice-journal," as it is now known to history, I read the last pitiful entry, evidently written in the morning, —

"*Oct. 30th, Sunday.* — One hundred and fortieth day. Boyd and Görtz died during the night. Mr. Collins dying."

So the close of the day that saw me finish and pack my sled at Tamoose doubtless closed the eyes and earthly career of the commander and remainder of as gallant a band of men as ever struggled against fate, or its cruel emissaries, ice, snow, hunger, and cold. The next morning, October 31st, was very cold, and a brisk breeze blew from the eastward, driving the snow in clouds and obscuring the faint glare of the sun, which had already settled behind the southern mountain range, not to show its face until the following spring. Old Vasilli, ready with his team of dogs, supplemented by recruits from Tamoose,

in company with some of the villagers and the occupants
of Kusma's hut, first paid his religious devotions before
the icon over the guest's couch, with all the elaborate
ceremony of the navigators of old when starting on a
long and perilous voyage. Bowing down even unto the
earth, on which he rested his forehead and which he
finally kissed, he arose, stood upright, and exclaimed
"Pi dome !" (Go on, or, We will go.)

The dogs had been hitched in harness for some time,
and were now restless and eager to be off. There were
eleven in our team, comprising a variety in size and color,
some being party-colored, though the red fox dog (that is,
a kind of dog much resembling the red fox in color and
shape) was in the majority. The rest were mongrels of
every hue and build, the largest weighing about forty-
five, and the lightest about twenty-five pounds ; and this
motley team had been making the icy air resound with
its discordant solos and chorus.

I seated myself sideways on the sled with my feet
trailing on the ground or snow, allowing room in front
for Vasilli. Composing himself, he seized the great iron-
shod staff with which he guides the sled and dogs, and
when in ill-temper, beats them too, and grasping the
bows of the sled gave it a gentle sway, shouting the while
to the team. Away we went, with the dogs in full cry,
all yelping, snapping, biting, and seizing each other from
behind, those in front turning round to fight back, until
some were drawn off their feet and dragged along at a
fearful rate ; Vasilli, yelling at the top of his voice, coaxed,
scolded, and anathematized by turn, until at length, by
dint of twisting and rolling over, the team became entan-
gled into one living mass of vicious flesh. To pacify and
disentangle the crazy canines, Vasilli leaped upon them
with his iron-pointed guiding staff, and the only astonishing thing to me was how the poor brutes could live under
such a heavy basting. It is true some of them, after re-

ceiving a severe blow on the small of the back, did drag their hind-legs for a few minutes, but in the end it did not seem to check their desire to bite and fight. Yet they were considerably more tractable after this, their first beating, and ran along at a more even pace, following the leaders, who in turn were guided and governed by Vasilli's word of command: "Tuck, tuck! Taduck, taduck! Stoi, stoi!" (right, right; left, left; stop, stop;) and a general chuckle of encouragement.

Directly the dogs had outlived their excitement and settled strictly to their work, they looked beautifully picturesque, with heads down and manes and tails up and wagging, while only an occasional yelp burst from their ranks as they scudded along the ravines and over rivers, taking the top of the hard snow at about six miles an hour. Approaching steep banks, the dogs are sometimes turned loose, and the sled lowered by hand; but when not too steep the whole force dashes down the descent, and, if great care is not exercised by the driver with his staff, sled and riders are rolled over on the ice or snow, and not always without serious injury. Such an accident occurred to us the first day out from Tamoose, in which my left arm above the elbow was so injured as to render it powerless for hours, and even at this late date the swelling remains.

The sleds of Siberia are from twelve to fourteen feet long, about twenty inches wide, and raised about ten inches above the runners, which are five or six inches wide, single-ended, and made of birch, when procurable. The uprights, of which there are usually five to each runner, are made long enough to extend as high above the deck or flooring of the sled as below it, in order to receive a rail; while joining the tops not only adds strength to the frail-looking frame, but also forms a guard for the load. The uprights are fashioned with conical ends which fit into corresponding holes in the

runner, and midway in the height of the upright, at a
swell given for strength, is a conical hole, with the larger
circle on the inside. Into these holes are fitted the
cross-bearer pieces, on which rests the deck or floor, generally made of one or two thin slabs of wood, smoothed
down, after splitting, with the blade of an axe used as a
jack-plane. The uprights rake aft a few degrees out of
the vertical, and are lashed down into the runners by
thongs which run up through and are counter-sunk or
let in beneath the runners, and pass through holes bored
near the bottom of the uprights.

The whole affair is lashed together, but left as elastic
as a willow basket, none of its parts being tightly wedged
or pinned, since, if made rigid, the rough travel for which
it is intended would soon break it to pieces ; whereas, if
any of the lashings should break or wear out, a ready resort can be had to the harness or long trace to which the
dogs are hitched. A large birch bow, one and a half
inches in diameter, bent in one sweep of nearly a circle,
binds the two runners together in front at the same
time that it wards off projecting pieces of ice, and to it
the lanyard or trace is also attached.

I said that our team consisted of eleven dogs of various breeds. They were hitched in pairs along the centre
trace at equal intervals of about four feet, with a leader
in advance. At a convenient distance from the bow commence the toggles, by means of which the dogs are fastened to the trace. The dog harness of Siberia is of the
kind known in this country as " Dutch harness," with
breast strap, etc.; and I do not consider it near so sensible or comfortable in arrangement for the dogs as that
in use by the natives of Norton Sound or St. Michael's.
The former runs up and chokes the dog around the neck
when not carefully adjusted, — something which the traveler in cold weather cannot always find time to do;
whereas the latter harness rests upon the back of the

dog's neck, and when he hauls settles down, bringing the load on his shoulders. It is made in the form of a figure eight, the head passing through one of the loops, and the other being long enough to pass under the fore-legs and up on the back, where a short trace takes hold and attaches it to the toggle. These toggles make it a comparatively easy task to clear a dog-team when tangled up in a fight.

I noticed that a peculiarity of the trained Siberian dog is that after being turned loose for any purpose, he will at once resume his place in harness again when called ; although a strange dog may sometimes require a little coaxing, which the natives do by playfully throwing up their mittens to attract his attention, and so protect him from the fury of the "old stagers." After a run of an hour or less the dogs are usually brought to a stop and permitted to rest; whereupon they roll around and rub the rime out of their eyes and ears, and from their heads, and then, stretching out, lick their paws, which soon become very sore from travel. A team can seldom endure more than ten days' continuous work, for, no matter how well fed, the feet wear out and bleed, and the dogs are shortly so enfeebled as to be almost useless. A native will not willingly drive his team two days in succession, the custom being to travel one day and rest the next.

Not so with me, however, for I insisted upon pushing forward as rapidly as possible, and when night fell we pulled up at a *povarnia*, about sixty versts from Tamoose. Here was congregated a mixed crowd of natives and small traders who were proceeding on their respective routes to secure the cream of the fall trade, all huddled together, men, women, and children, in the hut, which was about twelve feet square and four and a half feet high, with a fire-place in the centre of the floor, over which a dozen pots and kettles were cooking for the different parties. Making room for the new-comers, they all concentrated their gaze and conversation on us, while Vasilli placed my sleeping-bag in a corner of the *povarnia*, and put on his tea and cooking kettles, the latter containing a portion of a reindeer head. Then having staked and fed the dogs, and eaten our supper, we turned in for the night. Thirty persons in a hut not more than twelve feet square! After the day's journey I felt none the worse, save for the hurt I had received from the overturning of the sled; though the pains in my legs and feet, which had not yet entirely recovered from their former freezing, were terrible, and fresh blisters had formed on my heels and shins, and the toe-nails had turned black and begun to curl up like burnt feathers. Yet in a little while we were all asleep, now and then disturbed by the howling of the dogs without or the biting of the vermin within.

Day broke with a fierce wind, which, drifting the snow in clouds, caused the poor dogs to bay out their misery with all the strength left in their weak quivering bodies. For, on a journey, they are never housed at night, either when halting at a *povarnia* or elsewhere, but a stake is driven in the snow at the bow of the sled to retain it in place, the main trace is hauled well taut and made fast to another stake, and the driver's great iron-shod staff is finally hitched into the centre of the trace and elevated so as to barely permit the middle dogs to lie down. This is done to prevent fighting and consequent entanglement, and while thus confined they are fed, each dog voraciously devouring his fish or fraction thereof. The young and vigorous members of the team, with good, sharp teeth, quickly absorb their rations, and then endeavor to seize the whole or a part of their aged neighbors', which, menacing the enemy with sundry snaps and growls, manage after occasional frenzied bites to bolt their frozen meal, provided that the aged one is not assaulted from the

rear, either from "pure cussedness" by a co-laborer just
robbed, or by a youthful and vicious marauder urged to
the attack by his knowledge that the old and defenseless
one has lost his teeth and is fain to "gum" his food.
Often in an affair of this kind the attacking party fails
after all to secure the fish, for, while owner and enemy
are making war for its possession, a sly cur takes, lawyer-
like, a quiet hand in the affray and carries off the prize.

Our dogs could only be seen as trembling masses of
snow, and Vasilli did not wish to start until the weather
cleared or the wind calmed. Few if any of our fellow-
travelers cared to face the storm, and only those who
had a following wind gathered their traps together and
ventured forth later in the day, when the gale had slightly
abated. Vasilli then intimated his willingness to resume the journey, protesting, however, his fears for my
safety in such weather. I was still limping and miserable, and the cold wind and snow sifted through my tattered garments, setting me to shake and shiver, while the
natives, observing my condition, shook their heads and
muttered "Morose" (cold and hungry). But I urged
Vasilli on, and so with their blessings and crossings, and
a few presents of dried fish from the sympathetic natives,
began my second day's journey toward Belun.

The weather cleared as we proceeded, but the cold became more intense, severely cramping my frozen feet and
legs. During the day Vasilli halted the team about
every half hour, and while the dogs rested I thrashed
about in an endeavor to coax the blood to circulate to
my extremities; for it was out of the question for me to
run alongside of the sled, since the dogs, with lightened
load, quickened their pace beyond the power of one so
crippled as myself to keep up with them. Day darkened
into night, and we still staggered on over the bed of the
river, having left the mountains. It was here that I had
hoped to intercept Bieshoff, the Cossack commandant,

with his deer-teams ; and it was my intention to either turn him back to Belun, or start with him immediately to the north, if he knew the location of De Long and party from the information given him by Nindemann and Noros, or by the natives who had found them starving in the hut at Bulcour. The ice on the bed of the river was exceedingly rough, and thrown up in great heaps and ridges like windrows, forcing us to take a very devious course in picking our precarious way along the ice-shelves by the banks of the river, which we crossed and recrossed many times. I thought I should perish of cold before we reached Ku Mark Surt ; for I could do nothing but sit on the sled, beating my limbs to keep warm, while the cold blasts tortured and froze me. As it grew darker it grew colder, and kind old Vasilli kept encouraging me, saying, " Malinki, malinki, balogan " (a little, little way to the house); and as he chattered on and scolded the dogs, he would occasionally place his hands on me, as if to assure himself that I was there and alive, and then, with a cheerful word and laugh, would seem perfectly content.

Long after midnight, while we briefly halted to afford the team a breathing spell, Vasilli pointed ahead with his staff, saying, "Ku Mark Surt," and, at the same time, stretching forth his arms with the hands and fingers drooping and trembling, after the manner of tree branches, repeated over and over again, " Lis, lis, masta " (leaves, leaves, wood). Finally I caught the outlines of low, dwarf pines fringing the banks of the river, and at once understood that we had come to the place where trees were growing, or, in other words, had reached the limit of timber growth in that region. A pleasing sight to me, indeed, for it was the first standing timber I had seen for more than two years, and no matter how mean and stunted, I felt as though I had met in it an old friend.

Soon we detected the baying of dogs in a distant village, and our team, listening for a moment, answered the cry with interest, and then dashed forward with renewed vigor. In a little while we could see sparks rising from huts on the steep west bank of the river, and, shortly after, the villagers, comprising three or four families, aroused by the noise of the dogs, seized our team and assisted us up the bank and into a new, cozy, and warm hut, where a prosperous family, composed of a widow with three sons (one being paralyzed in the legs), two daughters, and an old aunt, and a blind stranger, lived in true Yakut luxury. They had a good hut, plenty of fish, fresh and smoked, some tea, and a very little salt. Vasilli told them our story all over again, and of course the neighbors were present to see and marvel at another of the queer beings who had apparently risen out of the frozen sea, — the *boos byral*, — the thought of which seems to fill them with terror, for I found them all ready and willing to perform any duty except venturing upon the sea.

CHAPTER XII.

AT BELUN.

I am Admired. — Deer-Sleds. — Buruloch. — Native Gossips. — Meeting with Nindemann and Noros. — Their Piteous Story. — The *Malinki Pope.* — An Unexpected Visit from Bartlett. — Back to Buruloch.

THE villagers, who had either seen or heard of Nindemann and Noros, proceeded to tell me all about them, how they looked, and what they and their companions had suffered. Then while all busied themselves in the preparation of a supper of hot tea and boiled fish, I was regaled with some raw, frozen fish-bellies, which melted into oil when I placed them in my mouth; and the natives were very much surprised to see that I preferred the other parts of the fish, cut thin and free of the oily fat; since with them the height of happiness and good living consists in a feast of fat things, and as raw, fat fish-bellies are the fattest things in Northern Siberia, they ate what I rejected, and doubtless wondered at my poor taste. Our meal also included some smoke-cured fish, and after eating heartily we all turned in and slept soundly until broad daylight.

When I awoke, a small wooden vessel, made in the shape of a butcher's tray, was placed on the ground before me, and a member of the household stood ready with a ladle of water, which he poured into my hands while I washed, the wooden trough catching the drippings. When breakfast, which differed from our evening meal only in the time of eating, was prepared, the whole household examined and admired my tin drinking-pot; and, as I had removed my outer clothing, and stood in my red flannel under-clothes, now faded, patched and torn, they all overhauled me, old and young, male and female, — even to the poor old blind man in the corner, who though unable to see my gaudy raiment was yet led across the hut that he might feel the texture and criticise the cut of the stranger's clothes. Every one was delighted with my sleeping-bag, and indeed it was quite an improvement upon the gear in use by the natives. They sleep on deer-skins, each sleeper being furnished with a long narrow covering of light cotton calico, quilted with white fox and rabbit-skins, the bottom of which is turned under so as to form a short bag or pocket, and into this he thrusts his feet and legs to the knees, tucking the sides of the loose quilt under him. When a couple sleep together, the bag is of course wider. The only difference in the native's mode of sleeping in or out of doors is that when housed, he strips to nudity, whereas when retiring in a snow-drift, he usually retains a portion, at least, of his clothing.

After breakfast, I was told by Vasilli that for want of dog-food he could go no further, but that the *starosti* of the village would conduct me by deer-sled to Buruloch, the next station; and in a little while I was again under way, having taken an affectionate farewell of my good friend Vasilli, who seemed as proud and careful of his charge as he might be, in his old age, of a baby. My sleeping-bag and small store of provisions were placed on a little sled about six feet long and twenty inches wide, with three uprights, and rails, bow and runners arranged as in the dog-sled, compared with which, however, it was a poor affair. My driver had a separate sled, and to each two fine young deer were hitched, fastened loosely at the head by a halter. A strap of rawhide, one and a half inches wide, passed across the shoul-

der and neck and under a fore-leg of one deer, the centre
running around the bow of the sled, and then back again
to the other deer, so as to bring an equal strain on each
animal. A long rein led outside from the forehead of
the right hand deer, and was held in the left hand of the
driver, who, perched on the front of his sled, goaded the
teams onward by means of a slightly tapering pole, ten
or twelve feet long, and about one inch in diameter, the
free end furnished with a button made of wood or deer-
horn, with which he punched the haunches of the deer;
my team being hitched behind the other sled.

Away we went over the smooth banks of the river;
along the land when practicable, at other times on the
ice-shelf, and occasionally my driver would lead the deer
by the halter over and around the broken, heaped up ice-
bed. At such times it was impossible to keep on the
sleds, since they were continually overturning, while the
driver sought out a path, none as yet having been made
on the river. Yet, whenever an opportunity presented
itself, we indulged in fast driving, though such distances
were very short and disagreeable, too; for when the deer
were driven at the top of their speed, they seemed to
labor painfully along with heads thrust forward, tongues
hanging out, nostrils distended, sides working at every
leap like a great pair of blacksmith's bellows, and the
noise of their breathing like the exhaust of a locomotive. Maintaining their fearful exertions for about half
an hour, they suddenly swerve to the right or left among
the trees, or up a steep bank, to avoid their tormentor,
or, dropping down in their tracks, bury their heads with
open mouths deep in the snow, and eat voraciously of the
cooling dust.

Before night we had arrived at Buruloch, a deer station on the east side of the Lena River, and eighty versts
from Belun. Nothing worthy of note had transpired
during the day, except the novelty of reindeer riding,

and I had learned the reason why dogs and deer cannot travel by the same road, which simply is that the dogs are so fierce they will attack and kill the latter. So, just before reaching Buruloch, when we sighted a dog-team approaching us in the distance, my driver turned our team up a bank, and conducted it into the woods, back of the road, stationing me with a huge stick to prevent the team from following him. But the dogs had seen the deer, and came howling on in hot pursuit, their driver doing his utmost to check them. Luckily the team was a very small one of only seven dogs, and, as they dashed into the path that we had taken, I struck the leading dog a blow across the head and back, which, ruffling his temper, caused him to turn round and attack his neighbor, and in an instant the whole team was embroiled in a " free and easy." Leaving the driver to restore peace, I rejoined my team five hundred yards or so to the rear, and shortly afterwards we drove into Buruloch.

Fire was leaping from the chimneys, and from the peculiar location and appearance of the huts, they forcibly reminded me of the cabins of the charcoal-burners. So soon as it was known that a stranger had arrived, all the inhabitants crowded to the hut of the *starosti*, crying out, " Tell me, tell me ; " that is, " Tell me the news." Two Yakuts meeting, and while yet a considerable distance apart, promptly start to hail each other with, " Tell me, tell me." In this manner, without knowledge of our civilized scandal-monger, the penny daily, do they pass the news along from one to another, and it is amusing to watch the sled parties encounter. Driving a little distance past each other, they leisurely stake their dogs or deer, meet half-way, come to a full stop, take off their hoods and mittens, gaze solemnly at each other for a moment, and then fall to kissing on cheek, forehead, or lips, according to age or kinship ; the men going through their

11

salutations before the women. They then replace their hoods and mittens, and, sitting down on the snow, draw forth their pipes, tobacco pouches, and flint, steel, and tinder bags, for a smoke, using a small quantity of tobacco about the size of a little green pea, which, indeed, is an equal mixture of tobacco and bark or powdered wood. The small pipe-bowl is either of brass or pewter, and lashed to a wooden stem-piece, taper, and made in two parts for convenience; since, having no means to bore a tube, they first fashion the stem on the outside, and, splitting it in two, cut a groove down the centre of each half and then lash the two together,—it being thus left easy to cleanse the stem. And when two old cronies meet on their journeyings, or when the native feels especially well-disposed towards his spouse, he unwraps his pipe-stem, separates the parts, and then the two sit down for a social chat, cleansing the stem of nicotine by licking it with their tongues, after which the pieces are put together and fastened for a smoke.

My reindeer driver repeated our story as he had heard it from Vasilli, and with a great show of surprise and sympathy I was, at length, fed and put to bed. There was present a Yakut *copert*, who very much fancied my rifle, and wished to empty a good portion of his pack for it, provided it had not been a breech-loader. He did not approve of the style, and thought the bore too large, but that the breech might be closed and a flint lock fitted on. I finally turned in to get rid of my trading tormentor, and was out early in the morning ready for the road.

Madam was present, but the proprietor of the deer-teams could not be found. The poor woman was anxious to see me off, for my driver from the last station had informed her that I must travel forward without delay, and she was alarmed at my impatient and persistent demands to *pi dome*, coupled with some forcible, if not polite, expressions in rude Anglo-Saxon. Presently a driver with sleds and deer appeared on the scene, and we were soon dashing towards Belun. I left my hostess suckling a hulking big boy about five years old, who had thrust his head beneath his mother's deer-skin shirt for another pull at the life-giving element. This is a custom among the Yakut women, to suckle their children until one pushes the other away from the breast, and very often two or three of different ages derive nourishment from the same source.

It was almost dark when we approached the village of Belun. My driver rested his team a short distance from the village, and then with a grand flourish dashed into it and up to the *starosti's*, or public *povarnia*, a series of low huts attached to the public building, which was of more pretentious dimensions. As we drew up, the word was passed that a stranger had arrived, and immediately the people flocked around us. My driver told some of the loungers who I was, and where I wanted to go, whereupon several of them bustled ahead of me, and opened an outer door, but refrained from touching an inner one which opened into the apartment where were my two comrades, Nindemann and Noros. Pausing an instant, I pushed open the door, which was covered on one side with deer-skin, and on the other with woolen felting to keep out the cold. I was clothed as I had been when last seen by my comrades, save for the addition of a light deer-skin shirt over my old jacket; and I remained silently standing for a brief spell in the apartment to see if Noros would recognize me. He stood up facing me, behind a rude table, not more than ten feet off, holding in one hand a loaf of black bread, which he was in the act of cutting with a sheath-knife when I entered. Nindemann was nowhere to be seen. A dim light straggled through an ice-glazed window in the rear of Noros, and to the left, around a fire in a small alcove, a number of Yakuts were cooking their supper. At my entrance Noros glanced up from his bread, but did not know me, and was about to resume operations on the loaf, when —

" Halloa, Noros ! " said I. " How do you do ? " at the same time advancing towards him with outstretched hand.

" My God ! Mr. Melville," he exclaimed, "are you alive ? " And then, Nindemann, hearing my voice, arose from a roughly made bed and cried out : —

" We thought you were all dead, and that we were the only two left alive ; we were sure the ' whale-boat's ' were all dead, and the ' second cutter's,' too."

As soon as I could control my feelings, I told them that we of the whale-boat were all alive and well, and had mourned our comrades of the first and second cutters as lost ; that I had been trying to reach Belun for the previous thirty days ; that it was my messenger, Kusma, whom they had met and who had carried their pencil message to me ; and that I had hurried forward immediately upon its receipt, in order to obtain information of the whereabouts of De Long and party. At this we all broke down, Nindemann and Noros declaring that it was useless to search for their companions, who had died long ago ; that they had parted from them twenty-five days before, and for several days previous to their separation they had had absolutely nothing to eat, having subsisted on their skin clothing, and the alcohol, sweet-oil, and glycerine from the medical stores ; each man receiving but a couple of ounces of alcohol per day, with a teaspoonful of oil or glycerine while either lasted ; and that finally, at parting, De Long had divided the alcohol equally around, and started them on a forced march along the west bank of the river to a settlement, distant, he supposed, about twenty-five miles.

They told me of their great and many sufferings ; how they had eaten their dog, and managed to crawl along a

1   Nindemann and Noros starting ahead in search of help.   2. The meeting betwe
Melville and Nindemann and Noros.   3 Deer teams.

few yards at a time; how Ericksen had died and been
buried in the river; and how De Long, seeing that his
party could not struggle on in a body, had selected the
best two men for travel, Nindemann and Iversen, but as
Iversen had complained the day before of frost-bitten
feet, Noros was detailed instead. Nindemann was instructed to journey forward as rapidly as possible, keeping the west bank of the river, but if he found succor of
any kind, to return to the party, who meanwhile would
follow in his footsteps. He left them camped on the
bank of a small river running northwest from one of the
main branches, and followed the west bank to a great
bay, when, true to his instructions, he continued around
it to the westward, and thence in an easterly direction to
the river proper, where it opens out from between the
mountains into the wide bay which finally brought De
Long and his party to a stand-still.

This and a great deal more they told me, beginning
again and again, from their landing on the shore of the
Arctic Ocean to the time of their arrival at Belun.
Nindemann had a short section of a chart which Mr. Collins had copied for him from the small chart in De Long's
possession, and which I recognized as similar in part to
my own copy. I then made from their description a
rough map of the region north and south of De Long's
probable position, as a guide for my search. It was out
of the question for either Nindemann or Noros to accompany me; for, leaving aside the difficulties of transportation and food, both were so sick as to be barely able to
walk, vomiting and purging violently, — the effects of
having gorged themselves with some decayed fish refuse
which they had found at a hut called Bulcour. They
complained bitterly of the treatment they had received
at the hands of the starosti and natives; having been
furnished with a very limited supply of black bread by
the young priest of the village, and with nothing but

smoke-dried fish, although there was plenty of fresh fish
and venison to be had.

I slept in the *povarnia* with my two comrades and a
number of natives the first night; but, before turning in,
I started a telegram to the Secretary of the Navy, a copy
of it to the United States Minister at St. Petersburg,
and another to Mr. James Gordon Bennett. This message I first wrote out in English, and then, after a manner, translated for the young priest, who finally wrote it
out in Russian. I then sealed it across the back with two
feathers to indicate that it must fly, and saw the messenger off with it. Next day I diligently busied myself in
adding to the comforts of my two sick companions.
Noros had discovered a couple of good vacant huts in the
village; so together we visited the young priest, who
said, however, that he was very poor, having given the
two men all the provisions he could spare, and that he
had no authority to compel others to do the same. He
accompanied me to the vacant houses, but would not
dare to enter them; so I told him I would do as I pleased,
the American government would pay, and that General
Tschernaieff, the Governor-General of the district, was a
soldier, and would not permit any soldiers of the United
States to suffer. Whereupon I placed my shoulder against
the door of a hut, and, bursting it open, invited him in.
At first he was a little alarmed, saying the hut belonged
to a rich copert, who might claim damages from him, but
I allayed his fears by saying that I would be responsible; and then calling upon the Yakut *starosti*, I told
him that I must be furnished at once with pots, pans,
kettles, and other housekeeping utensils for Nindemann
and Noros, together with plenty of bread and venison for
them to eat. I also ordered him to have a native haul
wood and keep the fire going for the two sick men.
Then, having attended to these and many other kindred
things, I finally, when the hut had been well warmed,

directed the men to take possession of it, and having seen
them properly installed, I left them to the tender mercies
of the women, who dropped in "to fix things," and accompanied the young priest to the house of the old
priest, who had prepared some dinner for me.

I told my story to him as best I could, but, owing to
the obscuration of his mind from a deep and long-con-
tinued debauch, my task was a very difficult and thank-
less one. However, he treated me kindly, and promised
a reindeer the next day as food for myself and men. I
now found myself becalmed, as it were, and unable to
move until the return of Bieshoff, the Cossack commandant, who alone had the authority to furnish me with the
necessary outfit for my search for De Long and party.
Nindemann told me that immediately upon his arrival at
Belun he had prepared a message which he wished to
have sent to the United States minister at St. Petersburg.
Bieshoff said, "Yes, yes," took the message and put it
in his pouch to carry to me. For Kusma had seen him
meanwhile and informed him of myself and party,
making in all eleven men; and since Nindemann had
spoken of De Long and party as a company of eleven,
here was a coincidence strengthened by Kusma's reference to me as "Captain," which had led Bieshoff to
confound me with De Long and believe both parties to
be one and the same. So he took Nindemann's message
and hurried away to my relief at Jamaveloch, where
he learned of my departure with the intention of inter-
cepting him on the road. On his journey he had found
so little snow on the mountains, that the reindeer teams
he took with him would not be available for the transportation of the party, and so he sent them back to
Belun.

Nindemann's letter he gave to Mr. Danenhower, who
apparently regarded it as a most important communication, for he dispatched Bartlett with it to me at once,

notwithstanding that I was then in personal contact with the man who wrote it.

I was dining with the priest, when an old woman entered the house in a state of great excitement, and announced that another American had arrived in the village. I started immediately for the *Balogan Americanski,* as our habitation was called, and there found Bartlett. To be sure I was pleased to see him, though disappointed to find none of the others with him, and particularly Bieshoff, whose assistance I so sorely needed in fitting out the relief expedition for my missing shipmates; although assured over and over again by Nindemann and Noros that they were dead, and that it would be useless and perilous to seek for their bodies before spring-time. But the reader may imagine my astonishment when, asking Bartlett what had occasioned his hot pursuit of me, he answered that Mr. Danenhower had forwarded him with Nindemann's old communication to the United States minister at St. Petersburg.

So I had some additional conversation with Nindemann, finishing my written description of his journey after leaving De Long, and had almost completed my chart, when the *starosti* of the village and the young priest entered our hut, the latter bearing a letter, which, he said, had been written by Bieshoff, directing him to order the *starosti* to furnish me with two deer-sleds in order that I might meet him, Bieshoff, on the morrow or the day following at Buruloch, where, in the mean time, he would fit out two dog-sleds and guides to convey me north on the search for De Long. I had seen two of the three natives who had found Nindemann and Noros in the hut at Bulcour. One of them, Tomat Constantine, was *Golivar Candidat* (head candidate) of North Belun, and consequently a man of authority among his people; and as he knew the location of Bulcour, and likewise of the hut described by Nindemann as the " place of the sleighs," that is, a hut where a number of sleighs was stored, I selected him as one of my guides.

I then prepared a letter to Mr. Danenhower, in which I directed him to take all the men as far south as Yakutsk, and there await my arrival; but at the same time I told Bartlett to remain at Belun until I returned from my northern trip, saying that I would direct Mr. Danenhower verbally to leave him behind when he started for Yakutsk, in order that I might have some one to look me up if I failed to return to Belun within thirty days. And then bidding them all good-by, I left for Buruloch, where I arrived late that night; it having taken me nearly twelve hours to traverse the eighty versts, whereas I had previously made the distance in eight hours. For the snow had deepened and the winter storms had set fairly in.

CHAPTER XIII.

SEARCHING FOR DE LONG.

Bieshoff. — Poor Jack Cole. — I start in quest of De Long. — At Ku Mark Surt again. — How to eat Raw Fish. — The Maiden with the *Lively* Tresses. — Bulcour. — "The Place of the Three Crosses " — Sleeping in the Snow. — Mat Vay. — A Clue.

WHEN I arrived at Buruloch, my feet and legs had swollen to nearly double their healthy size, and the new skin had arisen in large blisters filled with water. I emptied them, and the old woman greased my legs with goose grease. Next morning I was up and out in good time, anxiously looking for the arrival of Bieshoff and my party. An hour or so before noon the yelping and baying of the dogs announced their coming, and soon I was being introduced to Bieshoff, a fine specimen of Cossack manhood, very large of stature, of a commanding presence and quiet demeanor. We had a general handshaking all around, and breakfast together; and then, much to my sorrow, I learned for the first time that poor Jack Cole had become demented. He had apparently lost all trace of time and circumstance, asking me if I thought we would see the captain in a few days, and saying that he was tired of the strange, mysterious fellows who were in that country, and believed he would like to go and see the "old woman." At this time he was perfectly tractable, although Mr. Danenhower told me that he had been occasionally obstinate, and that on the way from Jamaveloch it was with great difficulty he could be kept on the sled, having once, indeed, dropped off quietly without any one missing him until they had gone a considerable distance, when driving back they found him lying in the snow. He had now become quite a charge, requiring one man's constant attention and care. I was not altogether unprepared for this, since I had noticed during the last few weeks that Jack had grown so foolish in the repairing of his clothes, needlessly wasting the thread and needles, that the sailors had taken those articles away from him.

I was delighted to meet again my good friend Vasilli Kool Gar, who had driven over one of the sleds from Jamaveloch; and I hastened to secure him as one of my own drivers on the search. I had learned by this time that the limit of a dog-team's endurance is about ten days; so Bieshoff set diligently about equipping me for that space of time, providing me with two teams of eleven dogs each, two drivers, and ten days' supply of food for all of us. With everything satisfactorily arranged, and bidding good-by to my friends, who were to pass the night at Buruloch and proceed next morning towards Belun, Verkeransk, and Yakutsk, I set out for Ku Mark Surt, where that night I arrived and rested, full of hopes and fears for the future, — hoping for the best, yet fearing the worst. From Nindemann's story, I had very little hope, if any, of finding my comrades alive; but, if dead, I might yet be able to prevent the destruction of their bodies by the few ravenous animals of these regions.

My intention was to follow the back-track, if possible, until I came up with the party, dead or alive; pursuing the west bank of the river to the northward, guided by the notes I had taken from my conversations with Nindemann, until I arrived at the point on the Arctic Ocean where, after landing, they had made a cache of their books, papers, chronometers, and other articles.

My old friends at Ku Mark Surt seemed pleased to
see me, and I was doubly welcome because of a little
salt that I brought from Belun. The old lady of the es-
tablishment earnestly requested me to examine her son's
paralyzed legs, in the hope that I might be able to cure
or relieve him; for she had been told that at Moscow,
where the Czar was, he could be made whole and strong
again. But I said he was now too old, I feared, the
misfortune having occurred in his youth. They all understood me very well, —

"Baranchuck, bar, bar; mooshina, soak;" that is, "If
a baby, yes, yes; but a man, no." And all the family
joined with the poor cripple in sighing out, "Mooshina,
soak." After a supper of tea and boiled fish, we crawled
into our berths, and were out of them early the next
morning. While the members of the family busied themselves in the preparation of breakfast, I had an opportunity of witnessing some of their internal household
economics. Such fish as were to be boiled were first
thawed out before the fire, and then properly scaled,
cleansed, cut in satisfactory lengths, and placed in a
kettle *near* the fire, where they might simmer, but never
be allowed to boil. Fish to be eaten raw are selected
for their delicate fatness, and while hard-frozen briefly
exposed to the heat and turned quickly so as to only
thaw the skin. Then by a few dexterous cuts of the
knife, the dorsal fin and a narrow strip of the belly-skin
are removed, and the skin raised at the tail and cut up
to or near the gills, when it is caught at the tail between the teeth, and, with one jerk, peeled to the head,
the opposite side of the fish sharing the same fate.
These skins are cured, dried, and worked up into water-
proof bags, wherein the natives store their tinder, shavings, and sleeping-gear, while traveling.

The tea-kettle was put on, and several gallons of boiling water made; the stranger's particular little tea-pot

being filled and a quantity of tea begged for the family.
The raw, frozen fish were then cut down in thin slices;
the fat back and belly-pieces, however, being carved into
short, thick morsels, and turned towards the guest as
choice tid-bits of savory sweetness; and with this dish
the meal began. While breakfast was in course of preparation, I had noticed with interest that the natives, old
and young, performed their morning ablutions by taking
a large mouthful of water, spurting it into their hands,
and then rubbing their faces, — the little chicks, four or
five years old, along with the rest. Afterwards, the
young lady of the household, aged fourteen or fifteen,
began combing her jet-black tresses. The natives manufacture their combs from the fossil ivory of the mammoth, managing to turn out very creditable work, considering their rude means — the ever-handy and useful
sheath-knife, made in the shape of a single-edged dirk.

I observed that the young lady, after letting down her
hair, set about arranging on her lap a circular wooden
platter or tray about eighteen inches in diameter, with
an edge raised around it like the lid of a Yankee cheese-
box. I likewise took notice that it was the veritable
platter which had been used to hold the raw fish on my
previous visit to the hut, and from which we had all
eaten our food. Hair and lid properly adjusted for the
combing, that operation began, and continued with an
expertness only attainable by long practice. A down-
ward stroke of the comb through the hair; a sharp,
short knock on the lap-board for the purpose of removing any obstruction there might have been in the teeth
of the comb, or of releasing, perhaps, any prisoner held
in its narrow meshes; a circular sweep of the comb
around the periphery of the lid towards its edge, with a
partial sweep to the right and left, which brought anything that might have been there to the centre of the
lid; then a quick, decisive thrust with the flat portion of

the comb, which was apparently an effort to crush or
blot out some creeping thing; and finally, the remains
being brushed into the fire where simmered our breakfast, the lid thus cleaned was ready for service as platter
for the frozen sliced fish. These were heaped upon it,
and all the male members of the family at once began
the attack. It is needless to say that although I furnished the salt, still, on this occasion, I did not care for
frozen fish, and held my appetite in check until the second course, of boiled fish; for although there had been no
lid on the pot, yet since the old woman had kept skimming off the scum which arose to the surface, it seemed
likely to me that in doing so she had managed to rescue
the hair or any other thing which may have fallen in.

With breakfast over, the dogs were hitched, and we
started on our journey in a blinding snow-storm. It was
fifty-five versts to Bulcour, where Nindemann and Noros
had been found; old Tomat Constantine knew the place,
and so there was no doubt of our ability to take up with
the first landmark on our backward track. It is impossible for me to adequately describe our progress along
the bed of the river; crossing and recrossing it to find
a path among the *chevaux de frise* of jammed ice;
tumbling over, upsetting, hauling, and pleading with the
dogs, which, in turn, plead with each other in such a way
as to make the fur fly. A strong westerly wind cut our
faces, and it was very cold. My feet and legs had again
begun to swell, until my moccasins grew tight to bursting.

They did not pain me, for all feeling seemed to have
forsaken them, but what troubled me most was that I
had lost all control of them, and, being unable to stand
up, much more to walk, I was forced to forego the exercise
that would have prevented the blistering of my limbs,
which comes of suppressed circulation, and consequently
the tighter my moccasins became with the swelling, the
more severe became the frost-bite, each additional chill

acting as a further check to the diffusion of blood. It was long after day when we arrived at Bulcour. The place consists of two huts and a store-house; one of the huts being a *balogan,* and the other a *palatka,* — which have this difference, that the former is a frustum of a regular pyramid, from four to seven feet high, covered with earth, and having a hole in the centre of the top for the escape of smoke, etc., while the latter is a regular pyramid, with a square frame-work inside near the top, against which the poles rest, and through which the smoke finds an exit. A small river makes out into the Lena proper from the northwest, between steep banks probably a hundred yards apart, so steep as to be very difficult of ascent, and this, too, at an abrupt bend of the Lena, where it swerves from east northeast to north. The *balogan* is pitched on the northwest bank of this little stream, the *palatka* on the northeast, together with the detached store-house, which can be more properly described as a square box of ten or twelve feet, raised slightly above the ground on stilts or blocks. These two structures are nearest the main river, but are set so high up and far back that it is a wonder to me Nindemann and Noros saw them at all; indeed, the men were on the point of returning to the "place of the sleighs," some twenty miles to the northward, there to lie down and die, when, going out on the bed of the river from under the bank, one of them descried the *palatka* and store-house, and, of course, they at once sought shelter there. This is one of the favorite fishing stations of the natives for certain runs of small fish. It was abandoned for the season, but the natives had left some of their nets and other gear behind; and Nindemann and Noros had searched everywhere, without success, for food. In the store-house, however, they came upon a lot of mouldy fish, which the natives had pulverized and heated for the purpose of extracting the oil for their lamps, and though this refuse was decomposed and covered with mould, yet there was nothing else to be had ; the men had been long without food, having only caught a lemming, or tailless rat, several days before, which they roasted and ate, hair and all ; so it was quite natural that they should try to satisfy their hunger with the fish offal, which would, at least, distend their empty stomachs.

Accordingly they devoured a portion of the mess as they found it, and then starting a fire in the *palatka*, brewed some willow tea and warmed themselves. Afterwards they placed several large flat stones over the fire, on which to cook or heat the putrid mass, but it was not as palatable as when cold or frozen. A scarcity of wood compelled them to burn considerable of the interior woodwork of the hut, as well as an old canoe ; and they tarried here for a day or two, trying to recuperate their strength, and feeding on the offal, which soon sickened them so that they were fearful lest they should be disabled from continuing their journey. Summoning, therefore, all their remaining powers of endurance, they filled their pouches and skull-caps with the rotten fish, and started forward towards the settlement which they had hoped to reach long before, but which was really fifty-five versts beyond. It was very cold, and a cruel wind was blowing, and, in a little while, Noros complained that he was so sick and weak from frequent purging, vomiting, and previous exposure, that he could then proceed no further, and begged Nindemann to turn back for at least one day more.

This they did, and while Nindemann was busy repairing his moccasins inside the hut, he heard a strange rustling noise without. Hunger had rendered their sense of hearing doubly keen, and Nindemann, thinking that a reindeer was at hand, seized his gun, charged it, and approaching the door was about to peer out, when suddenly it opened and he stood face to face with a Yakut.

Naturally his first impulse was to embrace his savior; but that honest worthy seeing the mere spectre of a man, ragged, emaciated, frost-bitten, smoke-grimed, his face covered with scabs, and a gun held at "ready" in his hands, fell back in dismay, and sinking on his knees begged Nindemann not to shoot. To drop the gun and embrace the friendly visitor was the act of an instant, and then they drew him into their hut, — his hut, for it was Ivan Androsouff, the proprietor, who had left his nets in the store-house, waiting for the ice to make, and had then come to carry them with him on a fishing excursion up one of the northwest discharges of the river.

This lucky incident saved their lives, for, instead of recovering, they were wasting their little strength while feeding on the putrid fish, the condition of their bowels being such that they would soon have been unable to crawl. The Yakut was not a little alarmed at his situation, for he mistook the two men to be escaped exiles, whom it was his duty to capture, under penalty of being flogged. However, they made Ivan understand that they were very hungry, exhibiting the fish refuse of which they had been eating, and proffering him some; but he turned aside in disgust, telling them to throw it away.

Parleying a while, he signified that he would go off and return soon with assistance; and ere they could collect their wits he held up three fingers and left them. When he had gone, Nindemann said he was in doubt whether the three fingers were intended to mean three miles, three hours, or three days; and so he upbraided himself for having allowed the native to depart. Still, in several hours Ivan returned with two companions and a couple of deer-teams, and carried the men away to a hut in the woods, where there was a woman with children. Here they tried to make the natives comprehend that they had recently left the captain and party in a starving condition to the northward; but with their own

12

sufferings and the remembrance of their comrades' sad lot, they completely broke down, and mingled their tears with those of the gentle-hearted natives, who, believing that the men were anxious to reach Belun, urged them to sleep first, and then, in the morning, they would all push forward.

Next day they started for Belun, the worthy natives loaning them clothes for the journey. Poor Nindemann did his utmost to explain that they must go back and save the captain and party, but to no purpose; the natives persisted in the belief that he desired to hurry on; and it was during their progress to Belun that they encountered Kusma, my messenger, returning to Jamaveloch.

I was rejoiced to seek refuge in the *palatka* at Bulcour. Our dogs were unable to scale the steep bank with the loaded sleds, so I crawled up on my hands and knees before the natives succeeded in elevating the teams. The wind had strengthened to almost a gale, and when the dogs had been staked on the leeward side of the hut, and fed, Tomat and Vasilli set diligently about, and soon had a fire blazing in the centre of the hut. They carried up ice from the river to use in boiling our tea and fish; for the natives believe, with whalemen, that snow - water breeds scurvy; and I have seen them travel more than a mile for ice, when there was an abundance of snow at hand. Soon we had our supper of hot tea and boiled fish, and I delighted my two companions by presenting each with a small lump of sugar, given to me by Bieshoff, and which neither used, but stored away for home use. As the hut warmed up, we hunted around among the ashes of the fire-place and found several little articles that had been left or lost by Nindemann and Noros; and these fully identified the place, and satisfied me that I was now on the right trail.

The numbness and insensibility of my legs shortly

gave way to the most excruciating pains. I withdrew
my feet as far as possible from the fire, and thrust them
into the snow which covered the floor of the hut, but
without experiencing any relief. I dared not remove my
moccasins, for the swelling had so increased that I feared
lest I should not be able to get them on again. So all
that night I rolled in agony on the floor, becoming so
sick that I lost my supper. The wind raged without,
and when morning dawned was still blowing so fiercely
that the natives, taking a look at the weather, said, —
"Pagoda, poorga ; pi dome, soak."

There was no use kicking against the pricks ; neither
man nor dog could face such a gale ; so we did the only
other practicable thing, and abided its abatement, the
natives lashing their sleds and repairing their dog harness. Thus the day was spent, and during the night the
gale went down, subsiding so far that by morning we
were again able to take to the road. The delay had not
been altogether a loss, for I had been afforded an opportunity to translate and talk over with the natives my
written account of Nindemann's journey; and now our
next objective point was the " Place of the Sleighs."
Neither Tomat nor Vasilli had ever been there, but from
my broken translation of Nindemann's description they
understood pretty clearly about where I wanted to go.
So, making an early start, we drove rapidly along, and
at length reached the desired spot. It was a very small
hut, devoid of door or cover to the smoke-hole, and consequently snowed in. I found fragments of the sleds,
which Nindemann had broken up for fuel, and searched
vainly about for a sign of any one having followed in his
footsteps. Again we pushed on, this time for the three
Bulchoi Crasses, three *Propaldi Yakuts* — the place
where there were three dead Yakuts in cases raised on
trestles, with three large crosses. The natives said there
were many, many *Yakuts pomeree* and *crass manorga ;*

but I made them understand that these were near an old hut on a high bank, and that there were two canoes (*vi-atkas*) and an old shed on the river.

And so we pressed forward. Night came and found us yet on the snow, till, at length, the natives dug with their shovels a pit about seven feet square, and standing the sleds to the windward, in order to break the force of the gale and cause the snow to bound over us, drew forth some raw frozen fish, on which, cut down in thin slices, we regaled ourselves. The dogs being staked and fed, we then crawled into the pit and crept into our sleeping-bags, the few dogs which had been turned loose nestling warmly on top of us. For an hour or two we slept fairly well, but long before daylight we were so chilled that, for my part, I felt as though I could never stretch myself again. Indeed, as was often my experience, when I first lay down I was very cold, but with my blood flowing freely and the heat of my body confined within the bag, I soon became quite warm and comfortable, save at the feet, where, to be sure, I never succeeded in inducing any heat. And so in a little while, overcome by the genial glow of my body, I fell into a deep sleep, dreaming of long, weary marches; and, as the snow sifted into the rents of my old battered sleeping-bag and thawed upon my neck or face, I brushed it off as though troubled by a persistent mosquito. But in the course of five or six hours, when camping thus, the limbs of the sleeper begin to cramp, his body is chilled, the snow has drifted up the sleeves and around the collar of his jacket, he grows restless, and finally awakes with a jump as though branded with a hot iron. For the snow has melted under his jacket, the bag and body are about freezing, the wet sleeve has indeed frozen fast to his bare wrist, and in his haste to remove the burning jacket from his irritated flesh, he tears off the blistered skin, leaving a raw spot to scab and fester and fill up with reindeer "feathers" (hairs).

So I was glad when day dawned. We leaped from the pit, shook out the snow from our clothes and sleeping-bags, and ate our morning meal of frozen fish, — the natives topping off with a smoke. Then when the sleds were packed, the poor dogs shaken out of their little heaps of snow and beaten, snapping and shivering, into line without any breakfast, another day's journey began. Keeping the " west bank of the river on board," we carefully scanned every object as we proceeded, halting at times to run at right angles with our course and look for a track or trace of any one coming from the northward, directed by the foot-prints of the two who had marched out of the jaws of death. Here and there we saw where they had plunged through the ice while it was yet young, but wherever the snow had fallen and drifted, there was no sign or trail to guide us. Darkness settled around us again, bringing with it another pit in the snow, and a second miserable night; and in the morning after our frugal meal of raw fish, washed down with lukewarm tea brewed with difficulty over a fire built in the snow-bank, we resumed our journey.

At one place we observed the footprints of two men crossing the *gooba* or bay, toward the east side and returning again. The imprints only showed where they had been made in the ice when it was fresh or leathery, though we occasionally saw them in the sand-spits, which had afterwards frozen hard ; and the snow, driven violently by the wind, had swept across such exposed patches and lodged under the banks or piled up against the huge hummocks of ice with which the river was cumbered. Towards night we arrived at the place of the three crosses, and discovered the two old huts, canoes, and fishing shed, as described, together with the dead Yakuts stored away on their trestles in mid air. I could find the foot-prints of two men in and around the huts, but no more. These structures, which were dilapidated and filled with snow, afforded not the slightest evidence of having been tenanted since the sojourn of Nindemann and Noros. I was now nearly famished. The previous two nights had given me but little rest, and no renewal of vigor; and even the cold fish seemed only to chill me. I asked the natives how far it was to Mat Vay; twenty-five versts, they said; and there we might have fire, shelter, hot tea, and hot fish; so, although it was far into the night, I gave the order: " Pi dome, Mat Vay."

I had now followed up their tracks to where Nindemann and Noros said they came upon the first huts they had found after leaving De Long; and as I had stuck steadily to the west bank of the river, there could be no mistake. I had therefore concluded to stop at Mat Vay with the intention of renewing on the morrow my search over the shoals which we had traversed all day, to the Stolb or Stolboi, one of Nindemann's most prominent landmarks. We hurried on through the dark, I knew not how, save that we still clung to the west bank, and long after midnight brought up at a hut. Then, so torpid that I could barely move or speak, I sat down in the snow until the natives had opened the door of the hut, when, crawling into it on my hands and knees, I cried out for *agoime, agoime !* (fire, fire).

Presently one was blazing warmly on the hearth, and then I noticed by its glare that though the door of the hut had been properly banked up with snow, yet the boards used to cover the smoke-hole had not been replaced, and, as a consequence; the hut was nearly filled with snow. But as the fire burned more brightly and lit up the room, I was particularly surprised at the unusual arrangement of the bed of sticks, to which I drew the attention of the natives, saying, " Yakut soak." They looked very grave, shook their heads, and iterated "Soak" (no); and then, pointing to the open roof and snow, added, " Americanski."

Now, indeed, was I charmed; for I felt that I had found the new trail. Nindemann and Noros had assured me that they had not seen or entered a hut until they arrived as far south as "The Crosses," nor slept in one till they reached the "Place of the Sleighs;" and the peculiarity in the arrangement of the sticks which attracted my attention was, that they had been removed from the ground around the three sides of the hut, according to the custom of the Yakuts, and arranged in the bed form adopted by the North American Indians, with the feet towards the fire, and a log for the head. So it occurred to me at once that a second party, in all probability Alexia, had been sent ahead by De Long to follow up Nindemann; that coming upon this hut, and being unable to open the door, he had dropped down through the smoke-hole, and here passed the night; and that when leaving he had neglected to cover the hole in the roof, an oversight of which no Yakut would have been guilty.

A further search revealed no record or other evidence; and so, when we had readjusted the sticks Yakut fashion, and supped, we turned in, — I to dream that I had found the first clue of De Long.

CHAPTER XIV.

LOSING THE TRAIL.

Treachery. — I make a Cannibalistic Threat. — Cass Carta. — A Late Supper of Deer Hoofs and Horns. — Benumbed. — Another Night in the Snow. — "Balogan Soak " — North Belun. — I dive into a Hut and "receive." — Two Precious *Bumagas*. — A Yakut Bride — Sleeping in Close Quarters. — An Ignoble Economy. — Definite Tidings.

NEXT morning I noticed that the natives talked a great deal between themselves, and upon my urging them to be more active, packed their sleds with considerable discontent and mystery; conversing the while in whispers as though fearful of my overhearing them. Finally, when we were about to start and I had left the hut, Tomat, who had returned to see that everything was taken from it, reappeared, his face aglow, and presented me with a leather belt, saying that he had found it in the hut. A glance at the great copper buckle of homely make told me that the belt had been made on board the Jeannette; so I reëntered the hut and instituted a more thorough examination, shoveling out all of the snow, but to no purpose. I was now, however, more fully convinced than before that a second party sent out by De Long had been there. So jumping upon the sled I ordered the natives to *skaree* (hurry up) !

They both looked at me for an instant, afraid to speak, and then Tomat found voice enough to say, — "Soak, soak !"

Turning to Vasilli, who stood back, I inquired, — " Kack soak " (why not) ?

"Cushat soak" (nothing to eat), both replied.

"Bar, bar, cushat manorga," I said.

But they insisted, "Ballook soak, olane soak; savaccas propaldi" (no fish, no deer; the dogs would break down).

I was astounded. Bieshoff had told me that he had furnished food for ten days, and here it was all gone in four. I instantly suspected that the Yakuts, losing heart, were playing me false, and so caused them to unload their sleds of everything; but no food was brought to light. Yet I had full confidence in Bieshoff, and so it occurred to me that the natives, afraid to proceed further, had robbed me of the fish and buried them at Bulcour, or one of the snow camps.

The very thought enraged me. To turn back now that I had struck a new trail was impossible. The two natives seemed stolid, while Tomat jabbered away at a lively rate. I seized his great staff and dealt him a staggering blow, whereupon Vasilli took to his heels, and poor Tomat, dazed and frightened, in trying to do likewise, fell down and retreated on his hands and knees, while I followed him up with the iron-shod stake. So crippled was I, however, that they soon distanced me; though I had succeeded in cowing them, and that was all I wished. Yet, as they started off, the dogs jumped in pursuit, but, luckily for me, were staked fast. Then fearing lest the natives might desert me altogether, I seized and loaded my gun, and shouting "Fintofki, fintofki!" discharged it after them. The bullet went whistling over their heads, and at the report both natives fell on their faces, then turning round on their knees, began crossing themselves in terror, and making such low obeisance that their noses dipped in the snow.

I beckoned them to approach, at the same time reloading the gun and seizing the staff. They returned very contrite, but beseeching, "Masta soak; masta soak" (no stick; no stick); and then fell to pleading with me on the danger of traveling farther. We had no food, either for ourselves or dogs.

"Olane soak, ballook soak, savaccas pomree, too pomree, kack pomree" (no deer meat, no fish, the dogs die, we all die, why die?). Then, too, "Car manorga, pagoda poorga, manorga" (snow and gales of wind in abundance). "And," they argued, "you can find your dead comrades in the spring-time when the snow is gone;" and placing their staves upon the ground they blew and scraped and covered them with snow, to show me how it had buried in my friends. Then catching me around the neck they cast themselves upon the snow and feigned to die; old Vasilli afterwards acting the wolf or fox which, he signified, would certainly come and eat us.

Meanwhile I had somewhat recovered from my heat at the sudden exhaustion of our provisions, and now inquired the distance to the nearest village. "Two hundred and fifty versts," they announced, after counting by tens on their fingers. In what direction was it? They laid down their staves, pointing them northwest, and said, "Sever zaputh." Then I firmly demanded, —

"Pi dome, skaree, Sever Belun, balogan Tomat" (hasten to North Belun, to the house of Tomat).

Almost stupefied, they looked at me again as though to assure themselves that I was not crazy, and then burst forth into remonstrances. "There was nothing to eat, and we would all die; I was lame, and trembling with cold, and must surely succumb." Sitting on the sled with my gun in hand, I only replied, —

"Savaccas cushat, cushat manorga, Kack too, Yakut cushat!"

At this they crossed themselves, glanced at each other, and, seeing that I was inexorable, prepared to start, murmuring repeatedly my words, "Yakut cushat, Yakut cushat." For they had only smiled when I said that I would eat the dogs, of which there were plenty; but

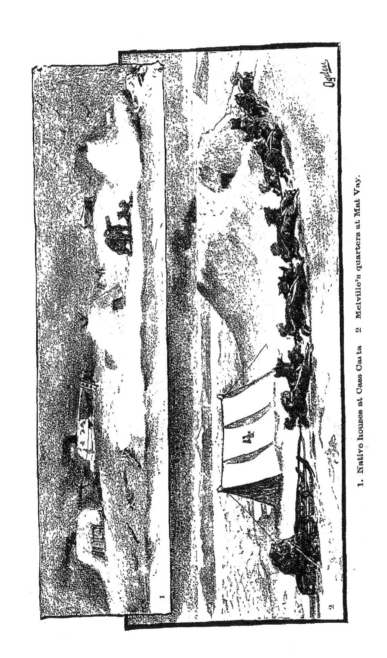

1. Native houses at Cass Cu ta   2   Melville's quarters at Mat Vay.

when I threatened to eat them too, they were entirely overcome, and eyed me askance.

But, setting out, I took a survey of the river and shoal in front of us, and then directed the natives to follow along the west bank of the main branch in Nindemann's reported tracks. But the river here takes a long turn to the westward, and there are numerous streams making out into the north around from the west by way of north, and finally to the eastward, where the largest branch is in sight, though there were more than a dozen headlands to be seen on the bay, each indicating the outlet of a great river. We had gone but a little distance when we came upon a few old huts, which I stopped to examine, but they yielded me no information; so we continued on our journey without interruption until late that night, when we halted at a place called Cass Carta (Goose Place), where there were two good huts and a couple of store-houses.

During our day's progress, I was convinced that we had left the trail, owing to the unusual number of huts we had seen; albeit Nindemann had informed me of the many old and new huts he had noticed while with De Long. Arrived at Cass Carta, however, I was assured beyond doubt of my having lost the scent; for, we found and ate a lot of offal in the huts, which De Long and party, had they come that way, would most certainly not have despised. This windfall consisted of some deer bones with tendons and a little ragged meat attached, — a few of the leg bones having hoofs on them, which we roasted in the fire, and, as the heels softened, pared readily with our knives and pronounced capital. We also found the antlers of deer, which had been killed while the horns were yet in the velvet, and these, porous as pumice stone, and filled with blood, the natives pounded with their axes into a meal, and this we ate and found pleasant. There was nothing for the dogs; but they had been turned adrift, and since Cass Carta in season is a good goose-hunting station, I noticed that they had all managed to pick up goose skins, feathers, wings, feet, etc., which had been thrown away the preceding summer. But they were very weak and foot-sore, some barely able to stagger along out of harness; and a few of these had, indeed, been cut loose and left to follow the team as best they could, or die by the road.

Our hot tea and warm hut were comforts, however, glorified by the fierce storm which blew in the night. I was eager to reach North Belun, where I expected to procure fresh supplies of dogs and food, in order to follow the main river as far south as the point from which I started; and, if I failed in my object, to then return to Belun and fit out an expedition to continue the search in the spring. I was now so badly lamed as to be utterly unable to stand up without assistance; but I urged an early start from Cass Carta, with a view to reaching Koogoolak that day. A veritable gale was blowing, fortunately across our path and not in our faces, and the drifting snow almost obscured the leading dogs. We moved along very slowly; the teams were completely fagged out, and when night overtook us I felt, as we floundered aimlessly about in the snow, that it made little difference to me whether I lived or died, if I could only sleep. The dogs crept on at such a snail's pace that I could readily lie down on top of the sled without danger of falling off; and should this happen Vasilli would certainly miss me.

Finally the natives came to the conclusion that they had better camp and get me into my sleeping-bag; for they seemed highly alarmed at my drowsiness, and soon I was cozy and warm and fast asleep in the land of snow and *tundra*.

We were up with the starlight and under way again. I had scarcely dozed, it seemed, as we struggled along against the furious storm. The natives' faces were blistered and sore; the dogs simply able to move, and no more; and our outlook was gloomy, indeed. I repeatedly asked Vasilli how soon we would come to the next hut, and he as often answered, " In a little while," pointing ahead with his dog-staff, as though the hut was just beyond the hill. But day faded into dusk, and darkness again intervened, and I almost despaired of living. At each of our more frequent halts to rest the worn-out teams, the natives would place their hands upon me and say something. I answered their calls by asking for a *balogan;* and they continued to assure me, " Yes, yes, a little way, a little while ; " until, at last, when they came to tell me of the imaginary *balogan*, I fiercely cried out " Balogan soak."

Doubtless, after that, they believed it unwise to contend with me, for they abandoned me to myself, apparently reasoning that there could be no danger of so obstinate a man dying, even of hunger or cold ; and, forsooth, they were somewhat warranted in supposing from my short, snappish " soak, — soak," supplemented by choice expletives selected at random from my own rich mother tongue, of which they knew nothing, but the essential import of which they eventually divined from the usual vehemence of my delivery, that I would yet keep for some time to come. Finally, long after midnight, Vasilli shook me, and exclaimed, " Balogan, mahor, balogan!" He received the usual rejoinder ; but, notwithstanding, tried to set me up on end, repeating, " Da, da, agoime " (yes, yes, fire), and the teams had stopped, so I looked up, and, sure enough, right beside me, flames were issuing from a snow-bank. So, at least, it appeared, for, although the huts are built upon the level ground, yet they are so covered and buried in by successive snows that before midwinter the dog-teams run directly over them, sled and all, sniffing the savory fumes

of goose or fish which float out of the chimneys. Hence
the ground-floor of the hut sinks farther out of sight at
each additional increment of snow, until it is anywhere
from four to five feet below the grade-level; and so it
becomes necessary, in order to keep the door-way clear,
and enter or issue from the *balogan*, to form or cut out a
series of steps in the snow.

At sight of the red cheerful sparks shooting up from
the cold, white waste, and the glare from the nearest
door-way, in which native women, alarmed by the barking of the dogs, stood holding lighted sticks or flambeaux
to guide us into the hut, my spirits revived, and I sat up
on the sled. Tomat and Vasilli helped me to my feet
with the intention of assisting me down the cellar-way,
but suddenly the dogs, not yet staked, seeing the bright
future ahead of them, bolted. Naturally, the natives let
go of me and sprang for their teams. Unpropped on a
useless pair of legs, I swayed back and forth for a moment, and then fell forward on my face and hands; but
the next instant began crawling towards the brilliant
door-way, where the women stood holding torches above
their unkempt heads and dark faces, while they gazed in
wonder at the strange object which approached them on
hands and knees. Arrived at the top of the snow stair-
case or pit, I first tried to turn about and back down the
steps in an orderly manner, but growing impatient at
the prolonged and cold proceeding, I all of a sudden shot
forward head first, and losing, of course, control of my
movements, rolled unceremoniously in at the feet of the
startled women, who drew back and surveyed me with
elevated torches, crying in Yakut. "Who, who, tell,
tell!" My reply in good, plain English only increased
their dismay, — "How are you, girls?" and then holding my hands toward them, I indicated my desire to be
helped into the hut.

Seeing that I was a white man, and spoke a strange

tongue, they stood back for an instant, and then, taking
courage at sight of my helpless condition and scarred and
frost-bitten face, two of them advanced, and, while others
lighted more torches or sticks, assisted me through the
low door-ways of the outer apartments into a large, well-
arranged, and comfortable *balogan*. It was floored, of
fair height, and around the three sides were the usual
berths or stalls, while a large fire-place, its chimney well
plastered with mud, stood almost in the centre of the
room, and flashed forth light and heat from its capacious
jaws.

I was conducted to the customary guest corner under
the icon; and the women then fell back for a better survey of the stranger. Meanwhile the news of my arrival
had spread through the village, and already the gossips
were pouring into the hut. The women at once busied
themselves in the removal of my outer garments, jacket,
trousers, boots, etc., since all articles of fur are kept in
another and cool apartment, in order to preserve them.
I myself had thrown off my mittens and cap upon entering; and now Vasilli and Tomat came in carrying my
sleeping-bag and other traps. They told the people who
I was, and informed them of my errand, seeming to ex-
cite a great deal of sympathy by the relation; for, gazing at my frozen and puffed-up hands, the natives, full
of curiosity and concern, broke into a clamor of exclamations and questions, — "Ah, oh! Is it possible!
Verily!"

Presently I began to feel the effects of the heat; my
sluggish blood began to circulate more freely, and went
pulsing painfully through my frozen limbs; and when the
women, ignorant of my condition, took hold of the top of
the moccasins to pull them off, they also partially peeled
the skin and scabs from my legs until they bled, and I
cried out in agony. The poor creatures started back as
though struck, while the rest crowded around and craned

their necks to have a look at the bleeding, hair-matted
sores; weeping, and saying, "Poor white man! he suffered from the cold." The women next used their knives,
and, ripping the outer seams of the moccasins, removed
them as gently as possible, leaving the hair of the fur
stockings sticking in the wounds; and then all held a
consultation over the/frozen limbs, which, after considerable controversy and wagging of heads, were given a
bath by pouring water over them; and, when dried, they
were finally smeared with goose grease, which was not at
all pleasant to the eye or nose.

By this time I was seized with an uncontrollable
drowsiness, and, notwithstanding the babbling of tongues
about me, fell into a sound sleep, from which I voluntarily awoke, at length, to find myself carefully covered up
with the furs of the natives. I had neither bandages,
nor material to make the same, for my legs and feet; so
I set the women at work on a pair of mufflers, and, not
knowing the proper name, called them " noogie recovit-
sas " (foot-mittens). The natives tendered me a pair of
reindeer coat-sleeves as temporary coverings for my legs,
and I then "received." A more motley or odoriferous
crowd of mortals I never saw packed within so small a
space. On a table placed in front of me, at the edge of
the berth, hot tea and raw frozen fish were served, amaz-
ingly to the delight of all of us, and afterwards we enjoyed a kettle of boiled fish and soup, or, rather, the hot
water from the fish, which, anyhow, was nourishing and
pleasant.

When supper was over, the crowd made way for a
rather fine - looking young man, who came forward to
where I was, and, bowing very low, said, " Drastie, drastie!" at the same time handing me a small paper (*bumaga*). Eagerly unfolding it, I read: —

ARCTIC EXPLORING STEAMER JEANNETTE.
At a Hut on the LENA DELTA, believed to be near
Tch-ol-booje, Lat. —, Lon. —

Thursday, 22d Sept., 1881.

Whoever finds this paper is requested to forward it to the Secretary of the Navy, with a note of the time and place at which it was found.

Here followed a brief statement of the Jeannette's voyage and loss, our march over the ice, and separation in the gale. The paper then related the landing of the first cutter, and continued : —

On Monday, September 19, we left a pile of our effects near the beach, erecting a long pole. There will be found navigating instruments, chronometer, ships' log - books for two years, tent, medicines, &c., which we were absolutely unable to carry. It took us forty-eight hours to make these twelve miles, owing to our disabled men, and the two huts seemed to me a good place to stop while I pushed forward the surgeon and Ninde-mann to get relief for us. But last night we shot two reindeer, which gives us abundance of food for the present, and we have seen so many more that anxiety for the future is relieved. As soon as our three sick men can walk, we shall resume our search for a settlement on the Lena River.

Saturday, September 24th, 8 A. M. — Our three lame men being now able to walk, we are about to resume our journey, with two days' rations deer meat, and two days' rations pemmican, and three lbs. tea.   GEORGE W. DE LONG,
Lieutenant Commanding.

When I had finished reading the paper, I turned to the young man, whose name, he said, was La Kentie Shamoola, and asked where he had found it. At a hut, he replied, called Ballok, on the east bank of the river (Oshee Lena), about fifty-five versts east of North Be-lun. Here I was in luck, and still more so when an old woman, who now came forward, searched down in the inner recesses of her bosom, and at last drew forth a second

13

paper, which proved to be another record left by De Long
at a hut called Osoktok, about 70 versts to the southeast of the village, and a little south of Ballok. It read
as follows : —

At a Hut, LENA DELTA,

About 12 miles from head of Delta,

Monday, September 26th, 1881.

Fourteen of the officers and men of the U. S. Arctic steamer
Jeannette reached this place last evening, and are proceeding to
the southward this morning. A more complete record will be
found in a tinder case hung up in a hut fifteen miles further up
the right bank of the larger tream.

GEORGE W. DE LONG,

Lieut. Commanding.

P. A. Surgeon J. M. Ambler,
Mr. J. J. Collins.

W. F. Nindemann,	N. Iversen,
H. H. Ericksen,	A. Gortz,
H. H. Kaack,	A. Dressler,
G. W. Boyd,	Ah Sam,
W. Lee,	L. P. Noros.
Alexia,	

The natives now told me that there was yet another
paper in the possession of the neighboring village, and
also a *fintofki*, or gun, which had been picked up still
further to the southward on the east bank of the river,
at a hut known as " Usterda," about ninety versts distant,
south southeast (*youke malinki ostok*). Here I produced
my chart, and by dint of much explanation hammered
its meaning into the thick heads of the wise men of the
village; getting them, at length, to agree that at Usterda
there was a branch of the river which ran east or east by
north. I then made them understand that at this place
De Long and party had halted four days, and then
crossed the river to the westward and traveled south to
a small hut on the west bank, where they camped for
several days, and buried one of their number in the river;

and the little old hut was about twenty versts Youke from Usterda.

To all of this the natives gave their approval, and traced out on the chart the course from North or Upper Belun to Ballok, Osoktok, Usterda, and thence across the river, and south to the *Malinki Starry Balogan* (Little Old Hut); and said that if I would wait a day or two the other paper (*bumaga*) and gun would be brought to me. They were horrified when I told them of Ericksen's burial in the river, and shuddered at the thought of the fish eating his flesh. The hut in which I was stopping belonged to Tomat Constantine, and was inhabited by his wife and children, including a married son and wife, and the customary quota of aunts, blind pensioners, etc., who filled up the berths, while many others occupied the sleeping space on the floor.

The young son had just been married; that is, he had but recently come into possession of his bride; for it is a custom in this section of Siberia for the bride and groom to separate immediately after the marriage ceremony and live apart with their respective parents until one year has elapsed. She had lately joined him in all the glory of her wedding outfit, made from the fine fur of the young fawn, checkered and strapped and covered with peculiar bead-work; interlaced and woven with curiously colored strips of leather; and patched around the bottom of the skirt and up the front with fur.

A funny leather belt with buckle confined the dress under her arms, for she had no waist; and her head was graced with a jaunty band bedecked with beads and spangles. She was pretty after her kind, plump and round, rather mischievous, and never idle a moment from teasing her young mate, who was very modest and shy, but apparently very fond of her. Tomat introduced her to me as the *jonah* of his *brat*, the wife of his son; but she quickly wriggled away from her father-in-law, and returned to torture her spouse; and presently ended her blandishments, coaxing, and love-making, by pulling off his boots as well as assisting him in the removal of other of his outer garments; and then, rolling the blushing youth into his little bed, she finally let fall the greasy calico curtain which shielded their love from our vulgar gaze.

It was nearly dawn before the other inmates of the hut settled to rest, all tarrying to see the stranger crawl into his sleeping-bag, when those without berths spread their skins upon the floor; and I then saw the old lady arrange the fire, and going out on top of the roof put several boards over the chimney, by means of which, to be sure, she kept the heat in the hut, but at the same time cut off the pure air from a mass of forty naked, unwashed and greasy sleepers of all ages compressed within a floor space of perhaps fourteen by twenty-two feet! And the atmosphere — directly it became indescribable. Nevertheless we all slept, and in the morning turned out to find the storm still raging furiously.

It was amusing to watch the natives yawning themselves into wakefulness. Many scenes were too ludicrous, if not peculiar, for me to recount to ears or eyes polite. The regular morning mouth bath was performed by all hands, after which they went through their religious drill before the icon. An iron dipper full of water was then brought by one of the sons, who poured the stream into my hands while I washed and afterwards dried before the fire. Breakfast was soon served, consisting of hot tea, sliced frozen fish, and a pot of boiled fish and venison. Each stranger, visitor, or member of the family, I noticed, had his or her separate little pot of tea and fish. The children were well-behaved; and the mother prepared the hot dishes, while the husband sliced the raw fish with his sheath-knife. And here is a noble economy: the body of the fish which remains after being pared to the bone from head to tail is invariably given to the woman, while his lordship eats of the fair and fat slices, the few scraps that may be left being divided between woman and children. Similarly, when the boiled fish has been eaten and the bones picked clean, woman and children go sucking over them again, rarely a profitable expedition; and if there should chance by some windfall to be "full and plenty" in the hut, then can the poor old blind mammy and granny who live behind the chimney hope for a first pick at the bones, but not before. It is here as in all barbarous countries: the woman is slave to the man; she does all the household work, carries the wood, makes and repairs the clothing, dresses the skins, assists in hauling the fish and game, and in making and repairing the nets, beside bearing the children.

The weather was so violent, that I had fears of the success of my messenger, who had been dispatched for the gun and record, as promised. While awaiting his return, I questioned the natives and made myself conversant with the state of the country, its people, game, etc. Tomat's eldest son, it appeared, had first found the gun and given it to the starosti of the village for safe-keeping. I made a sketch of it, much to his surprise and delight; and then to further identify the gun before its arrival, he cut with his knife a curling spiral shaving from a stick of wood to represent the spring which is fitted under the barrel of a Winchester rifle, and showed me how he had unscrewed the cap and the spring had jumped out of its tube. It was now well understood among the natives that I was searching for a party of twelve lost men, who in all probability had died of cold and hunger; and it was here that they inquired of me why De Long and men had not come to their village, inasmuch as it could be seen from the Oshee Lena with the naked eye, and very plainly with the aid of glasses;

the natives suiting the action to the words by looking
through their fists in imitation of field-glasses. It was
here, too, that Tomat told me of there being twenty-
three head of reindeer cached or staked on trestle-work
about sixteen versts to the westward of Usterda, where
De Long and party had camped, but on the opposite
side of the river, which, at this point, is in the neighbor-
hood of a thousand yards wide and five or six fathoms
deep.   Yet without means of crossing it, — their attempts at rafting having proved abortive, — it is doubtful if they saw the cache at all, since when pointed out
to me it was barely visible on the horizon; and even
though De Long did discern it, after the repeated disap-
pointments he had met with in visiting vacant huts along
the line of retreat, he would scarcely be justified in traveling twelve miles to examine into the nature of a black
object which looked as much like a hillock of earth as
aught else in creation ; not to speak of the fast-running
river which intervened.

The natives said the records and gun had been in their
possession about twenty days. When the ice had made
hard they had journeyed along the river bank towards
their homes, and had noticed the little sled-track and
many foot-prints in the snow, but were puzzled to know
by whom they had been made, fearing at first that some
ruffianly band of freebooters or fugitive exiles had come
that way. They found that many of their traps had
been torn up and used for fire-wood, and, at last, upon
arriving at the huts, they had looked in and discovered
the records and gun, along with some small articles of
clothing which had been cast off or lost. It was with
difficulty I could persuade them that poor De Long and
his poverty - stricken band knew absolutely nothing of
the location of North Belun, much less of the venison
cache. And, indeed, it is most pitiful to think how un-
consciously near they were to salvation. Alas! if some

misfortune, which would have proven such good fortune, had only detained them ten days longer in these huts, to be succored by the natives! It happened otherwise, however; and waiting four days until the river froze, they carefully picked their way across it, hauling poor Ericksen on a sled, and then keeping along the west bank they traveled towards the south, in the hope, as the record said, of reaching a settlement, — a hope that soon turned to despair and ended in the agonies of a cruel death.

It was during the progress of my interview with the natives, conducted by the combined means of pantomime, diagram, and what few Yakut, Tunguse, and Russian words I could muster, that the messenger made his appearance, accompanied by one of the roughest looking criminal exiles I ever saw. The *starosti* brought the gun and third record, which proved to be in regular sequence and of considerable importance. It read as follows : —

" *Saturday, October* 1, 1881.

"Fourteen of the officers and men of the U. S. Arctic steamer Jeannette reached this hut on Wednesday, September 28th, and having been forced to wait for the river to freeze over, are proceeding to cross over to the west side this A. M., on their journey to reach some settlement on the Lena River. We have two days' provisions, but having been fortunate enough thus far to get game in our pressing needs, we have no fear for the future.

Our party are all well, except one man, Ericksen, whose toes have been amputated in consequence of frost-bite. Other records will be found in several huts on the east side of this river, along which we have come from the northward.

GEORGE W. DE LONG,
Lieutenant U. S. Navy, Commanding Expedition.
P. A. SURGEON AMBLER,
MR. J. J. COLLINS," etc., etc.

This was definite and pleasing information for me. I

now had as a guide De Long's authentic narrative telling me the place of his landing on the shore of the Arctic Ocean, where he had cached his goods about three miles to the eastward of the main river; how he had visited the three huts in succession, of whose location I was fully apprised; and as for the rest, I could rely upon my written notes of Nindemann's story.

So I immediately determined to first visit the shore of the ocean and secure the log-books, chronometer, navigation box, sextant, and other articles of value belonging to the expedition, then return by way of Ballok, Osoktok, and Usterda, crossing the river where De Long did, and finally follow the west bank until I came up with the party dead or alive.

CHAPTER XV.

A TRIP TO THE ARCTIC SHORE.

Ballok. — The " Boos Byral." — The Cache. — I am Hoodwinked
again — Topographical Revelations. — *Myacks.* — Our Despised
"Mock " and " Tas " — A Coveted Demijohn. — Phadee Achin.

THE women had made a large pair of deer-skin mufflers for my feet, which, as well as my legs, were entirely
too sore and swollen to allow of my resuming the moccasins. I had also contracted for a new deer-skin coat and
pair of trousers, which materially added to my comfort;
and I now engaged three dog-teams with drivers, and arranged for a ten days' supply of fish. In the morning
I limped out to the sleds, and watched the process of
counting the fish for our journey, mistrusting the honesty of the natives by reason of the imposition practiced
upon me by Vasilli and Tomat. I then returned to the
hut, and as soon as my skin clothing could be donned,
bade good-by to my friendly hostess and her neighbors,
and set out with the three sleds, driven by La Kentie
Shamoola, Tomat, and Kerick; old Vasilli, discharged
with thanks, having gone home.

The weather was propitious, the dogs fresh and strong,
and with a light breeze on our backs we fairly flew along,
passing a small cemetery containing some forty or more
graves marked with crosses; and by dusk we had arrived
at Ballok, resting in the first hut which sheltered De
Long and his men. I found in the ashes a knife-blade,
some pieces of vials or broken glass, and other little arti-

cles attesting their presence. The hut was partially
filled with snow, which the natives cleared, and building
a fire cooked our supper of fish; when we all turned in,
and were out the next morning bright and early. I then
read and explained to my attendants what the first record said, that to the northward on the shore of the ocean
we would find a cache, over which a great pole (*bulchoi
masta*) was erected as a landmark; and the natives marveled greatly that not having been there I could yet
know all about it.

Following the east bank of the river to the north, we
at length came upon the green, heavy massed-up ice of
the ocean, and the natives, imitating the rolling of the
sea, cried " Boos byral ! " and handed me a lump of the
ice to taste, saying "Tooshe, tooshe ! " (salt, salt.) I
then turned to the eastward, and after running for nearly
an hour, finally espied the tall flag-staff, and pointed it
out to the natives, who could scarcely contain themselves
in their anxiety to see what was buried there. Arrived
at the cache, I had it uncovered, and then exhumed from
the snow every article it contained, much to the wonder
and delight of my drivers, who had never before seen
so much plunder in one heap, and who were especially
tickled with the two guns. I loaded all the relics on the
sleds, save one long, heavy steering-oar, and the flag-pole,
which I left standing. The boat was nowhere to be seen,
though I looked for it carefully along the shore ; but since
the ice was so jammed up, I reached the conclusion that
it had been crushed and submerged, or, perhaps, snowed
over. The discovery of it was of no consequence whatever, so far as finding the party or relics was concerned,
but then I wished to remove every vestige of their landing, in order that no future searchers might be misled or
hampered in their progress. For this reason I carried
away all the old sleeping-bags, clothing, etc., with the
intention of destroying the useless articles, or giving

them to the natives at the village, with instructions to keep them there.

I returned to Ballok, well satisfied with my day's work, and, after a supper of hot tea and boiled fish, fell asleep, in the hope of following on the morrow the track as far south as Osoktok. I slept soundly in my bag, reclining on a bed of soft snow, and when morning broke was ready and eager for the road; my feet and legs drying and healing rapidly, save where the deer hair had matted under the sores, for my loose mufflers were soft and warm. I now observed that the natives were having another secret consultation, and, at length, when I was ready to start, and told them to *pi dome*, designating the course we would pursue, they promptly refused to go. "Why not?" I asked. "There was no food," they said. This dumfounded me, for I had certainly seen a fish supply for ten days loaded on the sled. Old Tomat had been involved in a scrape of this kind before, and now edged past me towards the door of the hut. I saw his movement, and seizing a billet of wood began to belabor La Kentie, and the young son of the *starosti*. The former was a great, dignified sort of person, and unaccustomed to this kind of treatment; but as the blows fell thick and fast he beat a precipitate retreat, scrambling all over me in his haste to reach the door. I followed quickly, gun in hand, shouting, "Fintofki, fintofki!" in fear that they might desert me altogether. Old Tomat, well pleased at his escape from punishment, stood in the middle distance, laughing heartily at the discomfiture of his friends. La Kentie was sullen, hurt in feelings as well as body, and did not relish the fun which Tomat was poking at him; while Kerick, the *starosti's* young son, stood rubbing in dismay that part of his person which he had presented to me as a target when leaving the hut on all-fours. I called them to me, but as La Kentie exhibited some unwillingness to obey, I lowered

the gun and fired as I had before at Mat Vay. It had
exactly the same effect. They all dropped on their
knees and faces, going through their religious exercises
at a great rate. Old Tomat took refuge behind La
Kentie, and lied abominably, saying that " La Kentie
had stolen the fish, not he, who, indeed, knew nothing at
all about it."

However, I summoned them to me, promising not to
beat them, and then learned that the villagers had filched
the fish from our sleds, and returned them to the different store-houses ; because, they explained, I had appropriated all the fish in the village, and there had been a famine in that section of the country, eighty of their dogs
having starved to death, and that if I had carried off the
fish the women and children would have died; whereupon
they compressed their stomachs and made hollow their
cheeks. What could I do but swallow my vexation and
return to the village?

We started back in a driving snow-storm, which luckily blew from the eastward and against our backs. The
dogs were already weak from hunger, although they had
been out but three days, yet, as I have mentioned elsewhere, the natives never work their teams, if they can
avoid it, two days in succession. So we staggered along
at a snail's pace, and it was night before we reached
North Belun, having consumed nine hours on the journey, whereas on our outward trip we had accomplished
the same distance in six. I was surprised at the number
of rivers we had crossed in going from Upper Belun to
Ballok; for on my chart (a copy of Petermann's, by far
the most reliable known at the time of the Jeannette's
sailing) only three main branches of the Lena were laid
down. So on our way back I counted the rivers, inquiring of La Kentie as we crossed each frozen stream, " Kack
Oshee ? " (river) ; or " Ku mark ? " (dirt) ; to which he
would reply, " Da, da," or " Bar " (yes) ; or " Soak " (no),

as the case might be. We crossed thirteen streams, several of which were as wide, though perhaps not as deep, as the main branch, along which De Long had marched; so it can be imagined of what value our charts were to us, which located within this space of forty miles but two streams. While at Ballok and on our way to the ocean, I had diligently inquired of the natives the situation of "Sagastyr," but they knew nothing of such a place. They told me, however, of Barchuck, and the many old huts at Barkin, but said that no one had lived at the latter place for years. Still they were confident of the non-existence of "Sagastyr" or "Signalthorp." I was so particular about establishing this fact because Nindemann had informed me that when Erichsen was buried, there had been a signal station in sight, which De Long believed to be the "Signalthorp" marked on Petermann's chart. But the natives took the trouble to show me a dozen or more of their signal stations, which they erect for the purpose of guiding them when benighted or lost in storms. As they journey across the Delta, they halt at every *myack* (finger-post or pointer) and examine it, recutting the post-marks, when necessary, with their knives, adjusting a new pointer, or setting up a triangle which has tumbled down. The triangular *myack* consists of two short sticks supporting a longer stick which either indicates a point of the compass or points toward a particular hut or village. These pointers have certain marks cut in them, the significance of which is generally understood by the natives, and I have seen them, lost in the snow, drive aimlessly around almost in a circle, until they found a *myack;* when, taking a fresh start, guided by the direction of the wind or furrows of snow, they would successfully reach their destination.

The whole village turned out to witness our dejected return, and many hands carried the treasure-trove into the hut of Tomat Constantine. I selected every article

of any value to the expedition or government, and gave
the residue, which comprised a lot of old sleeping-bags,
clothing, an old cook-stove or fire-pot, and· some useless
rope and canvas, to Tomat and La Kentie Shamoola as
part payment for their services. Among the packages
which I made up to carry back with me to Belun was a
tin box of nearly a cubic foot capacity, filled with rock
specimens, mosses, etc., from Bennett Island, and when
I set it carefully to one side I saw the natives first peer
into the box, then pick over its contents, and after some
chattering among themselves, finally burst forth into a
loud guffaw at the idiocy of a man who, upon the point
of starvation, proposed to incumber himself on a long
journey with a load of worthless stones. I could plainly
hear their contemptuous comment on the *mock* and *tas;*
until, at length, to make sure that he understood me
aright, Tomat inquired a second time if I really meant
to carry them to Belun, and upon my replying in the
affirmative he impatiently cast the box among the other
articles with a look of supreme disgust, admonishing me
at the same time that the dogs and sleds would assuredly
propaldi (break down).

Amongst the things I had brought away from the cache
were a demijohn and breaker, containing some alcohol.
The natives soon learned that I had the spirits, and all
congregated around in the hope of having a spree. But
not knowing how the devil would act if I turned him
adrift among such crude material, I flatly refused to listen to Tomat's entreaties for "just a little." "It was
only good for fire," I told them, showing how it burned
in the alcohol stove, but still they coaxed and begged,
until I saw a young man seize the demijohn and bolt
with it. I caught him before he reached the door, and
snatching the demijohn from his lips struck him with it,
spilling the alcohol over the floor, whereupon he quickly
got down on his stomach and eagerly lapped up the pre-

cious fluid. I exhibited considerable anger at the young
fellow's forwardness; and then emptied the contents of
the wicker-covered vessel into the fire-place and among
the ashes of the hearth, where it took fire and burned
for a long while, greatly to the sorrow and dismay of
poor Tomat and his friends.

Before retiring that night, I arranged for the necessary
teams to convey me to Belun. La Kentie and Kerick
had all the experience they desired, while Tomat, though
obliged to return to Belun, had not enough dogs for the
journey, and yet, as I was compelled to find drivers in
order that the teams could be returned to Upper Belun,
Tomat became a passenger on my hands. A bright
young chap named Geordi Nicolai (George Nicholas) volunteered his own services and those of a fine team of dogs
which he had managed to gather together. Geordi had
been decently reared for a native (and I always found
that those who had been well-fed and nurtured were infinitely superior to their groveling fellows); he was intelligent and thoroughly acquainted with the road, and, as
it transpired, I liked him so well that on my second
search for De Long I hired him again. My next driver
was a half-breed Tunguse, named Phadee Achin, square-
jawed, square-headed, and resolute. There was not a scin-
tilla of nonsense in his composition; he ate his fish, bones
and all, and digested them, too. His face even then was
covered with blisters and sores, his cheek-bones were raw,
his complexion was a peculiar livid blue, and his lips
were black. Full-chested, square-shouldered, and clean
in the flanks, he was taller by far than any Yakut I had
seen on the Delta.

. Yet I was not so much struck by his face, however
comical, as by the peculiarity of his name when he introduced himself as "Phadee." "Good," said I, "Paddy,
you've got a first-rate name;" and I certainly thought
it a queer coincidence, for he had all the air and *tout en-*

semble of a large-boned, stalwart, but dark-skinned Irish-man; and I became morally convinced of the fact that in the long ago some adventurous Celt had forced his way into these lonely, frozen regions.· So I engaged Phadee, who talked in monosyllables. How many dogs had he? Eleven, he replied, holding up his fingers. When could he start? Now. Had he any food for himself or team? No. How did he expect to live? I asked, using the words "Cushat soak?" His only answer was a repetition of my query, "Cushat soak" (nothing to eat). "And so," I thought, "my man, if you can stand it, so can I."

But the two sleds, it was plain, were not enough to transport all my baggage; and so it became necessary to hire a third team to assist us part way, at least, and return when, from fatigue or want of food, it could go no farther. To fill this place an old fellow named Starry Nic-olai (Old Nicholas) was recommended to me, as a man who had walked all the way from the shore of the Arctic Ocean to Belun without anything to eat, and in the dead of winter; the narration of which exploit he indorsed as correct by simply adding "Verna" (truly). He was very poor as well as old, and had no team, but the vil-lagers promised to furnish him with seven dogs; and so with everything in readiness, — except, indeed, the rather important item of food, which the natives assured me would be provided in good time, — I at last sought my berth. The wind had almost increased to a hurricane when we turned in for the night, and long before dawn was howling in fury. I dreaded that it might prevent our departure, but nevertheless dressed and prepared to start, eating a hearty breakfast of raw and cooked fish. Tomat came to me and said, "Pagoda, poorga, pi dome soak." But I had been outside and was not so certain; the wind nearly carried me off my feet, to be sure; but then it was from the northwest, and, pointing in that direction, I told him so. Still, as he shook his head and persisted in repeating "Pi dome soak," I could do nothing but acquiesce.

Yet I was very anxious to be off, feeling confident that if I could rightly strike De Long's trail, I would shortly find him and party, doubtless dead, in some hut or crevice in the river bank. But then I must search at once before the valleys became entirely filled with snow; for it was only possible that the party would or could erect flag-poles to attract the attention of rescuers or passers-by. I scarcely expected to find them alive, my only hope being that they had fallen in, like Nindemann and Noros, with natives; yet they had journeyed along about midway between Arii, the nearest village to the southeast, and North Belun, the nearest one to the northwest. At least, I might be in time to rescue their bodies from the mutilations of wild beasts, and to secure our valuable records; for the face of the country clearly showed me that if I was delayed until spring all trace of my unfortunate comrades would be swept away by the floods, which at that season of the year completely submerge the Delta, and leave as driftwood great logs as large as ship-spars, some deposited on *tundra* beds forty feet above the river. When breakfast was over, Geordi Nicolai came in and firmly said, —

"Poorga, periscomb soak; sarsun" (gale, walk no; to-morrow).

But presently "Paddy" made his solemn appearance, armed cap-à-pie for a battle with the storm; head-gear, gloves, dog-stake, and all. He had already come some little distance, and was, I feared, about to sanction a postponement of our departure; so I hastily addressed him, —

"Pi dome, Paddy?" said I.

"Pi dome," he assented, without changing countenance.

14

Tomat put in a vigorous protest; but as Starry Nicolai had meanwhile arrived in evident preparation for the journey, though, it is true, strongly inclined to belie his looks and take sides with old Tomat, I felt, reinforced by Paddy, considerably in the majority, and so gave the peremptory order to "Skaree, pi dome" (make haste, go on).

CHAPTER XVI.

STRUGGLING WITH BOREAS.

WE set out in a tempest of snow, laying our course for Osoktok, where De Long's second record had been found. Had it only been snowing our discomfort would have been comparatively slight; but the wind blew fiercely, veering from northwest to north, and ere long to east, full in our faces. The dogs were poor, the sleds overladen, and old Tomat kept croaking dismally that it would storm for ten days. I rather regretted having forced the natives out, for they disliked to face the wind, as did the dogs, which refused to follow the course, lowering their heads and turning clear around to avoid the cutting blasts. This consequently made trouble for the drivers, and I almost despaired of pulling through; the dogs howled in unison with the storm, and the natives kept time with their cudgels. "Paddy" graphically described the situation to me, as he clambered upon the sled, after clearing the harness and hauling his team into line for the hundredth time, —

"Savaccas a mooshina, poorga booda" (dogs and men are alike), said he, placing his forefinger at the bridge of his nose, or where it meets the forehead; "poorga pomree." By which he meant that the wind striking them between the eyes would kill both. And he was right; a cold gale first produces pains in the head, then drowsiness, and lastly the sleep which knows no waking.

We toiled patiently on, and long after the time when we should have reached Osoktok brought up at a little old hut, or rather its ruins. The natives had driven to it merely as a landmark, and to take a breathing spell. It was impossible to get inside, so we all sat down under its lee in the snow. The natives smoked and then lunched on several raw fish, laughing at my refusal to join them; for I looked forward to a hot meal at Osoktok, where, considerably past midnight, we arrived, almost exhausted by our tiresome journey. The dogs were so fatigued that, as soon as staked, they coiled up and fell asleep without even looking for their accustomed fish; and, indeed, it was just as well, for there was nothing to give them. I was pleased to notice upon our arrival a tall staff with an arm attached, pointing towards the deer cache which one of Tomat's sons had described to me.

The door and roof being far from tight, the hut was partially filled with snow. This we leveled off, spread our beds, cleared the fire-place, and in a little while softened before a kettle of hot fish and a pot of tea. All this time I was supposing that the natives had brought fish or food of some kind along for themselves; but, jolly devils that they were, my bag of ten fish had, according to their custom of sharing supplies while on the road, been the only store from which they drew; and as they had previously eaten two raw, and now helped themselves to two more while the pot was boiling, only three fish remained, and this was our first day's journey. Yet I then imagined that but one fish had gone into the pot as my contribution to the meal, and so ate on in peace, as also did they. I afterwards searched the hut for any small articles which De Long's party might have abandoned or lost; but found nothing beyond some deer bones, the hide of which, one of the natives told me, had been picked up.

A restful sleep, and betimes in the morning we were out and on again. I observed that my dusky companions refrained from their frozen fish, and that our kettle of hot fish was smaller than before; yet as there was plenty I did not care. The wind and snow still raged and drifted, and the miserable dogs yelped and shivered from hunger and cold. They seemed more like wild wolves than domestic animals, in their mad impatience to be off, though some were almost too weak to stand up. Following the bed of the river, we pushed on against the storm, now fairly in our teeth, and so thick that it obscured the leading dogs; which at length, utterly overcome, lay down, and howling wearily refused to move. The natives then took turns at putting a drag-rope over their shoulders and pulling the leaders along, while the rest of the teams were beaten into motion. I could do little else than sit on my sled and shout out encouragement. Our situation was serious indeed; four hundred versts intervened between us and the nearest succor, Ku Mark Surt; and the natives assured me that such storms continued for ten days or two weeks. If this was no exception, we would certainly be snowed in; for should our twenty-nine dogs entirely succumb, the natives could not possibly drag the sleds, since even now, with both laboring at the harness, we could barely worry along. At any rate we must make the distance of fifty versts between Osoktok and Usterda as one day's journey; and so we did, reaching the hut long after midnight. It was pitched on a point of land between the Lena and a branch running up into the northeast, called Obi Bute Yaisia; but when the natives looked in they found it full of snow, so we kept on to Macha, a hut about a mile below or just beyond the mouth of the northeast branch.

This was comparatively new, weather-tight, and warm, and soon we were seated around a good fire sipping our hot tea. The natives, I noticed, went without their usual feed of frozen fish, and were dilatory in preparing supper; so I told Tomat to put on the kettle of fish.

"Ballook soak," said he.

"What!" I cried in astonishment. He only shrugged his shoulders, and extended his open palms, repeating with genuine sorrow, "Ballook soak."

Knowing they were all right as long as my fish lasted, and with no concern for the future, the rascals had hastened to consume my provisions without giving me a thought; and now when I accused Tomat of the theft he simply pointed to Paddy and lied, saying he believed them to belong to him, whereupon Paddy passed the lie to Geordi, and so it went around, all fibbing in the fatuous manner of children instead of manfully ascribing their duplicity to a pardonable necessity. And there the matter rested. I curled myself up in my sleeping-bag and went supperless to bed; but not to rest; for, though not very hungry, I was very tired, and the old pains, revived by the heat of the hut, were again biting and gnawing at my legs. No new blisters had appeared, and the sores were healing, but yet my agony was terrible, and I tossed about craving to lie in the snow, until, at last, I fell asleep from pure exhaustion.

After our hot tea in the morning, the natives dug around among the heaps of rubbish in search of offal, but there was not a scrap or sign to be found. The half dead dogs were then dragged from their beds of snow, and back we started towards the hut at Usterda; for it was my intention to cross the river where De Long did, and then, as he had done, keep along the west bank, in the hope of reaching a settlement. Accordingly, I searched about the hut, and then followed the foot-prints of the unfortunate party, and the track of the sled on which they had hauled Ericksen. This was quite plain on the ice of the river, for the fierce gales had swept it clean of snow, and when the party had crossed over the runners had cut deeply into the soft young ice. I also saw where one of their number had sounded the ice, or had punched holes in it with a pike or staff; and again where more than one had broken through, and in their flurry to get out had plunged in again or oftener, and only escaped by retreating.

Once over the river, I turned to the southward, rounding the bend, and thence on to a high bank, where the party's foot-prints had been plainly frozen in the soft snow. I now proposed to pursue these tracks until I came to the little old hut, which, Nindemann had told me, they reached after a slow march of two days; the place at which Ericksen had become too sick to be moved, and where, waiting until he died, they had buried him in the river. It should properly be distant about twenty versts, and when I explained to the natives where I wanted to go, they said they knew its location. So we struggled along all day, and at length came to a hut which agreed with Nindemann's description; at least, in its distance from Usterda. He had said, however, that when Ericksen was buried they had cut an inscription in a board, which had previously served them as a table, and had placed it over the door of the hut, leaving, beside, a gun and some ammunition there. I made a thorough examination of the place, casting out the snow, but neither board, gun, nor any evidence of the party's presence there could I find.

Evidently I had lost the trail, but *how* I could not understand. There was no doubt of my having followed the main branch of the river, and certainly I had stuck to the west bank; so where was the mistake? Shortly after leaving Usterda I had noticed that the river took a great bend to the westward, and had then inquired of the

natives if that was the Oshee Lena and whether it led to
Mat Vay ; and upon their assurance that such was the
case, I had kept on, finding subsequently that the main
branch turned again to the southward, and, farther on,
to the southeast. I now asked the natives if they knew
of any other huts on the west bank. Yes, but they were
a long way off, or far to the westward of the river. Yet
they did know of one close at hand on the *east* bank.

Now it occurred to me as quite probable that Nindemann had become confused and mistaken, or forgotten
the exact location of the hut ; for the whole Delta is
nothing but a congregation of islands ; and it was on one
of these, in sight of a tall signal staff, and some trestle-
work which they had taken for a signal station, that
Ericksen, according to Nindemann's account, had died.
This was a clue for me ; but as soon as I took the natives
into my confidence, they straightway found me a dozen
signal poles. I therefore concluded to try the hut on
the east bank of the main river. Slowly we crept on,
the poor dogs staggering with weakness, and the natives
trudging along without a murmur; while chilled to the
bone, I sat on my sled in a kind of dreamy torpor, with
no other feeling than that of hunger. We halted at
last before a hut, and looking above the door I saw there
was no board, and knew that we had only been straying
farther from the trail. The natives, dropping inside
through the smoke-hole, hunted about but found nothing, and as the hut was charged with snow we could not
sleep there.

It was not far, the natives said, to a place called
Sister Ganak, so thither we directed the weary teams.
Paddy's face grew longer at each step, as he coaxed and
cudgeled the dogs by turns, and occasionally paused to
quarrel with old Tomat for not working as diligently as
himself. It seemed an interminable journey; the tempest strove to overwhelm us before we could reach

shelter; and I, benumbed and half-famished, saw nothing towards the end but a wild white sheet dashing incessantly at us, until, finally, a hut loomed up, and shortly afterwards I was seated within it before a roaring fire. The place was roomy and comfortably free from snow, and warmed up by our hot tea and the cheerful flames we soon forgot the day's misery. Searching around, the natives discovered some fish entrails, a number of dried fish heads strung on a reed for use as bait in fox-traps, and some reindeer bones with shreds of meat or tendons left on the legs and haunches. These things satisfied me that De Long's party marched in another direction, for certainly they were not in a condition to reject even such offal. We roasted the bones and fish heads in the fire by means of a stick, and I would have relished them had it not been for the vile odor they gave forth. The natives put on their kettle and made a compound soup of all the refuse they could find in the ash heap; and they seemed as happy as though the pangs of hunger had never assailed them. Not so our wretched dogs, which howled piteously throughout the night; though Paddy set free his two leaders and permitted them to scratch in the ash heap for whatever bones or bits we had (?) overlooked.

Would the storm never cease! I asked the natives how far it was to the next station or hut on the river, and they said forty versts; from which to Mat Vay it was seventy more. A glorious prospect, indeed: nearly three hundred versts to the nearest settlement; remorseless weather; not a mouthful to eat; ourselves half-frozen; and our dogs, upon whom alone we could depend for rescue, already on the verge of death. When morning broke in wrath, it seemed that the tempest had only rested during the night to gather fresh force for the coming day. A pot of tea for our breakfast, and then, though the natives begged hard for a day's rest, I insisted upon continuing the journey forty versts to Qu Vina, the next hut.   They demurred; the dogs could not possibly proceed further without food or rest; but here I explained to them the necessity of pushing on, promising to stop and recuperate at Qu Vina for one day; and so at length they yielded an unwilling consent.   Again we faced the storm, weaker and hungrier.   My legs, however, in their loose warm mittens, had so healed that I was gradually regaining the use of them, and could manage, when the dogs went slow enough, to stumble along at the side of the sled with my hand on the rail.   The day's march differed in nowise from the preceding ones, unless in the increased feebleness of our poor dogs, which limped painfully on, exacting more and more aid from the drivers.

The snow was now so deep that at times the dogs were buried and almost useless, wallowing helplessly about in their harness.   They would labor for a few minutes after a long breathing spell, and then, when the sleds stuck fast, would lie spiritlessly down, howling and yelping as if in expectation of a beating.   And so we struggled and rested, and struggled on again, each time as though it were our last pull for life; and it seemed to me that the terrible journey would have no end.   Now and then, discouraged, I would decide to cache the relics at the first safe place we came to, returning for them when I could ; but after a moment's reflection, recalling how persistently we had clung to these treasures, — the records and valuable accumulations of our two years of toil and suffering, — and setting my teeth against the storm, I would swear a new oath to carry them through, let come what might.

From our limited knowledge of the country, as well as from my lost faith in our charts, I still had hopes of finding my lost comrades dead or alive.   The natives, with whom I now could converse quite intelligibly, told me that the river which we were following led to Mat Vay, and was the west branch; hence I felt assured that even if De Long had abandoned it he must eventually have returned to it again, as I had followed Nindemann's track as far north as Mat Vay, and there found the Jeannette waist-belt; for I did not suspect that there were a dozen rivers to the east of me along which he might have marched; and yet that he had really left the branch I was pursuing was evident alone from the amount of offal which we had discovered in the huts; and then, too, we had come upon no trace of him after leaving Usterda. Nor was this made clear to me until five months afterwards, when, returning on my second search accompanied by Nindemann, I first conducted him across the river at Usterda, and then on down the west bank to where the river takes a great turn to the westward, and was surprised to have him assert, " Here, sir, we recrossed the river to the eastward, and stood away down to the southward and eastward."

And this was the cause of my losing the trail, though, of course, I had no means then of knowing it; but now in looking back, I can readily understand how easy and natural it was for De Long to make such a mistake. The river veered to the westward. He did not wish to go west, but south, and consequently recrossed it. Then again on his chart, as indeed on all charts, just about where he imagined himself to be, is marked a large branch, which flows almost as far west as the river Ala-nek, and he doubtless believed this to be the main western branch, and that in recrossing it he was on his way south to a settlement.

The days were very short, the sun having forsaken us when I quit Jamaveloch to go to Belun; so it had been pitch dark for many hours ere we drew up at Qu Vina. This was an old hunting lodge belonging to Tomat, and was in a bad state of preservation, yet I cannot remember that the sight of any dwelling has ever filled me with half as much joy as my first glimpse of it. For I was exhausted to the verge of fainting, and frozen to complete numbness or dumbness, since I was conscious only of acting mechanically. I was awake and aware of all that was transpiring around me, but had lost all feeling and power of speech, and existed like an animated dead man. I lay on the sled until a fire could be started, and meanwhile, as the natives went back and forth, I saw their jaws working, and knew from their unnatural manner of talking that they had found something to eat. Crawling in on my hands and knees I secured a position near the fire, the snow which covered the unboarded floor forming a soft bed for me, and, although there were a few berths raised around the sides of the hut, I fell asleep where I lay.

When the natives had housed our sleeping-gear and arranged matters for the night, they awoke me to partake of the supper which they had prepared. As the hut was located on one of their deer runs, there was a goodly quantity of offal inside and hung up about the door. The fur of the deer legs is peculiarly adapted for the making of boot-tops, and is instantly cut from the slaughtered animal and handed over to the women to be dressed. After the skins have been removed, the legs are roughly stripped of what little eatable tissue there is on them, and then, as the sewing-thread used by the natives comes largely from this source, they are generally hung up to dry, and the tendons removed at will. Of these there was quite a number in the hut, mostly with hoofs attached, which, indeed, when heated or softened in soup, are well worth eating.

I drank a pint pot of tea, feeling considerably refreshed thereby; and then emptied a wooden bowl full of the odorous, if not savory, juice which the natives skimmed from a large kettle containing the shanks, hoofs, and various bones of the deer. When the kettle had been boiling for some time, the natives drew forth from it and gave me portions of the putrid, stringy mess, and, swallowing some of this together with the soup, I finally closed my eyes in a much needed sleep.

The dogs had been fed on a few of the scraps, the leaders getting the remainder of our hot mess, and the weakest were brought inside the hut to thaw out. Next morning we awoke to find the storm as vigorous as ever ; but as Mat Vay was only forty versts distant I determined to accomplish the journey at once, since from that point there was no place of refuge short of Bulcour, and no succor nearer than Ku Mark Surt, fifty-five versts still farther to the southward. The natives again entreated me to tarry, saying it was *Vos Crusina* (the Sabbath Day). "But," said I, "bless your Sabbath, we must go on or perish." An enervating dysentery had attacked me, and now besides cold, exhaustion, and slow starvation, I had a fresh enemy to struggle with.

I was now satisfied that I had done all that was possible for me to do at that season of the year ; if De Long and party were alive and in the hands of natives, they were certainly as well off as myself ; if dead, and of this there could be scarcely any doubt, then the natives were wise in admonishing me that I should die too if I persisted in searching at that season of the year for a lot of corpses, which I could find with safety in the early spring. So I urged an immediate departure, saying that to delay meant to die of starvation ; but the natives looked incredulous, and pointing to a heap of refuse, rich in fish heads, skins, and bones, goose wings with feathers attached, deer-skins, and other scraps, only smiled with an air of jollity, and answered, "Cushat manorga," (plenty to eat).

Yes, thought I, a profusion, but heavens ! how hungry a man must be before he would even acknowledge such carrion as food. Much of it was alive with worms, which, indeed, the natives did not bother themselves with removing. And here is where I saw my friend Paddy do justice to his digestion. His heavy lower jaw went up and down with the regularity and power of a pair of shears for cutting iron, crushing with ease the soft bones of fish and birds; and Paddy was good to me, pulverizing with his hatchet some small bones, which he extended to me with the assurance that I need not fear starvation; they would keep hunger from our door, so to speak.

They further argued that the dogs must be rested and fed, after which we would start and go directly on; that a short distance from Qu Vina, they knew where a quantity of venison was buried, and that if I would wait over *Vos Crusina* they would stop and get it, and we would then have a sufficiency of food to carry us to Ku Mark Surt. I was by no means loath to recuperate, and eat of our bountiful supply of offal; and, besides, in doing so and resting the teams we were doubtless making haste slowly; so I made a virtue of necessity; telling the natives, however, that if the weather improved, we would not halt at Mat Vay, but would keep right on and camp in the snow, for a native will never willingly abandon a hut in bad weather, if provided with food.

Monday morning dawned calm and clear, but intensely cold; and the stars glittered gloriously in the heavens. Running in an easterly direction for several hours, we then drew up, and the dogs being staked, Tomat stretched out my sleeping-bag on the snow alongside of the sled, and told me to crawl into it and *spee* (sleep), until they came back. For about an hour I lay there comfortably warm, and then the dogs bayed out their return. They had secured a lot of bones which had been interred the summer before, and were now covered with the black earth of the *tundra;* for it is a custom among the Yakuts to cut the meat from the carcass of their game for convenience in transportation, and then bury the bones with an eye to future use in case of famine, or an emergency such as had befallen us.

Strapping the bones on top of our sleds, we made another start, keeping along a branch of the river bed to the eastward of the one which we had previously followed; passing a couple of huts which I had the natives overhaul; and, at length, coming in sight of Mat Vay, far off to the southward.

The weather had remained clear, and the traveling was good, save for the sand-spits in the river, which cut the sled-runners and worried the dogs. I had now but one object in view; that of reaching Belun, and thence hurrying on to Yakutsk, where I could fit out another expedition to continue my search in the early spring, before the floods could carry off the dead and their effects to sea. So, agreeably to my expressed intention, I passed by Mat Vay, albeit the natives wistfully eyed the hut and longed to hear me sound a retreat. But on we went, pursuing the main channel of the river, between great blocks and hummocks of ice, which, lodging there the previous spring, had outlived the heat of summer, and now lay fixed and immovable where they had grounded.

Huge masses of solid ice were there, as large as ordinary dwelling-houses; and what surprised me most was how such monsters could have formed in one season. They were evidently the growth of a single winter, cast out, upon attaining their maturity, in the summer time, and were now in a proper position to stem the spring floods and act their mischievous part in the great ice-jam which yearly gorges the Lena outlet, and plays such sad havoc with the natives on the Delta, who must needs flee to the mountains for refuge. As we entered the mouth of the main river, where these giant blocks of ice lay stranded like so many monuments of the Druids, the wind began suddenly to pipe and blow in our faces. We had sought an entrance where the Lena proper debouches from between the mountain ranges which guard its banks, and where a vast natural funnel had been formed, through which the cold winds, collected by the high hills, were driven with the force of a fan-blast.

Awed and charmed at sight of these enormous masses of ice, I was absorbed, as we toiled slowly and laboriously along, in a study of their peculiar structure, oblivious of the wind and the monotonous " Yap" and " Tuck " of the drivers, when Geordı Nicolai rudely awoke me from a dream,—wherein a battle of the giants who could so promiscuously hurl such massive missiles was being waged, — to say, —

" Bulchoi balogan, Mahor!" (Big house, Major); aptly referring to the size of the ice-blocks. To be sure, they must have had their origin in the fall, when the young ice running and telescoping formed into a gorge, and the water percolating through the masses cemented them into solid bergs.

But, aroused from my reverie, I looked around and saw that the natives were engaged in an unequal struggle with the storm ; for —

> " Boreas had burst his dungeon, armed with ice,
>   And snow, and hail, and stormy gust and flaw."

Standing on smooth patches of ice, it was impossible to resist the strong gale of wind, which, by sheer force, swept sleds, dogs, and men to leeward. Ah! it was cold! The blast seemed to go clear through me ; and presently, seeing that it was almost impossible to make any headway against it, I told the natives to camp behind one of the blocks of ice. But no ; they combated my order in a body, possessed of a healthy dread of the river, even in winter-time ; for they had known it to overflow its banks and carry destruction to the valley in an hour.

STARTING OFF ON THE SEARCH FOR DE LONG.

So we labored along among the boulders of ice for a the than a mile, and then opening out, at length, into the clear bed of the river, we turned to the west bank, where there was a sufficient snow-fall; and, digging a square hole, as before, in the drift, set our sleds up to windward, and crawled, cold, supperless, and altogether miserable, into our sleeping-bags, beyond the fiercest fury of the storm. And it was an unspeakable comfort, indeed, to lie there stretched out horizontally in the snow, and feel the warm blood slowly coursing through our veins, until, glowing with heat, we sank into a sudden and refreshing sleep.

Elsewhere I have portrayed the alarming vicissitudes of a night in the snow; how the first stage or flush of pleasant heat, engendered by extra clothing, gradually cools, and the warm moisture or perspiration chills, until suddenly the sleeper awakes — awakes if it is extremely cold — with a start which robs him of considerable skin, and renders him the more painfully conscious of the beautiful snow which has taken advantage of his slumbering to drift under his garments. To this I may add that it is very arctic weather when the sleeper's nose is frozen, and then his thumb, when he righteously tries to thaw out his nose by holding it in the palm of his hand with thumb extended and exposed; which thumb he, later on, thrusts into his mouth to thaw, and so on *ad nauseam*. Thus we lay through the night, and when day came it brought no lull in the storm; and, since we could not hope to push forward, we remained all that next awful day, without food, cramped and motionless, with the poor dogs cuddled shivering on top to keep us warm.

But when the second morning arrived, and the gale had sunk to a gentle breeze, it was absolutely necessary that we should creep out of our bags and again start our stagnant blood into circulation; and to do so was no mean effort, for we all experienced considerable difficulty

15

ᴏraightening out, as it were, the crooks in our backs.
ᴛhe natives tried in vain to build a fire with the water-
soaked and frozen driftwood which they picked up about
the camp, until, finally, in despair and impatience to be
off, I bade them *pi dome*, trusting to the near future for
a good warming and a cup of hot tea. But here my extra dog-team, driven by Starry Nicolai, must leave us.
He had come, according to agreement, as far as he could
without food for himself or dogs, and would now return
to North Belun. Yet, before we parted, it was no more
than right that he should be given a repast of the best
that our sleds could afford ; and so, when his load had
been shifted upon the other two teams, and I had exchanged our poorest dogs for the best of his, Tomat drew
forth the skeleton ribs of venison which had been disin-
terred the day we left Qu Vina.

With an axe the rib pieces were soon severed from the
back-bone, and then from the inside of these the natives
cut strips with their sheath-knives, and handed me a
chunky morsel from the loin, as breakfast. I bit into it
without any ceremony, while the dogs clamored frantically for a share. So long as it remained frozen the meat
did not exhibit the vile extent of its putridity ; but directly I had taken it into my mouth it melted like butter,
and at the same time gave off such a disgusting odor that
I hastily relinquished my hold upon it, and the dogs captured it at a single gulp. The natives first stared in
genuine astonishment to see me cast away such good
food to the dogs, and then burst forth into hearty laughter at my squeamishness. But I was not to be outdone,
much less ridiculed, by a Yakut, and so ordered some
more, perhaps a pound of the stuff, cut up into little bits.
These I swallowed like so many pills, and then gazed on
my Yakut friends in triumph ; but not long, for in a
little while my stomach heated the decomposed mess, an
intolerable gas arose and retched me, and again I aban-

doned my breakfast, — my loss, however, becoming the dogs' gain.

At this the natives were nearly overcome with mirth; but I astonished them by my persistence, requesting a third dose, albeit the second one had teemed with maggots; and, swallowing the sickening bits as before, my stomach retained them out of pure exhaustion.

And now Starry Nicolai was ready to begin his return trip of almost two hundred and eighty versts to the northwest. I broke off and gave him about a quarter of a pound of my brick tea, but beyond this he had not an ounce of food for his long and lonely journey. He begged hard for one of the tin cups; but I was obliged to refuse him the simple thing, fearing lest it might lead some other searcher astray; and my heart misgave me as I watched him, with his miserable little team of dogs and rickety old sled, trot slowly out of sight. Yet were my fears happily groundless, for I hired him again the following spring, and he continued to drive teams for me until I left the Delta.

CHAPTER XVII.

END OF MY FIRST SEARCH.

Forcing the Journey. — "Oo, oo." — Bulcour. — Rough Repairing. — "Paddy" despairs. — But quickly revives — And performs Wonders — Ku Mark Surt. — Buruloch. — My Deer Train. — On to Belun. — Tedious Progress. — Ajaket. — Belun. — Epatchieff.

AND now began the most difficult and distressing stage of our journey; for we had scarcely set forth when the wind rose again, and, gathering up the sand and snow from the bed of the river, assailed us with fresh fury, cutting our faces and filling our clothes with the coarse rough grains. The dogs, unable to keep their feet on the glassy ice, were borne hither and thither by the cruel wind which showered sand in their eyes, and so utterly demoralized them that they fell down and howled in terror. Then, too, the sled-runners were greatly impeded by the driving pebbles; and when the natives guided the teams along the shore where the sand lay in bare spits, the dogs were unable to pull their loads. Moreover, the sled-runners had now worn off to such an extent that some of the lashings which held them to the stanchions had parted, and instant repairing was in order. But the natives were determined, if at all possible, to accomplish this long march to Bulcour in one day, and when I suggested to Paddy that we camp at nightfall, he answered sturdily, —

"Soak, pomree; poorga manorga."

The natives saw that I was very sick from eating the

decayed venison, but they were really afraid to stop without shelter or fire. At length one of the runners gave way; and turning the sled upside down, a piece of driftwood was found, chopped into proper shape, filled with the necessary number of holes, bored by means of the fiddle-bow drill instrument, and then lashed on with thongs cut from the dog-harness or trace; and soon we were off again. But lashing after lashing broke, until it looked as though the sleds were entirely going to pieces. I tried then to induce the natives to cache the loads, hasten on to Bulcour for the night, and send back for them the next day. Tomat was perfectly willing, but Paddy hung on like a bull-dog, telling me constantly, "a little while, a little distance," until I became utterly helpless. We passed the "Place of the Sleighs," and Paddy only said "a little while," and would not consent to stop at the hut. So I lay on top of the sled writhing in agony, until at last, and it seemed an age, the teams stopped and I heard the natives speak of water (*Oo, oo*), striking the ice with their iron-shod staves. Looking up, I saw that we had reached a stream of comparatively warm water,—for the ice had hitherto averaged a thickness of four feet,—and that the drivers were stretched out on their stomachs eagerly sucking up the water through holes which they had punched in the thin ice. My efforts to follow suit were not rewarded, for the water would not rise to the surface, and my suction power was too feeble to draw it up. Afterwards, particularly between Belun and Verkeransk, I found many such springs bursting forth and flooding the ice for miles.

It was almost morning when, after a succession of accidents, we arrived at the foot of the hills on which are pitched the huts of Bulcour. The dogs, too weak to draw the sleds up the bank, were turned loose, and the natives then unloaded the sleds and carried them into the hut for repairs. While they were thus engaged I

made strenuous endeavors to scramble up the bank, but
it was so steep that I as regularly slipped back, and accomplished nothing but the pounding of a shelf in the
snow. So I decided at length to wait until the natives
had time to assist me, and sitting down I was soon fast
asleep, warm and comfortable from my fruitless exertions. And there I remained until the natives, missing
me, instituted a search and set up a great shouting, which
finally resulted in their discovering me quite at home in
the cozy pocket which my struggles had formed in the
snow. Sliding me down to the foot of the bank, they
seated me on one of the sleds, and so drew me up to the
hut; and I had no sooner crept into my sleeping-bag than
my eyes closed again. Hot tea, which they aroused me
to partake of, was all I had for supper, since, forsooth,
there was nothing else, save the putrid reindeer bones,
and of them I had already eaten more than enough.

When we awoke, the natives busied themselves in making good the damage to our sleds; but they did so in such
a rude manner, piecing and splicing the runners with
rough short sticks of driftwood, that I remonstrated with
them, asserting that the sleds would surely break down
before going any distance. They only looked at me quiz-
zically, as though they would say, " Go away, child; what
do you know about sleds or venison ? " and Geordi Nicolai playfully offered me a piece of the latter which he
was eating. So they had it their own way, and late in
the forenoon we succeeded in making a fresh start. I
tried again to prevail upon the natives to store all my
heavy weights at Bulcour, and send for them from Ku
Mark Surt, now only fifty-five versts distant; but Paddy
was tenacious in his grip, and confident of pulling them
through, so I did not quarrel with him. Yet we had
barely driven a hundred yards, when one of the sled-
runners gave way; and I had consequently the laugh on
the natives, telling them that though they could eat rot-

ten meat, still I could mend a broken sled. But even after this, successive breaks occurred, until I feared a total dissolution of our teams.

The snow had deepened considerably since my passage north, some twenty days before; the dogs were worn down to mere skeletons; and the distance which, in going, I had made in less than eight hours, and on my return had trusted to accomplish in ten or twelve, it now looked as though we would never cover. Ku Mark Surt contained all that was necessary for our comfort and recuperation; but it was evident that we had forced the distances and overworked the dogs. The natives were worn out, and even Paddy threw down his stake in disgust and cast himself on the snow, cursing bitterly, I imagine, in Yakut. Yet it always ended in a smoke of the pipe, and then, when dogs and men were rested, the journey was once more resumed. Thus the day died out, and, at length, long after dark, we all became so weary that I proposed that we camp over night in the snow, and finish the short distance to Ku Mark Surt in the morning. Paddy was discouraged, and had dejectedly taken a seat by his sled; so I ordered Geordi Nicolai and Tomat to make camp, which they were proceeding to do, when Paddy suddenly revived, and stopped them with the remark that it was only a little way to the village. Then stepping to the front of his team he seized the leading dog by the back of the neck, pulled him out of the snow, shook him vigorously, and then, pointing his own nose to the stars after the manner of the dogs, began howling like a wolf.

At first I thought he had gone mad, but presently the dog caught the spirit of his master, and elevating his head howled dismally. This seemed to be precisely what Paddy desired; he eagerly encouraged the other dogs, and in a few minutes both teams were making night hideous. With his hands to his ears, Paddy then stood in a listening attitude, while the discordant chorus grew louder and louder, until, finally, during a brief lull when the dogs paused to catch breath, he triumphantly exclaimed, —

"Savaccas, savaccas!   Ku Mark Surt savaccas!"

I bent eagerly forward but could not hear the dogs at the village, for the howling of our own; but Paddy assured me it was *da-loca, da-loca* (far, far away). And now I noticed a restlessness among our teams; they barked and snapped and leaped in their harness; for they had detected the far-off wailing of their fellows, faint, but certain, in the frosty night air. Paddy warned me to be quick, and I had just time to fall on my stomach across the sled, when the now thoroughly aroused and excited teams dashed forward at full tilt, vying with each other in their wild ardor to reach the village. On we flew, but Geordi Nicolai and Tomat had not been nimble enough, and were accordingly left to trudge on in our wake laden with snow-shovels and bedding; nor did they overtake us, for Paddy's trick was a brilliant *coup*, and the answering howl to his Lochiel cry inspired our dogs with a new life and strength — loaned them wings, and when they halted panting at the foot of the steep embankment there was a host of canine friends yelping out a welcome to them; while from the chimney tops the flames leaped cheerily, and the women flashed their flambeaux from hut to hut preparing the way for our reception.

Soon the sleds were hauled to a place of safety, the dogs liberated and staked in their harness, and later on, when they had rested, given a feed of fish heads. We were ushered into my old quarters, the principal hut of the village, and home of the young lady who had so adroitly combed her flowing and *lively* tresses. Here I was entertained with a disjointed recital of the game of "hide and seek" between the espravnick and myself; I

had taken to the roads, they said, and the espravnick
had followed me all around with a bag of bread, — which
seemed to amuse them greatly. Some of the natives had
been to Belun and seen the sailors, and consequently
must tell us all about their peculiarities : how one had
lost an eye, and another who had lost his mind wished
constantly to box with the Yakuts; and him they mimicked to perfection like so many monkeys, distinctly pro-
nouncing the words " Jack Cole." They chattered about
the guns and hatchets and other riches of the white
men; whom they plainly considered a great people and
very wealthy, notwithstanding that we lived solely upon
their bounty, and were beggars, indeed.

I need not tell how completely I enjoyed our supper of
fish, or the refreshing sleep which followed. Before turning in I had removed my outer clothing, and bathed my
feet and legs in a tub of hot water prepared by the women; they then greased my limbs with fish-oil or goose
grease, which softened and loosened the clotted blood
and matted hair. My toe and finger nails, I found, had
turned black, beside curling up, and were painful to the
touch. I slept soundly in the consciousness of having
done all I could for the relief or discovery of my lost
comrades ; and, however much I regretted my failure to
find them, still was it not something — nay, a great deal
— to have recovered our valuable records and relics ?

Early next morning I was out again, but could procure
no deer-team as I had managed to do before, and so we
were compelled to have recourse to our dogs, reinforced
by a few recruits from the village. A comparatively
good road now lay before us, tracked out by the many
teams which were coming and going along the east side
of the river, the entire distance being marked out by
stakes and tree branches set up within sight of each
other. Arrived at Buruloch, the deer station, I discharged Geordi Nicolai and Phadee Achin, giving them

all the tea and small articles which I could spare, together with the assurance that they would be fitly paid for the valuable services which they had rendered me. After our parting they refreshed themselves, and coolly set out on their dreary return to Sever Belun.

Buruloch is the first regular reindeer station on the road, and here the dog and deer teams are forever meeting. To protect the deer, which are allowed to roam about in the woods and feed on the moss which abounds in this particular locality, several large corrals of tall poles are erected to confine the dogs. The owner of the station complained to me of the great loss of deer from wolves in that vicinity, and he had consequently some difficulty in gathering together enough animals to transport my party and effects to Belun. After a night here, including a supper and breakfast of boiled venison, which in its best condition is infinitely worse than very poor mutton, I watched the process of arranging my teams. First in order was a leading driver with two deer hitched abreast to his sled, to the rear of which my team was attached. Tomat came next, driving his own deer, his sled laden with part of my treasures; then followed a driver conducting two teams as before, of two deer each, bearing the balance of the goods; and lastly came two relief teams, making sixteen deer in all.

From this the reader may partially appreciate the complex difficulties of Siberian travel, even when conducted under the most favorable circumstances.

A team of deer is not supposed to haul more than five *pood*, that is, two hundred pounds, or one passenger, whose provision box, not to exceed the same weight, requires a second team, and his driver must have a third; so that one traveler must needs employ six deer in his transportation. The provision box is a necessary part of his outfit, since he cannot rely upon securing any provisions while *en route ;* and should, indeed, take the precaution of purchasing and slaughtering a reindeer for his own use. As he retreats from the sea, fish in any quantity cannot be procured, and bread or meal, to the north of Verkeransk, is utterly out of the question. True, the traders and coperts carry kiln - dried black bread made of unbolted rye meal, but they use it very sparingly, and consume large quantities of venison, fish, or beef; whereas to the southward of Yakutsk black bread becomes literally the staff of life; and among the natives between Verkeransk and Yakutsk, the principal article of food is boiled milk with a little rye meal stirred in to thicken it; though, to be sure, venison and beef, when procurable, — as, indeed, the flesh of any horse which may die from disease or be killed in harness, — are all eaten with thankful avidity.

We made an early start from Buruloch, and I hoped to reach Belun within ten hours as I had easily done before. The snow was deeper, but I had plenty of deer and two good guides; yet we had not journeyed far before I learned that in proportion to the number of teams so is the difficulty of traveling. For when the first sled went slow, the second one was sure to foul with it and upset, and this it did with provoking persistency. Then, too, the guides lost their way (and who ever had a guide that did n't); we recrossed the river in search of the road, which we could not find, and so climbed along for a spell over the heaped-up masses of ice on the broken river-bed, blindly returning at length to the west bank. Time ran on apace, the air grew colder and more blustry; and it was all we could do to keep our sleds upright on their runners. I now found myself becoming more cold and numb than usual, and could not understand it, for I had been well-fed and rested during the past two days and nights; but it suddenly occurred to me that the manner in which I had boiled myself at Ku Mark Surt and removed the scabs from my feet and legs had much to do with my increased susceptibility to the weather. And this was the only reason I could give myself. The natives, too, complained bitterly of the cold, and, after wandering vainly across the river once more, drew up at a hut on the east bank. Here we rested and warmed ourselves, and then the drivers, fortified with some fresh information, again crossed over to the other side, and scaling a thickly-wooded declivity drove to a wild, romantic spot in the woods, where a few huts were congregated. We halted at one and made hot tea, the natives stuffing themselves with raw fish, of which I did not partake. Though half dead from cold and exhaustion, I was very anxious to be off, and roundly rebuked the guides for losing the road and idling away their time in the huts.

This place, I learned, was Ajaket, only a short distance from Belun, but a long way back from the river; and when I remonstrated with the natives for having gone so far out of their course simply to drink tea, they explained that the night was much colder than usual and they were afraid I would die. But the truth was and is, that a Yakut will travel forty versts in order to drink tea with his neighbor, the more gladly if the neighbor supplies the tea. Our road now lay through the low woods, across deep ravines, and over hills until we came upon the river bank again, when a brief run of less than an hour brought us to Belun.

The village was slumbering peacefully, and it required no little rapping to arouse the Commandant, whose welcome, however, was cordial in the extreme. Yet this last short ride had severely taxed my remaining modicum of strength. I felt as though all my vitality of body and mind had deserted me; and had many more hours been added to my arduous journey of twenty-three days, I certainly could not have outlived them.

The natives had gone to the Balogan Americanski and apprised its occupants of my arrival, and Bartlett and

Nindemann were soon greeting me. Swallowing some hot tea and refreshments, I then returned with the men to their hut, where they were comfortably quartered, and found that Mr. Danenhower had been unable to secure transportation, as I had ordered, for the whole party, but had furnished five of the men, beside himself, with fur-clothing and a bountiful supply of food, and started, in the best of spirits, for Verkeransk, leaving the others behind to await my return.

Bartlett then informed me that Kasharofski, the es-pravnick of the district of Verkeransk, had dispatched his assistant, Epatchieff, to the Lena Delta, to learn who and what we were. The people of Eastern Siberia had never heard of the Jeannette expedition; so when my tel-egram passed through Verkeransk, the espravnick, re-marking the strange characters in which it was written, and receiving at the same time an explanatory letter from the Malinki Pope, sent for an exile, M. Leon by name, who could write and speak French, German, and English, and had him translate it, — forwarding it imme-diately after by special courier to Yakutsk. He then sent his assistant at once to Belun, armed with a box of medicines prepared by Dr. Buali (White), another exile, and a letter, translated at his dictation by M. Leon into French, German, and English, inquiring who we were and what he could do for us.

I now decided to wait until Epatchieff returned from his protracted search for me; and meanwhile I urged Bieshoff, the Cossack commandant, to exert himself to procure reindeer clothing, and moccasins, mittens, caps, etc., for all my party to use on our sledge journey to Ya-kutsk.

I arrived at Belun on the 27th of November, 1881; and though I had not succeeded in finding the bodies of De Long and party, yet I had diminished the distance to be searched to the space, north and south, between

Usterda and Mat Vay, which, as the crow flies, is less than one hundred miles. And here I may mention the surprise which Noros occasioned me upon laying claim to the belt which I had found at Mat Vay and had supposed, on the strength of his and Nindemann's story, to belong to one of the remainder of De Long's party. During my journey I had traveled in straight lines 1,140 versts; but considering the devious course, the turnings and wanderings, the distance may easily be computed as so many miles; a verst being sixty-six per cent. of a mile. I also fully informed myself of the character of the natives, the location of their villages, and the resources for supplies of food and dogs, and, in fact, all things which would insure a successful search in the coming spring. For I had now made in addition a very correct chart of the topography of the Delta, and had De Long possessed it, none of his party need have perished.

Epatchieff, the assistant espravnick, arrived in Belun on the 29th, and I at once had an audience with him. At first our progress in comprehending each other was painfully slow, but Bieshoff, with whom I had had two days of profitable conversation, acted as interpreter, and presently our jargon and pantomime grew upon us, until Epatchieff was able to say, "Very good," when either made a particularly good hit in the other's tongue. So that, indeed, by the time we were ready to start for Verkeransk we had become quite chatty together, and while on the road, when we halted at *povarnias*, we talked politics, religion, and anything else we wanted to. To be sure we would now and then go astray in our conception of each other's arguments, as when we were talking of the relative cost of steamboats (*parahotes*) on the Lena River, some of which are made of wood and others of iron,—wood being very plentiful and iron very scarce, so scarce that Epatchieff emphasized its value by saying it was worth its weight in gold. At this, I told him of the great rarity and consequent value of iron in Central China (*Keti*), remarking, also, that I had seen a boiler weighing thirty tons cut up into little pieces and forwarded into the interior on the backs of men. To make it plainer I had reduced the weight to pounds, 60,000, and called the boiler, *kottle*, or kettle. This was all right, but in describing it, I had called it *golatz* (gold) instead of *jalazia* (iron); whereupon Epatchieff looked very mystified and then incredulous, until finally showing his finger ring and tapping one of the dinner-knives he besought me to tell him whether I meant *golatz* or *jalazia*. Presently we understood each other and laughed heartily.

With everything in readiness, I set out December 1st in company with Epatchieff for Verkeransk. Before leaving I first saw that the remainder of my party were all properly equipped and provisioned to follow me, and, putting Bartlett in charge, I directed him not to depart from Belun if any man complained of his clothing or outfit, but to promptly see the commandant, who would attend to all their wants. I started in advance in order to prepare the way for their coming and save time, the better so accompanied by the espravnick, since he has the right of way in everything. Then, too, a party of six with equipment cannot always be furnished with transportation at one time, the law limiting the number to three; but I succeeded in making special arrangements for the whole party, and thus avoided much suffering and delay, for although we were all enjoying fairly good health, still our former privations had left us very weak, and our feet and legs were yet tender from the effects of frost-bite.

CHAPTER XVIII.

FROM BELUN TO VERKERANSK.

My Record of the Journey. — Minus 45° Réaumur. — A Russian's Views of America and her Institutions. — Kasharofski. — M. Leon. — My Letter of Instructions. — A Visit to the Hut of the Exiles. — Leon's Case. — Life at Verkeransk. — Facts about the Political Exiles.

HERE is what my journal says of the journey to Verkeransk: —

December 1st. — Traveled ninety versts and halted at a *povarnia* for tea.

December 2d. — Traveled night and day, sleeping on top of my sled while in motion, with my sleeping-bag under my head for a pillow, but with no other covering than my deer-skin clothes. Much troubled by the upsetting of the sled, and the terrible pains in my feet The deer go too fast to permit of my running alongside the sled for warmth and exercise, so I must only grin and bear it. Stopped at six A. M., and had tea. Made eighty versts; stopped again for two hours to rest and feed the reindeer and breakfast; then made seventy versts before dark, halted for tea and fresh teams of deer, and started across the *tundra* one hundred versts.

December 3d. — On the go all night, losing our way on the *tundra*, and wandering aimlessly about. Ours was the best driver to be had, but he became hopelessly confused. Epatchieff pointed out the road, at length, remembering a small ravine which we had crossed. I had a compass with me, but no one knew the course over the *tundra.* We should have accomplished the distance in ten hours, whereas it took us fifteen hours and a half; and when we halted, our feet, hands, and faces were swollen from the intense cold.

Sunday, December 4th. — Traveled day and night, arriving (one A. M.) at the first *stancia* (station). I received two notes from Danenhower, saying that he secured venison and frozen milk here. We changed teams and drivers and were off again. Before leaving the station, we had a dinner of venison roasted on the coals, and "noodle-soup," Epatchieff having been so thoughtful as to lay in a store of "noodles" for our journey. Made sixty versts, and then another sixty, hauling up at a *balogan;* and finally halted at five o'clock, having made one hundred and forty versts between the hours of four A. M. and five P. M.

December 5th. — Made ten versts; stopped to rest the deer and dine, and then on again to a station, with eighty versts more added to our list of magnificent distances. From noon we had eighty versts ahead of us, and this point is two hundred and twenty versts from Verkeransk. Hauled up at eight P. M. for tea, having covered the last eighty versts in six hours and a quarter.

December 6th. — Drew up at three A. M., with sixty versts to our credit, in five and a quarter hours. Froze my left foot again, and it is bleeding all over. Cold!! Epatchieff's Réaumur says minus 45°. Made sixty versts and drank more tea. On again with only thirty versts to go, and these we made by six P. M.

Nine hundred versts in five days and eighteen hours. A fair showing indeed. But we traveled night and day without once sleeping in a hut or *stancia.* During our stoppages I had a number of pleasant talks with Epatchieff, who is a Russian, born in Yakutsk, sociable, intelligent, and withal a very fine fellow. He had many queer questions to ask concerning America, the " Great

Republic," and her constitution. He lives in the hope
that Russia will some day have a constitution, as many
unhappy people would thereby be saved from exile. He
has no faith in the criminal classes, believing, and rightly,
that they should be punished. He told me all about the
assassination of President Garfield, but said he had been
stabbed ; and he drew the inference that too much freedom was the death of presidents as well as czars ; a republic could not be all good, else we would not have
killed our president. He is a sturdy Greek churchman,
but has strange views on religion ; a man born a Russian
and brought up under the influence of the Greek church,
has no right to change or alter his religious belief ; and
those persons who do, he thinks, are very properly exiled.
A man born and educated a Romanist or Lutheran (for
he knows no other of the Protestant faiths) may, however, retain his religion and be true to his state ; but it
seems impossible to him that a Russian could abandon
the national religion and still remain loyal to his country.
He is very much interested in our marriage and divorce
laws, and particularly inquired if it was necessary to
have both the civil and religious ceremony performed ;
appearing somewhat surprised when I informed him that
each was equally binding, though certain persons made
assurance doubly sure by marrying three times, — once
by means of a state official, and twice by means of clergy-
men of different denominations. As to divorce, that was
purely a civil process, and could only be compassed for
cause. He had a very mistaken idea in regard to this
matter, having somehow or other arrived at the conclusion that men and women in America married and di-
vorced themselves at pleasure. I plainly told him that
Siberia was the only country I had ever been in where
every man had apparently a family, and few, if any, had
wives. This may have been putting it rather strong, but
throughout Eastern Siberia I found as many men with

concubines as I did with wives, and, moreover, met hosts
of these concubines at evening parties, and they mingled
freely with the wives of other men, and it seemed all
right.

Immediately upon my arrival at Verkeransk, I was
driven to the residence of Kasharofski, the espravnick,
who received me with great kindness. Danenhower and
the first section of my party had been nicely entertained
here, the men having been quartered at the house of a
widow lady, where they were well fed, and furnished
with plenty of tobacco and a very limited amount of
vodki; and, as a matter of course, enjoyed themselves
hugely. Jack Cole, poor fellow, though *non compos
mentis,* was not factious, but jolly and full of all kinds of
nonsense. But he became so peculiar, at length, that
Mr. Danenhower found it necessary to place him under
the care of a Cossack; and then after a few days of rest
and preparation they all set out for Yakutsk accompanied by a Cossack, who vouched for their expenses on
the road. The journey was a pleasant and merry one,
for they had been well-provisioned at Verkeransk, and
though it was very cold no one suffered greatly, since the
stations were located at easy distances. They had not
left Verkeransk when a courier arrived from General
George Tschernaieff, with five hundred roubles for the
use of the party, tendered by the General from his private purse. Mr. Danenhower took two hundred of the
roubles and left the remainder for me, but as I had no
need of any money I placed it in the hands of Kasharofski, and had all my bills carried forward; this I did to
avoid a complication of accounts.

Directly after my arrival, the espravnick sent for M.
Leon, one of the political exilès in his keeping, to act as
interpreter between us. M. Leon came and introduced
himself as the gentleman who had written the letter for
the espravnick to me while I was at the Delta; and

now, while he interpreted for us, he managed to inter-
polate for my benefit a part of his history. We had a
capital dinner together of teal, duck, snipe, and other
game which Kasharofski kept frozen in his ice-cellar the
whole year round. Leon told me that he had never before eaten at his table, though often, at first, invited to
do so; for Leon was a very bitter Nihilist, and would
not fraternize with his keepers. Kasharofski had a
son, whom I named the *Malinki Soldat* (Little Soldier),
greatly to the delight of both. Mrs. Kasharofski was
a pleasant, fair-haired, good-looking woman, seemingly
quite content to spend her days in this remote wilderness of snow, cooking and caring for her house, and
spouse, and little son. She waited on us at table, and
did not sit down until after we were through. It was
the first good meal I had eaten since leaving San Francisco. There was some red wine on the table, called
melivki, which is made of dilute *vodki* and wild red ber-
ries, pleasant to the taste and not very heady; and then
there were also cognac, and plain *vodki*, which is nothing
more than unrectified rye whiskey of about sixty per
cent.

Our conversation was prolonged until three or four
o'clock in the morning, for, of course, I had the story of
the Jeannette to tell in all its mournful details.

Leon informed me that no news had ever reached them
of our party; that they had recently heard of a German
Expedition, but knew nothing whatever about us or our
cruise until we actually came up out of the Polar Sea to
visit them. I could see his eyes brighten when I spoke
of the facility with which I could navigate the coast of
Siberia in a vessel as small as our whale-boat. Kasharofski was anxious to learn all about the clothing and provisions necessary to undertake journeys such as we had
been making; and as my narrative passed the lips of
Leon, the young exile drank in every word, and his face

flushed with hope and joy while I opened to his glistening eyes a glorious vision of escape from an odious imprisonment.

Next morning Leon came again to breakfast with us and to continue in his office of interpreter. While at table he told me that in putting Kasharofski's questions he would take the liberty of asking a number of his own. When we had finished our meal I requested Kasharofski to send an order at once to Bieshoff, the commandant at Belun, directing him to continue the search until my return, or the arrival of some other American officer on the spot. The following is my letter of instructions, which Leon translated; and it was instantly dispatched to Belun by special courier, a copy being forwarded to General Tschernaieff.

VERKERANSK, *December* 7, 1881.

ESPRAVNICK, ETC. :

SIR, — It is my desire and the wishes of the Government of the United States of America and of the projector of the American expedition that a diligent and constant search be made for my missing comrades of both boats. Lieutenant De Long and his party, consisting of twelve persons, will be found near the bank of the Lena River, west side of the river. They are south of the small hunting station known among the Yakuts as Qu Vina. They could not possibly have marched as far south as Bulcour; therefore, be they dead or alive they are between Bulcour and Qu Vina. I have already traveled over this ground, but followed the river bank; therefore it is necessary that a more careful search be made on the high ground back from the river for a short distance as well as along the river bank. I examined many huts and small houses but could not possibly examine all of them, therefore it is necessary that all — every house and hut, large and small, must be examined for books, papers, or the persons of the party. Men without food and but little clothing would naturally seek shelter in huts along the line of their march, and if exhausted might die in one of them. They would leave their books and papers in a hut if unable to carry them

farther. If they carried their books and papers south of that
section of country between Mat Vay and Bulcour, their books
and papers will be found piled up in a heap, and some prominent
object erected near them to attract the attention of searching
parties ; a mast of wood or a pile of wood would be erected near
them if not on them. In case books or papers are found, they
are to be sent to the American minister resident at St. Petersburg. If they are found and can be forwarded to me before I
leave Russia, I will take them to America with me.

If the persons of my comrades are found dead I desire that
all books and papers be taken from their clothing and forwarded
to the American minister at St. Petersburg, or to me if in time
to reach me before leaving Russia. The persons of the dead I
wish to have carried to a central position most convenient of
access from Belun, all placed inside of a small hut, arranged
side by side for future recognition, the hut then securely closed
and banked up with snow or earth, and remain so until a
proper person arrives from America to make final disposition
of the bodies. In banking up the hut have it done in such manner that animals cannot get in and destroy the bodies.

Search for the small boat, containing eight persons, should be
made from the west mouth of the Lena River to and beyond
the east mouth of the Jana River. Since the separation of the
three boats no information has been received concerning the
small boat ; but as all three boats were destined to Barkin, and
then to go to a Lena mouth, it is natural to suppose that Lieut.
Chipp directed his boat to Barkin if he managed to weather
the gale ; but if from any cause he could not reach a Lena mouth,
Lieut. Chipp would continue along the coast from Barkin, west
for a north mouth of the Lena, or south for an eastern mouth
of the Lena River. If still unsuccessful in getting into the
Lena River, he might from stress of weather, or other cause,
be forced along the coast toward the Jana River.

Diligent and constant search is to commence now, in December, and to continue until the people, books, and papers are
found, care being taken that a vigilant and careful examination
of that section of the country where Lieutenant De Long and
party are known to be is made in early spring-time, when the

snow begins to leave the ground and before the spring floods commence to overflow the river bank. One or more American officers will, in all probability, be in Belun in time to assist in the search, but the search mentioned in these instructions is to be carried on independently of any other party, and to be entirely under the control of the competent authority of Russia."

Leon now urgently invited me to visit his comrades in exile. I asked Kasharofski if he had any objection, and he said, "Oh no; he did not believe a Nihilist could hurt a Republican; but dinner would be ready at four." So he sent me to the hut of the exiles in his sleigh, which returned again for Leon and myself at dinner time.

M. Leon was a slenderly built, dark, and cadaverous-visaged young man with a Jewish cast of countenance; though when I asked him, he said he was not a Hebrew. His hair was black and long, reaching to his shoulders. He had been a student of law, he told me, and was arrested in a students' row on the streets, and afterwards brought before three separate tribunals, not one of which, however, had been able to find anything irregular in his habits of life; and so, indeed, his final commitment papers expressly stated. While *en route* to Siberia he asked the Cossack officer of his guard, a good-natured fellow, to permit him to look at these papers. His request was granted, and he learned that he had been sent out, after the different tribunals had acquitted him, on what is known as an "administrative order," — a remarkable instrument which closed with the following piece of legal logic : —

" *We can prove nothing against this man, but he is a student of law and no doubt a very dangerous man.*"

And he was accordingly banished for life. Leon had preserved a copy of his commitment papers, which he exhibited to me with a great deal of merriment at the peculiar philosophy of the administrator.

In the hut I found four other young men, Messrs.

Loung, Zack, Artzibucheff, and Tzarensky, all political exiles; the oldest twenty-seven, and the youngest eighteen years of age. They were all professional men, and spoke French fluently; some, German, too, and others a little English. All were earnest Nihilists, though several said they had not been so until after their banishment. Each had his sad and sorry story to tell, and all looked upon me as a most curious phenomenon. They came from different parts of the empire, had known the interior of Russian prisons all the way from Archangel to the Crimea, and were finally sent to the frontier to insure their safety. They were eager questioners in regard to the navigation of the Siberian coast, having in their possession a number of charts and maps, and they had often talked and dreamed, they said, of attempting an escape, but two thousand miles of coast-line and more than one thousand miles of river navigation had seemed an impossible feat until we had accomplished it, and risen before them like a pillar of hope.

With Kasharofski's permission, I visited them daily while awaiting the arrival of my men from Belun. In the evenings several little parties were given, where I met the *élite* of Verkeransk. At these affairs the people sang, played, ate, and everybody seemed to gamble, drink, and smoke. The women had separate apartments wherein they did all these things; and I dumfounded the assemblies by telling them that I never played cards, not even in my own country. Leon, who was present, said:

"They will suspect you of some evil, for they argue thus: This is a queer man who neither gambles nor drinks: he must be always thinking, and a man who thinks much must have some evil thoughts — so banish him at once!"

But this was the speech of a poor exile, whose life was ruined because by reading and reflecting he had learned to speak the truths of moral and political science, yet had

unwisely spoken them too loud, and so convicted himself
as a corrupter of the truth. He was fully acquainted
with the works of our modern philosophers and political
economists, John Stuart Mill, Richard Cobden, Herbert
Spencer, etc., and longed for a supply of English books;
for at the hut, though they had French, German, and
English dictionaries, they had no reading matter of any
kind in our language, and so implored me to give them
the Bible or any other English print I had in the navigation box; but as these were relics of the expedition, I,
of course, could not part with them.

My stay at Verkeransk was both a pleasant and profitable one. I made a copy of the Russian chart of the Lena
Delta for future use, and had frequent interviews with the
political exiles. Their hut was a miserable affair, built
in the manner of the Yakut *yaurta*, of vertical timber
covered with mud. There were the usual outer and
inner apartments, a kind of weather porch, and a kitchen
which contained their fire-place and cooking apparatus.
In the inner apartment they lived and kept their books,
beds, clothing, etc. The walls were overspread with
Russian picture papers, but the room was so low and
dark and musty that it was a very disagreeable place to
visit, much more to live in. At midday it was necessary to burn candles for light; and they were compelled
to this manner of living by their poverty. The government allowed each one for all his wants a monthly sub-
sidy of twenty-five roubles, — a sum equivalent to about
twelve dollars and a half in our money; a stipend on
which they must feed. clothe, and house themselves,
beside procuring fire-wood and service. And this, too,
in a locality where rye meal costs five roubles per *pood*
of forty pounds Russian, or about thirty-six pounds avoirdupois, as the Russian *pfeund* is about fourteen ounces
avoirdupois as against the American pound of sixteen.
Sugar is valued at one rouble per pound; but venison,

beef, horse-flesh, and fire-wood are not very expensive.
Yet all the necessaries of life, — everything which re-
deems it from a primitive state, — are costly in the
extreme. Some of the exiles had wealthy relatives who
sent them money, but such sums could not exceed three
hundred roubles at one payment, and the mails are very
irregular, often but once in six months, though there
may be a mail at odd intervals throughout the winter,
whenever a copert or some official has occasion to go
over the road. Yet no sealed package of any kind is
suffered to go or come from an exile; everything must
be opened, and read or examined by the espravnick of
the district, or the post or police master, and perhaps
appropriated.

My coming filled them with the wildest hopes, for
heretofore it had been considered as impossible to effect
an escape by the ice of the Arctic Ocean as to cross a
living sea of fire; and doubtless for them it would be, as
there was not a sea-faring man in their number, or one, I
suspect, who had ever seen the rolling ocean. Yet before
I left they told me that they intended to make the
attempt, and I ardently hoped that it might be crowned
with success. For here I saw youth, intelligence, and
refinement immured for life in an Arctic desert, with no
companionship of books or cultivated society, surrounded
by filthy and disgusting Yakuts, who were partly their
keepers. For the natives are held strictly accountable,
under penalty of the dreaded knout and imprisonment,
for the escape of an exile, since it is utterly out of the
question for any one to travel a great distance into the
country without their aid or knowledge. As a guest of
the nation and a continuing recipient of its succor and
hospitality, I could not honorably abet the exiles in their
plans for escape; yet as a Republican I am free to say
that all my sympathies were with them, — the oppressed
for speech sake. For it was one of these young men

who told me that all they asked and strove for was a
constitutional form of government, let the constitution
be what it might. They only wanted the *privilege* of
being imprisoned, and hanged, if needs be, under a Russian law and constitution; and not driven like a herd of
sheep by the police master of a town or city into prison
or exile, without the benefit of trial before any tribunal,
or if a mock hearing could be had, as in Leon's case, yet
not before such an administrator, who on his very commitment papers would record himself a judicial ass.

Still, Leon, in his character of interpreter, obtained for
himself and companions the full benefit of my recountal
to Kasharofski of the Jeannette's cruise and equipment;
our retreat, supplies, clothing, and line of march. The
youngest of the exiles, called the "Little Blacksmith,"
had been a polytechnic scholar, and seemed to be the
physicist in general of the party. He gazed fondly on the
sextant in my possession; for with it he could find his
way across the *tundra* and the ocean. They had watches
and compasses, but no means of determining latitude, or
tables for computing longitude. So this earnest young
Nihilist began the construction of a sextant, and had
already his navigation tables in course of preparation,
using a Russian almanac to find the sun's declinations,
etc. It was their intention to build a boat on the Jana
River, near Verkeransk, and attempt a passage of one
thousand miles to the sea-board, and then a voyage of
nearly two thousand miles along the coast of Siberia to
East Cape or Behring Strait.

I afterwards learned with regret that they had indeed
essayed, but unsuccessfully, to carry their bold project
into effect. Eluding their pursuers, they succeeded, after
many difficulties, in working their way down the Jana,
past a large village near its mouth, to within sight of the
sea, and could then have accomplished their escape with
comparative ease; but the rolling waves paralyzed them

with terror and tumbled into the boat, which was over-laden with its freight of thirteen exiles; and when they ran ashore it swamped and soaked their provisions. One of their number was a young woman, of whom more anon; but even she was made of sterner stuff than the two others who, frightened at their situation, straightway surrendered themselves to the authorities at Oceansk, who soon after captured the rest and sent them all into worse exile, if possible, than before. Leon was forwarded to the river Kolyma, and others were removed from the settled districts, and placed among the Yakuts. And what else could I do but admire them and their pluck, whose greatest offenses had been boyish indiscretions, rows in the streets, for none of them had yet become master of his profession? And so, in the eyes of every American, born to believe that free speech and a free press are absolute and indefeasible rights, must the overwhelming and horrible punishment meted out to these exiled youth appear shamefully despotic and cruel.

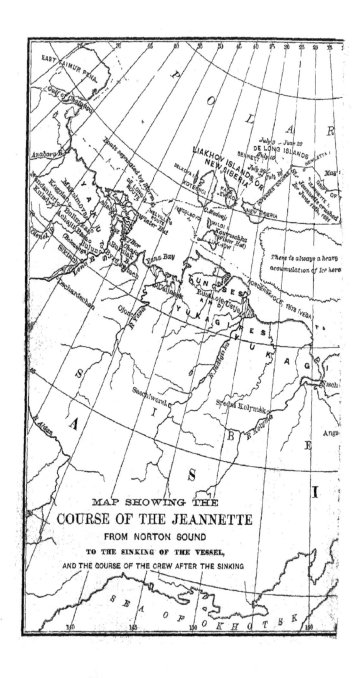

MAP SHOWING THE
COURSE OF THE JEANNETTE
FROM NORTON SOUND
TO THE SINKING OF THE VESSEL,
AND THE COURSE OF THE CREW AFTER THE SINKING

CHAPTER XIX.

FROM VERKERANSK TO YAKUTSK.

Penavitch. — Doctor Buali. — His Sad Story. — Those Terrible Nihilists. — " A Dead Nihilist and a Dead Czar." — Happy Lovers.
— I take a Remarkable Russian Bath, and a Very Bad Cold. —
Off to Yakutsk. — Siberian Scenery — The Horse and his Trou-
bles. — A Queer Predicament. — Kangerack. — Crossing the Divide. — A Dangerous Descent. — A Tunguse Home.— Exasperating Yamshicks. — A Sickening Sight — Native Grist-Mills. — A
Learned Exile. — The Scaups.

DURING my sojourn at Verkeransk, there arrived from
Yakutsk a police agent named Penavitch, who was likewise one of General Tschernaieff's secretaries. He and
Kasharofski were warm friends, and while in Verkeransk
he stopped at the house of the espravnick. I made arrangements to ride back with him to Yakutsk, and so
awaited with increasing anxiety the arrival of Bartlett
and party.

Penavitch was a great, jolly, good fellow, and he took
me to visit some of the coperts in the town, who, after a
manner, were also choice spirits; ever ready to make a
rouble, but good-livers and hospitable to a degree beyond
their means. They all wear a peculiar dress prescribed
by their guild, which consists chiefly in allowing their
long woolen shirts to hang outside their trousers, broadly
belted in, however, at the waist, and I saw the embryo
Yakut copert sailing about, in imitation of his white in-
structor, with his colored flannel shirt flying to the
breeze.

I also visited another exile, named Dr. Buali (White), who lived apart from his companions, and was then performing the duties of the government surgeon, who had become blind from a cataract and was about leaving the district to have it removed. Dr. Buali had been very kind in his attentions to Danenhower and Leach, and it was he who had prepared the box of medicines forwarded to us at Belun. His life held more of sorrow in it than that of most men. He had been a practicing physician in one of the small cities of Little Russia, or the Crimea; had neither committed any crime or ever belonged to any secret society; and he seemed to believe that his only offense had been the marrying of the lady whom he loved, in a neighboring village; for he knew not that he had a rival.

The story of his capture and removal to Verkeransk was tearfully translated to me by Leon, his friend and comrade in misery. It seems that for a week he had been attending the little daughter of the police master, who at length so far recovered that he erased her name from his visiting list. But one morning he was surprised to receive a call from the police master's Cossack, who said his presence was demanded at once at the house of his master, for the child was seized with another attack of illness.

"I imagine it is not serious," Buali had replied; "I will be there after breakfast."

But the Cossack insisted upon instant attendance, and so he bade him wait until he got his coat, but here again the Cossack interposed, saying that it was not necessary, the house was only a little distance off, and that the police master intended that he should breakfast there. So, dreading lest the child might indeed be seriously ill, he hastened off at once, but on the way the Cossack said they must first go to the office of his master. He followed wonderingly, but never for a moment dreamed that anything was wrong; and so he was ushered into the police office, and later into an anteroom, where the police master told him he was a prisoner.

"A body of exiles," said the heartless wretch, "are ready to start for Siberia, and you will go with them."

Buali laughed — it was a good joke. but the police master assured him it was a solemn fact; and then the poor man, completely overwhelmed, prayed gaspingly for a delay in his transportation. Why was he deprived of his liberty? Who was his accuser? What was the accusation? No answer save " Administrative order."

But could he not revisit his house under guard, and secure certain necessary articles of clothing? *Could* he not at least return and say farewell to his young wife? The brutal officer denied him everything; "And then," he said, "I wailed in anguish, but they placed me in solitary confinement to await the departure of the draft, and in twelve hours I was on my way to Siberia."

Of course, he was half-crazed. What would become of his young wife — what would she think of him? Doubtless, that he had deserted her. A thousand thoughts and suspicions racked his mind, and he had already lived through an age of torture, when, hurried on across the frontier by rail and steamboat, he chanced at a certain railway station to recognize, from the cage-carriage in which he was confined, an old merchant friend. Calling to him eagerly, he briefly told the great wrong done him, while the train halted at the station, and supplicated his friend to visit his wife and relations and inform them of his fate.

And here let it be known that immediately upon his sentence an exile loses his identity — John Brown ceases to be John Brown, and becomes " Number 101;" his estate is administered upon as though he were dead, and apportioned, with the usual forfeiture to the state, among his heirs. So that it is almost impossible for any one, save the authorities, to trace out and discover his where-abouts.

When Dr. Buali had arrived at Irkutsk he was detained there for some time, and meanwhile his friend, the merchant, true to his promise, had hastened to tell the circumstances and conditions of his banishment to the poor young bride; who, quickly as she could, set out to join him in exile. With womanly wit she managed to apprise him by letter of her coming, and a light broke in upon his grief. Daily, hourly, constantly he looked and longed for her, and just when he knew she *must* come, they sent him on to Yakutsk, and from thence to Verkeransk.

She, poor creature, filled with sweet sympathy and expectation, hoping from his delay at Irkutsk that he would be permanently detained there, arrived two days too late. Picture her anguish — when, having traveled with wifely devotion 4,000 miles over the worst country in the world, she learned the cruel news. Still 2,000 miles away; and even then would she find him! It was too much for the poor heart; she lost her reason; raved for a little while in a madhouse, and died. He received the sorrowful tidings, so different from what he had expected; and when I saw him he was just recovered from the effects of a suicidal attempt by poison.

This is the sad story of one of the friends whom I met at Verkeransk, as told by himself and interpreted to me by Leon. Dr. Buali was not a Nihilist, or at all intemperate in his political views, and consequently was not held in the highest favor by Leon and his companions. He was on pleasant visiting terms, however, with the espravnick and others, who liked him cordially for his own sake; yet he was not permitted to practice his profession for profit, but only to perform the duties of the blind old incumbent at the post, from whom he might receive little or no recompense, just as that worthy official saw fit.

Indeed, no exile is allowed to carry on any business, teach school, till the soil, labor at a trade, practice a profession, or engage in any work otherwise than through the medium of a master. If I wanted any service, an exile would sometimes come and offer to perform it, but I would have to pay his master, upon whose bounty he must depend for remuneration. This is a monstrous mistake. Russia has been striving in vain to populate Siberia for a thousand years, and she will never succeed as long as she continues in her present policy of converting the land into a vast penitentiary, wherein the prisoners are prevented from making an honest livelihood, and so driven, if criminals, to a further commission of crime. Beyond doubt there are rogues of the very worst type in Russia and Siberia, but certainly it is plain that their mode of punishment will never tend to elevate or reform them; and it is utterly impossible that Siberia, under its present system of government, should ever be populated or improved, as have been the penal colonies of the French and English.

The ignorant Yakuts are very fearful of the whole class of exiles, for they are told exaggerated stories of the atrocities of the Nihilists; and the police officials are ever on their guard for an outbreak or revolt. I was much amused at Kasharofski, who told me that he lived in constant terror lest some of the exiles should assassinate him. He showed me a long knife and a revolver, which always went to bed with him, he said; and every night a Cossack slept in the outer apartment. Leon, too, assured me of all this, saying that he and his comrades found a never-failing source of fun in the frightening of the police master, the Cossack guard, and the trading merchants, who would sell them goods at a ruinous discount, in order, as indeed they would explicitly state, to curry their favor and avoid their vengeance.

17

"But," said Leon and his friends. "why should we kill these poor creatures? What good would it do? To be sure, if their death might secure our freedom, we would not mind killing a thousand, but such is not the case."

Another episode in the life of the exiles at Verkeransk, and I will abandon them to their sorrows. I have noted that the walls of their miserable habitation were decorated with illustrated papers; but in addition to these there were two prominent pictures: one a photograph, and the other a wood-cut or print from some journal. They faced each other on opposite sides of the room, and I stood looking at them, struck by their similarity; for the wood-cut I recognized as a portrait of the dead Czar. He lay in state near a window, dressed in his cerements, his hands holding a crucifix and resting on his breast.

One of the exiles, Artzibucheff, observing my silent comparison of the two pictures, approached and said: —

"The two men are very much alike, are they not?"

They certainly were; faces sharp and peaked in death, hair and beards arranged the same; I thought both were likenesses of the Czar, and said so. The exile smiled: —

"No," said he; "the photograph is one of my brother, who perished of cold and hunger in the horrible dungeons of the fortress of Petropavlosk, on the Neva. His body was photographed on his bier near one of the gun-ports, which has the appearance of the palace window wherein is laid out the corpse of the Czar. My brother was murdered in the fortress; my friends murdered the Czar in his palace; 'things that are equal are equal in all their parts;' a dead Nihilist and a dead Czar!"

He laughed, and added that because of his brother's death he had been arrested and sent into exile; that he had a sweetheart to whom he was affianced, and that she, too, had been banished as far as Archangel; but as

it was no remittance of her sentence to be transported to
Eastern Siberia, she had been granted permission to join
her lover at Verkeransk. He was a typical Nihilist, as
portrayed in our comic papers, with long, bushy black
hair, combed out until it resembled a mop-head; dark-
skinned, and fine-cut in feature and figure, with eyes
and mind as bright as the Dog-star. He told me gayly
that he expected his sweetheart every day, and that if I
failed to see her at Verkeransk, I should certainly meet
her on the road. And sure enough, I did, on the afternoon of my departure. She was young and attractive,
of medium height and excellent form; her eyes and hair
were light, her nose slightly *retroussé*, and her mouth as
pretty, pouting, and cherry-red as one would care to see.
She had with her a number of French books, of which,
she said, she intended to make translations. She spoke
French and German fluently, but knew very little English; and at this time I only saw her for a few minutes,
but met her afterwards when going north and returning
from my second search.

On the evening of December 15th, Bartlett and party
arrived safely from Belun, and I instantly set about making the necessary arrangements for their transportation
to Yakutsk. They were capitally quartered at the residence of the Cossack commandant, where they had plenty
of good food and enough *vodki* to make them merry.
Before their arrival I had a sufficient supply of bread
baked, and beef killed, cut up into proper sizes, and
frozen for their journey. I was now suffering from a
severe cold, the first I had caught since leaving the
United States, and it occurred in the following way.

Upon my reaching the house of the espravnick, he
doubtless suspected my cleanliness, as well, indeed, he
might, for I had suspicions in the same direction myself;
so, when he proposed a bath, I gladly consented. He
then ordered a Cossack to prepare the bath, and brought

forth some clean under-clothing and a suit of gray cloth for me.

" You are acquainted with the bath ? " he inquired.

" Oh, yes," said I, for I could not think of any intricacies in the operation which might not be learned at a glance.

So, headed by the Cossack, who carried my clothing, towels, etc., I set out for the bath-house, which was located about one hundred yards from the main dwelling. I found it to be a square box of a building, perhaps eight by ten feet, and seven feet high ; the door was covered with an ox-hide and felted, to keep out the cold ; the floor was earthen ; and in a corner stood the stone-furnace and chimney. The furniture of the room was composed of one stool and a small table ; two large tubs filled with water, one hot and the other cold, with cakes of ice floating in it ; two shelves, one about two feet from the ground, the other about five feet, and both broad enough for the bather to lie out upon ; several small wooden vessels ; an iron dipper ; and then, beside my under-clothing and towels, a couple of sheets intended for wrappers. I noticed a large hole in the side of the furnace from which the flames and gases were rushing into the room, for a board had been placed on top of the chimney as a damper.

And now the Cossack told me to undress. I did so. He next filled the dipper with water, and asked me if I was ready ; I said, " Yes," and he then cast the water through the aperture into the glowing furnace. Instantly a thick volume of steam burst forth, and the Cossack, looking at me a moment, said, " More ? " I assented, and he threw in another dipperful, whereupon the upper part of the building filled with steam. He glanced at me askance, and asked again — " More ? "

" Yes, yes," said I, impatiently ; " heave it in, manorga ! "

Quick as thought he dashed two or three dipperfuls of water into the furnace, and then, dodging his head, bolted out of doors as though he had hurled a keg of powder into the fire.

Two candles were burning in the room, — one on the table, the other on the upper shelf. This latter was extinguished in an instant. I relighted it at the other, and, apprehending my plight, set them both on the floor, where they burned with a blue light. Meanwhile the scalding hot steam was lowering nearer and nearer to my head. I crouched down, but it followed me. The candles flickered, and were going out; evidently I could not stay in the dark and be smothered or scalded to death. So, without thinking of the sheets, I threw myself against the low door and shot forth into the open air and snow, the dense steam, literally in hot pursuit, pouring out after me.

The Cossack had fled in dismay to the house; and there I stood abiding the exit of the steam, in nothing but my skin, dancing up and down in a temperature of about sixty degrees below zero. It was not long, however, before I could see a current of cold air rushing in beneath my inflamed enemy, and I crept back in its wake, and, when things were cooled off, closed the door and leisurely bathed in one of the tubs, tempering the water to suit myself. When I at length found my way back to the house, and told Kasharofski of my escapade, he said the Cossack was under the impression I had deliberately boiled myself; and, indeed, I was so badly affected that Bartlett's first exclamation upon meeting me was, "Why, what have they been doing to you?"

Perhaps it was the change of clothing, but at any rate the cold clung to me until I again began living in the open air, and slept once more on top of a sled.

The time of my departure from Verkeransk was set for the morning of December 18th, but the espravnick had a large mail to send, and, of course, postponed preparing it until just before I was ready to start, so it was midnight when we at length set out. Before leaving, Kasharofski told me that he had received information from the Kolyma River that nothing had been heard of the second cutter or her people; for as soon as my telegram had passed through his hands he had sent word to the espravnick of the Kolyma at Kolymsk to keep a bright lookout for any stranger that might turn up on the coast.

My journey to Yakutsk, although made for the most part by reindeer, was not so rapid as that between Belun and Verkeransk. Yet in winter this trip is one of wondrous beauty: cold, forsooth, but the deep dark woods, with their little glimpses of sky; the dashing along under the low-reaching arms of the evergreen trees, league after league of forest bowed down to the very earth, and, in places, prostrated with its white weight of snow; the weird ride over hill and mountain, skirting ravine and precipice; the breaks along and across the numerous water-courses, over rude bridges, or along deep gullies where rough wooden guards preserved the teams from disaster, — with this quick succession of scenery wild and strange was I kept constantly awake and charmed.

At the *stancias* we met the traveling merchants with their long trains of goods, hauled by reindeer or packed on the backs of horses. Five *pood* is the regulation load, and all packages are put up in drums, bound with raw-hide, and so strapped that they can easily be transported by the pack-horse, which carries a half load on either side of a saddle-tree prepared for the purpose. Many of the merchants employ horses and reindeer the whole year round for the transportation of their goods. The *stancias* are let by the government to agents, who sublet them to the Yakuts or other bidders, who, in turn, keep them in repair and transport passengers and freight

at the rate of three copecks per verst for a passenger,
and three copecks per verst for each five *pood* of freight.
Of course there is an endless wrangle going on between
the station-keeper and the traveler or merchant.   At
times we would encounter a great train of pack-horses
hitched to each other's tail, with a conductor riding
alongside or in front, and another horseman in the rear
to look after any stray animals or lost merchandise.

In this section of Siberia there are a great many
draught-horses and cattle, which latter are housed during the winter months, and generally under the same
roof with their owners — often in the same apartments.
The horses, however, I noticed, were not stabled even in
the severest weather; excepting, of course, those of the
rich, which are only used for carriages or sleighs.   The
poor animals are compelled to dig down through the deep
snow in search of grass or any herbage, like the reindeer
after their moss.   The grasses though coarse are sweet
and nourishing, for the short hot summer has scarcely
brought them into life, when winter comes quickly on,
freezing and preserving their nutritious juices; yet the
Siberian horse, like the Spanish mule, does not limit his
diet to grass alone, but can apparently eat hickory hoop-
poles.   As he stumbles and staggers along under heavy
loads, I have seen him crop the twigs and branches of
the birch, yew, and scrub pine.   The *stancia* horses, indeed, receive better treatment, for they are fed on hay,
cut and cured during the summer; but north of Yakutsk
it is very rare that a horse receives any attention or shelter.   They are to be seen as far as Verkeransk, and many
wild horses roam the snowy plains; yet of these I saw
but few.   The horse-driver carries a wooden sword, in
one edge of which is inserted a roughly-toothed saw, and
with this he scrapes off the rime and snow from the
horses when they halt to rest or arrive at a *stancia*.

We traveled the first part of our journey, as far as

Kangerack station, by deer-teams. In crossing one of
the tributaries of the Jana River, which scatters itself
into a series of ponds and lakes, we suddenly found ourselves in a queer predicament; for although the temperature was down to minus 40° Réaumur, still there were
from ten to fifteen inches of water on the bed of the
river; and, ere we could grasp the situation, the *yamshicks*
had driven directly into it. The ice beneath us was very
slippery, so the deer were unable to keep on their feet,
and floundered hopelessly around in the water, which had
here spread itself over several miles of ice. The deer
attached to the sled of Penavitch fell down, and the
drivers, who had dismounted and were splashing about
in water over their boot tops, carried my big friend to
the bank, where he had a lonely walk of a mile or two to
perform. Meanwhile, the natives managed, after considerable difficulty, to get and maintain the prostrate deer
in an upright position; and my *yamshick* succeeded, by
wading and leading our two teams, which had not entirely lost their equilibrium, in reaching and scaling the
steep bank. This flooding of the ice, which the Yakuts
consider very dangerous, is caused by the hydraulic pressure up stream, which raises the ice-bed and finally bursts
it open; and the water continues to overflow until the
pressure is relieved, when it freezes in again.

But here we were at midnight, thirty versts from a
stancia or *povarnia*, with the temperature so low that I
shiver to think of it. Luckily the Yakuts knew of a hut
in the woods, and thither we went, where they built a
fire and dried their boots and clothing; and Penavitch
won their best regards by giving each a drink of *vodki*.
At daylight we were off again, threading our way around
the flooded ice; and Penavitch, who, I noticed, as well as
the *yamshicks*, had been very much excited, took occasion
to tell me that these overflows are very dangerous; that
they sometimes occur with so much force that travelers

are drowned; and whole teams, deer, drivers, and passengers, have been found frozen to death, where the waters suddenly rushed upon and soaked them.

It was dark on the 24th day of December when we arrived at the Kangerack *stancia.* Here we met a fine, fat copert just from Yakutsk, who knew my traveling companion, and was full of good nature and other good things; and the enthusiastic fellow expended his time and treasure in giving a swell dinner to the first American he had ever seen. We were obliged to rest here, as it is the last deer-station on the road toward Yakutsk, and is located at the mountain divide between the districts of Verkeransk and Yakutsk. Next day, however, we got under way about ten P. M., and traveled all night, crossing the divide about twelve o'clock.

It was severely cold — ah! ferociously so, from minus 40°–45 Réaumur; but the soft, clear moonlight was gorgeous and glorious. We were about 4,500 feet above the level of the sea; the mountains were grandly wild; and, stripped for the purpose, we toiled up their steep ascent on foot, with the teams plodding slowly on in front. Above us on either side, the gigantic peaks lifted their hoary heads far into the blue vault of heaven; silent, frigid, and white. Ah! what grandeur! I rejoiced that it was night, and so cold and still; for they filled me with an awe, those snowy summits bathed in the silver radiance of an Arctic moon, such as I had never known before. And though I twice recrossed the divide, yet the spell was not upon me as on that wonderful night, and the splendor I then saw can never depart my memory.

Arrived at the summit of the gorge, we halted for a while, and then the four sleds were lashed together, two abreast, with a driver sitting on each bow, and the reindeer hitched behind. When everything was ready, the natives worked the sleds to the edge of the divide, and over they flashed. I expected to see them roll in a heap;

but no; they steered with their feet, while the deer held
back, and kept safely on for about one hundred yards,
when they stopped in the deep snow, until the frightened
animals had quieted down; and then sped on again for a
mile and a half. The incline was so steep that it was
with difficulty I could stand upon it; consequently, I sat
down, with a stick in my hand, and at once shot off like
a sky-rocket. In vain I tried to control my speed by
jabbing the staff in the snow between my legs, but it
only slewed my body around, the heavy part taking the
advance like the ball of a shuttle-cock, and away I went,
sliding, tumbling, and rolling, until I at length brought
up near the sleds, considerably confused at my rapid descent. And I had rather a chilly time of it, stripping in
the open air to rid my clothing of its burden of snow,
which, to say the least, was very unpleasant next my
flesh; not to dwell upon the peculiar sensation produced
by a pool of cold snow-water in the convex section of my
trousers.

Towards daylight we came upon a band of wandering
Tunguses, who were camped in a ravine under a ragged
tent made of birch-bark and reindeer-skins. The lower
part, for a height of about three feet, was carried up ver-
tically as a cylinder, and then the skins were stretched
around poles which inclined to a tapering cone. The inmates were very miserable; two or three women and a
litter of children, covered with tattered furs, lay around
the floor, and a sickly fire smouldered in the centre of
the tent. At this we prepared our tea, and the women
brought forth their kettle to brew our leaves over for
themselves. Soon we were on again, and next halted at
a *povarnia*, to find a poor woman who had just given
birth to a little Yakut. Our drivers built a fire, warming
up the hut, and we made tea and gave her some. She
seemed happy and healthy, and had her baby stowed
away in a wooden bowl.

The Yakuts were becoming more squalid and filthy as
we journeyed south from the ocean ; those living closest
to Yakutsk being the most disgusting in their appearance and habits, and apparently devoid of any moral
sense. They all live under the same roof with their
cows, some, however, with a partition of bars between
their apartment and the cattle stalls. A wealth of patience is here required to cope with the exasperating lazi-
ness of the *yamshicks*. When the team is hitched and
the traveler is about stepping into the sleigh, his driver
drawls, — "Just one minute ; I have not smoked ; "
which usually means he has not dined, or drunk his tea,
or smoked his pipe, or *teaed* again, until he has protracted the delay to an hour or two. I could not hurry
them up, try as I would. If necessary, they stole away
to a neighbor's and drank tea, and then more tea, while I
vainly scoured the woods and ransacked the stables in
search of them.

At one place where we stopped for several hours, the
natives had a dead horse in the hut, where, I think, he
had died. The carcass was intact, save where they had
partly turned back the skin from the stomach to the
hock-joints of the hind legs, and had cut and eaten the
flesh from the haunches. The animal was not even dis-
emboweled, and the stench which arose from it was so
intolerable that I crawled into my sleeping-bag and lay
on top of the sled, in preference to staying in the hut,
while we waited for a relay of horses. But the natives
sliced off the meat and cooked it without wincing at the
sickening sight and odor. And yet there is a large quantity of beef raised for the Yakutsk markets and the gold
mines to the southward, though, it is true, the natives,
when they have paid their taxes, have little left of anything, and the tax-gatherer in Siberia is inexorable.

Nearly all first-class huts are furnished with primitive
grist - mills, in which the natives grind a few grains of

rye at a time, and bake the meal on the end of a stick or stir it into the hot milk. These mills are made of two blocks of wood cut from a poplar or other large tree trunk. One section or block is set up on three legs, with a pin in the centre to receive the upper and revolving millstone. Around the periphery of the nether block runs a bull's-hide shield, which catches the meal as it is ground, and delivers it at an inclination or dip into a receptacle placed on the floor. In the top stone is inserted an upright handle for one or two persons to turn; and occasionally a hole is bored which receives a staff suspended from the ceiling, and two women sitting opposite to each other twist this around, one of them from time to time dropping a pinch of grain into the orifice of the top block. I was at first a little surprised at the capital manner in which these rude machines performed their work; but upon lifting off the upper half I found that the wily Yakut had set at a proper depth small cubes of flint stone into the faces of both blocks. The meal is coarse and unbolted; but then the husks go to fill the abhorred vacuum, which at the Delta in times of famine I have seen filled with wood. When the grist is ground it is kneaded upon a stick, usually in the shape of a large cucumber, and stuck in the ashes and slowly turned before the fire; though, at times, it is kneaded in the shape of a fan, or baked on a board set up at an angle in front of the fire.

Truly the Yakuts lead a wretched existence, and the Yakut women especially. They all beg, lie, and steal; are ragged, diseased, and unclean. As we approached Yakutsk I noticed that the number of blind old men and women did not decrease, and that the mode of ablution by squirting water into the hands and then applying it to the face, thus transmitting the syphilitic virus or lymph from the mouth to the eyes, was common to all. In these miserable huts I found, now and then, one or

more exiles, political or criminal, quartered upon the
natives. Among the exiles, too, there were many Jews,
known to both Yakut and Russian as *Judes ;* who, true
to their instinct, were all strenuously striving, though
poor as church mice, to do a little trade.

The exiles ! A witching theme, and one upon which,
with space and leisure at my command, I could love to
dilate ; for I saw and heard so much of them that I am
sure would interest the reader. Fancy a poet and *litté-
rateur*, one of those rare Russian souls whose wonder-
working effusions must ultimately enlighten and enfran-
chise the people — a Turgenieff — immured for life in
this snowy desert. Yet there was such a one, and even
the savagery of his surroundings could not dispirit him
or cool the ardor of his genius. From his prolific pen
flowed a ceaseless stream of learning and of light; he
wrote and wrote, and in the writing forgot his wrongs
and sorrows. The authorities were overjoyed to see him
in this mood; they fostered his rich whim, for his fame
had gone before him, and they established him in better
quarters where he might lay his golden eggs for them to
seize and sell, and they gave him servants, who might
watch and see that none of the eggs were lost; and even
the bishop of the diocese could find it in his heart to appropriate and make capital of the learned exile's translation of the Bible.

But soon he saw all this, and came to realize how
precious he had grown in the eyes of his captors, and so
he shrewdly sought to use his talents for his own advantage. A Cossack, who bore a marked resemblance to
him, — his double, — was diligently searched for and
finally found in the empire. To secure his services ; to
transport him to Siberia ; to train and cut his beard, edu-
cate his manner and accent, and, in fine, model his person upon the exile's until, in appearance, they were one
and the same, and the deception was all but perfect ;—

to accomplish this the exile labored with superhuman
powers, and the fecundity of his brain increased under
the stimulating, the intoxicating hope of freedom, — one
blessed hour of which is said to be worth a whole eter-
nity of bondage. And at last the glorious day arrived
when, leaving his well-trained double to act his part and
cover his escape, the joyous exile, his liberty achieved,
set forth upon his hazardous journey, — a journey which
he never ended; for, alas ! the fickleness of fortune and
of friends as well! there came a slip in his plans, a cog
was missing, his hopes were dashed to the ground, — he
was apprehended, and buried alive again, this time beyond the possibility of resurrection.

The best clad and happiest of all the exiles whom
I saw in Siberia were those known as the " Scaups."
They are a religious sect, whose doctrines of late years
have widely spread throughout the empire, and whose
votaries seem to defy the efforts of the Russian government to crush them out. A peculiarity of the sect is that
it can only acquire new members by recruital, since both
sexes so mutilate their persons that they can neither be-
get nor bear children. They do not live apart, however,
except in the manner of the American Shakers, with
which people they seem very well acquainted, and de-
nominate them as the " Wet," and themselves as the
" Dry Scaups." They study all the economies of nature,
and neither drink spirits nor eat flesh ; they live in communities under the police, are invariably farmers when
permitted to be, and, like the Shakers, send every variety of produce to the markets. Somehow or other, the
women seldom leave the communities, and the men cannot ; though I saw one woman who had apostatized, and
borne a child to her new spouse; but owing to the mu-
tilation of her mammillary glands, she was unable to
suckle her babe.

The men, of whom I saw very many along the banks

of the Upper Lena, were all large, fat, bloated fellows, devoid of color or very sallow, and beardless as a rule, for the beards of those who might manage to grow any would slowly fall out. I found them for the most part intelligent, but not at all bright; they were stolid and flabby like stalled oxen. In conversation with a group of them who came on market-day to dine at our hotel at Yakutsk, I was asked if we had none of their sect in the United States. No, I told them, but we had in Utah their antithesis, the Mormons. They apparently saw no joke in my allusion, and said, " Yes, they had heard of the polygamists, and thought they were very sinful." Yet they seemed to be the only rich and prosperous people in the vicinity of Yakutsk, for they are sober, frugal, and industrious, and General Tschernaieff informed me that previous to their advent every pound of flour used in Yakutsk was imported from the southern provinces, while now they are exporting grain, meal, beef, butter, and vegetables. The general believed that the one object of the Scaups was to accumulate an abundance of this world's goods, and that their religion was merely a means of shirking the responsibility of raising families; and so it certainly seemed; but I presume the mainspring of their prosperity is no other than their total abstinence from intoxicating drink, which is a clog and a curse not only to Russia but to all the world beside. I had many dealings with the Scaups, and found them uniformly upright and honest; something which I cannot say of any other people I met in Siberia, except General Tschernaieff, Epatchieff, the assistant espravnick of Verkeransk, and Carpuff, the lieutenant of police.

CHAPTER XX.

AT YAKUTSK.

The "Balogan Americanski." — General Tschernaieff. — How He received me. — Mr. Danenhower and Party set out for America. — Instructions from the Department. — *Praesnik.* — Preparations for my Second Search. — Yakutsk Society. — New Year's Eve. — Nova Goat. — The Bishop receives. — Masquerading. — Bulky Money.

I ARRIVED at Yakutsk about an hour after noon of December 30th, 1881, the journey of 960 versts having consumed twelve days, or more than double the time required in going from Belun to Verkeransk. The last two hundred and fifty or three hundred versts had been especially tedious, and the Yakuts, living in their cow-stables, incredibly disgusting. The *stancias* were close together, not more than twenty or thirty versts apart; and we passed through several deserted villages of from twenty to thirty *yaurtas.* I asked Penavitch why they were deserted, and he said the inhabitants had all died of small-pox.

I was driven direct to the Balogan Americanski, where Mr. Danenhower and the sailors were lodged. The balogan was a government house, for the use of which I paid a small weekly sum, and was located across the way from the Guestnitsa Hotel, kept by Madame Lempert, who fed the party at the daily rate of one rouble apiece. I found all the men enjoying themselves greatly, dressed in tight-fitting boots, white shirts, and choker collars. They seemed comfortable and happy, and were already on visiting terms with the inhabitants. Many, too, had

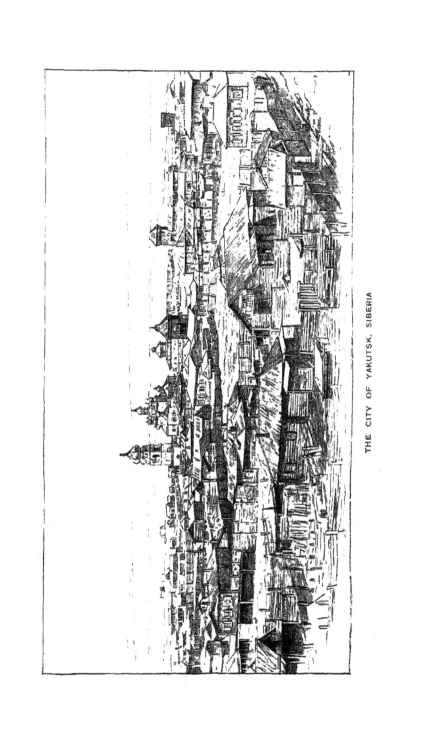

THE CITY OF YAKUTSK, SIBERIA

sweethearts, and, I fear, had they stayed much longer some would have had wives. Poor Jack Cole, I was grieved to see, had lost his mind entirely, but he was in the best of spirits, and told me that he had a body-servant now, and intended to marry Queen Victoria.

I learned from Mr. Danenhower that when they reached Yakutsk, they were first driven, as is customary upon the arrival of strangers, to the police station; but he informed the police master who he was, and demanded an audience of General Tschernaieff, which was granted at once; Dr. Capello, the district physician and inspector of hospitals, acting as interpreter. The general received him cordially, provided him with everything he wished, or that the place could afford; quartered the men at the government house, and placed them to board with Madame Lempert, beside loaning money from his private purse for the use of the party. He had also insisted that Mr. Danenhower should dine with him every afternoon at two o'clock; and as I had arrived at one P. M., I promptly prepared to pay my respects to the general. His sleigh came as usual for Mr. Danenhower, and together we repaired in it to the gubernatorial mansion, which, compared with the balance of dwellings in Yakutsk, is a house of imposing dimensions; built, like all others, of wood, part square and part round timber; and located at the corner of a street, surrounded by a fenced inclosure containing stables, servants' houses, and other buildings.

The general is a bachelor, sixty-two years of age, — twenty-one years of which he has spent in Siberia. He is two inches taller than six feet, straight as a spear-shaft, and rather spare; with full flowing white hair and beard, large aquiline nose, handsome face and carriage, a very soldierly air and bearing; and he was always dressed in uniform which fitted him neatly. His reception of me could not have been warmer had I been his own son.

18

Heaiing the door open, he came out of his cabinet across the dining apartment into the reception room to welcome Danenhower, when, seeing me clothed in skins, and my face frost-scorched, he gazed for an instant in surprise, and then, before Danenhower could introduce us, had caught me closely in his arms and kissed me on both cheeks. He called me his *brat* (son), and with tears rolling from his eyes he pressed me to his breast again and again. So apologies for my appearance were not necessary; he was a soldier, and appreciated the exigencies of the service, and, consequently, of the occasion. He and Dr. Capello had been old campaigners together, and were still constant companions; and the doctor, who spoke French fluently, now interpreted between Danenhower and the general.

We ate a capital dinner of soup, fish, beef, and game, Yakutsk potatoes, and a variety of canned vegetables, all washed down with the wines of the country, *vodki*, claret, madeira. and the *quass*,[1] a favorite beverage with Dr. Capello. We finished with a bottle of champagne, and after several hours of conversation, my first reception by the governor was over; yet before leaving he exacted a promise that I should dine with him daily during my stay in Yakutsk.

The following day he returned my call, and for the balance of the week I could scarcely do aught but receive and return visits. Yet immediately upon my arrival I began to arrange for the departure of Mr. Danenhower

---

[1] Subjoined is a translation of the *quass* recipe as written out in French by Dr. Capello : —

To make fifteen bottles of quass one must take twenty bottles of boiling water, six pounds of black rye bread, and one ounce of English mint (*folia menthae p. per lac.*), and boil for twenty-four hours. Then pass the contents through a sieve. The residuum must be left twenty-four hours longer, after one has added two spoonfuls of yeast and two pounds of sugar. When the forty-eight hours have elapsed, pour it into bottles and cork. — ED.

and all the sailors, save Bartlett and Nindemann, who were *en route* for Yakutsk, and were beyond doubt the best men in the party to assist me in the early spring search. Two months had now elapsed since I started my telegram from Belun, and still no answer. General Tschernaieff offered to advance government money for our transportation and supplies as far as Irkutsk, but would not hearken to my solicitations for means to renew the search for my missing comrades. However, as a mail was hourly expected, I postponed my departure for Irkutsk, whither I purposed going in order to put myself in wire communication with the United States.

January 1, 1882, I gathered a few bottles of wine, some *vodki*, white bread, cakes, and cold meat, for our table, and along with the men kept open house. The sailors had made numerous friends, who all paid New Year calls, and the day passed quickly and merrily.

As soon as possible I forwarded a mail and telegram to the Navy Department, in which I requested orders to remain in Siberia with two men and continue the search for De Long in March.

Meanwhile I had purchased sleds and provisions, and prepared everything for the comfort and safety of the party on their journey to Irkutsk. General Tschernaieff detailed a Cossack to take special care of Jack Cole; and he advanced me 6,000 roubles, 5,000 of which I transferred to Mr. Danenhower, with written orders to proceed at once to Irkutsk, and thence to the Atlantic sea-board, communicating from time to time with the Secretary of the Navy and apprising him of his progress. On January 6th, Bartlett, Nindemann, and the rest of my party arrived at Yakutsk, and on the 9th, with everything in readiness, Mr. Danenhower started homeward with the nine enlisted men.

The governor and half the population were out on that blue frosty day to see the Americans set out for America;

and there were many exiles there of all grades, who hungrily eyed the travelers and envied them their journey to the blessed land of liberty.

Within my short life I have seen a most respectable, affluent, and sensitive people driven like cattle into one section of a great European city, and the gates then closed and locked upon them — all and only because of their religion. And so it is to-day as it was eighteen centuries ago — over the greater part of the so-called Christian world, man is abused by man for conscience sake; though, thanks to the aggressive leaders of modern thought, much of this intolerance is dying out, yet not so rapidly in Christian Russia. And I pitied the poor exiles gazing wistfully on our little band of sailors, as though they were so many happy spirits bound for the mythical heaven, without the necessity of undergoing the imaginary horrors of death.

When my comrades had left, with the hearty well-wishes of all, I began to make instant preparations for my second expedition to the Lena Delta, in obedience to the following telegraphic order from the Navy Department: —

" *Washington.*

"Omit no effort, spare no expense, in securing safety of men in second cutter. Let the sick and the frozen of those already rescued have every attention, and as soon as practicable have them transferred to a milder climate. Department will supply necessary funds.

" Hunt, *Secretary.*"

Upon the receipt of this telegram General Tschernaieff said I could have anything I wanted, for I now had the whole Russian nation at my back; but unfortunately this was a *praesnik*, a holy day or holiday — in either case a day on which it was impossible to have any work done, for every store was closed, and even those farthest removed from the church refused to trade. Indeed,

throughout the whole empire, everybody was engaged
in the Christmas and New Year festivities; and I really
believe the governor thought I was rude or had gone
mad, from the energetic way in which I aroused the merchants and mechanics, and importuned him to interpose
his authority to compel the people to work and attend to
business. For, albeit the entire town was drunk during
the holidays, I nevertheless succeeded in gathering together materials and supplies for six months, all bagged
and put up in rawhide packages, ready for transportation
by pack-horse, deer-sled, and dog-sled across the mountain range (four thousand five hundred feet above the
sea level), and onward two thousand miles to the Arctic
Ocean, in a section of country where I have seen the
mercury fall in Fahrenheit's scale to seventy degrees below zero.

The governor forwarded my plans in every particular,
and assisted me greatly by his valuable advice. I engaged the services of three interpreters: Captain Joachim
Grünbeck, a Swede, formerly of the Nordenskjöld expedition, and commander of the steamer Lena when she
plied on the river as a freight and passenger transport,
who spoke Russian and English very well, and in fact,
with the exception of a few talented exiles, was the only
person in Yakutsk at all acquainted with our tongue;
Captain Constantine Bubokoff, an ex-officer of cuirassiers,
stripped of his titles and sent to Yakutsk for cause, who
spoke Russian, French, German, and Yakut; and Peter
Kolinkin, a Cossack sergeant, and the special body-servant of General Tschernaieff, who spoke Russian and
Yakut. Thus my company numbered six men, three
search parties of two men each, all so well provided with
languages that we could not help but understand and
be understood.

Our outfit was complete and satisfactory, including
tobacco for the party and for presents or payment to

some of the Yakuts on the Delta, together with a number of small articles suitable for trade and presents. With an eye to our possible failure to finish the search by early spring, whereby we would be compelled to stay all summer and return in the fall, I arranged with General Tschernaieff for the sending of an additional six months' supplies to Belun. I drew money on government requisitions, and kept Bartlett and Nindemann busy all the days long purchasing the small articles for our use.

The governor gave his own personal attention to the laying in of the bread, beef, flour, tea, and other heavy supplies; while the plates and numerous copper platters, pans and kettles were all procured through the police master, whom the governor, however, watched closely. The Scaups furnished us with many of our provisions, and the dried beef and butter, composed of equal quantities of butter and suet, were to be prepared at Verkeransk; where also, to save transportation, we would obtain our bread, which was first baked in large loaves, then cut while soft and hot into two-inch cubes, and finally kiln-dried in ovens and called *sucree*.

In the evenings we generally visited our friends, of whom there were not a few. As at Verkeransk, so here, and indeed everywhere in this cold, dreary climate, the people of all ages and classes have but one means of whiling away the long dark evenings — the card party, at which everybody smokes, drinks, or gambles; and I was again stared at as a most curious person when I confessed to never playing cards. However, I partook of their *cacuska* (liquor, raw fish, caviare, pork, etc., spread out on side-boards), and drank sparingly of their *vodki*. Yet the people were equally astonished, knowing of the abundant supplies I had purchased for the expedition, to learn that I had not included in the list a single drop of alcohol.

During my sojourn in Yakutsk, I had many pleasant

insights into the social life of the Siberians, a great
number of whom are free Russians, merchants of the
first class, who have emigrated there for the advantage
of trade. Others are the free children of exiles, and
then there is quite a host of government officials, military, Cossack. civil, police, revenue. church, etc. On New
Year's Eve the governor, officials, and all persons of rank
congregated in the public assembly room to welcome the
birth of 1882. Ladies and gentlemen mingled, drank
and conversed together, or danced to the lively music of
a large orchestra ; the gambling tables were set, and all
the *élite* of Yakutsk had evidently turned out, and, as
the lieutenant-governor remarked to me. on this night
as on no other, every man had his own wife at his side,
instead of some other man's. Let this be as it might,
however, all seemed cheerful and happy.

The governor was playing cards, and, as the rest of
the party could not understand my Russian without great
torture, I lapsed into silence, and my mind reverted to
distant scenes of ice and snow.

As the clock ticked out the last seconds of the old
year, and the critical moment drew nigh, every one became silent, many standing with bottles of champagne
in their hands, the cork strings cut, ready to fire a salute.
Suddenly the bell began to ring, and the governor aris-
ing announced the arrival of the *Nova Goat* (New Year).
Then, amid a volley of popping corks, the life and health
of the Czar and all good Russians were drunk ; and after
considerable hand - shaking, congratulations, and wine,
the party dispersed. The morrow was a notable church
day, on which it was the duty of the governor-general,
followed by all the populace, to first call upon the bishop.
I was present at the ceremony. A grand procession was
formed of the clergy in their robes, carrying crosses,
crosiers, books, etc. Chanting, they approached and surrounded the bishop, who sat in a large arm-chair, and

each one then advanced in order, with his hands outstretched, the left resting in the right, and both palms
turned up. The bishop graciously placed his hand in
each extended pair, and had the back of it humbly
kissed by the honored recipient of his blessing. When
the clergy had performed, the governor and all his official household fell into line according to rank, and did
likewise; after which the people followed suit, many,
however, refraining from the kissing act. Later on in
the day, the bishop with several of his aids visited the
governor, armed with their crosses, crosiers, and other
insignia of office. I was on hand by request, and conversed for a few moments with the bishop. He shortly
left, and then the household icons were blessed amid
prayers, chants, genuflections, etc., the service winding up
with the presentation, by the governor, of paper roubles.

Next day the bishop returned my call at the Balogan
Americanski, and seemed very much interested in my
work and people. All that day I saw numberless pro-
cessions of the clergy visiting alike the houses of rich
and poor, who in return for the blessings their little gods
received made liberal presents to each squad of religious
visitors. And as this sort of thing continued from early
morning to night, I have an idea that the purses of the
people must have been wofully depleted. In the evening,
the streets were filled with masqueraders, in parties of
half a dozen or so, who carried their own music, generally a little accordeon, and called promiscuously at all
the houses and danced without invitation. Every family
was ready, however, to receive and refresh them with
vodki, cognac, tea, etc., the inevitable raw salt fish, cut in
little pieces, smoked salmon, sliced sausage, white and
black bread, dried cakes, etc., and so as night drew on,
not only the masqueraders, but most of the clergy, too,
were gloriously drunk.

There is a regular military and Cossack force stationed

at Yakutsk, aggregating two thousand men. These military and Cossack forces are entirely distinct commands, and live in separate barracks, which are finely appointed; and the soldiers are all as well-clad and fed as the average denizen of Yakutsk. The city has been fortified for three hundred years, and some of the old Cossack towers are still standing. A government bank is located here, in which the revenues are deposited, and it does a bulky business, since many of the taxes are paid in fox-skins instead of cash. I watched the clerks counting, sealing, and baling up these skins like so much paper money, for their transportation to the East, where they are sold by the government's agents at Nijni-Novgorod, or some other fur mart, for the benefit of the Czar. This arrangement affords an opportunity for peculation on the part of the espravnicks of certain distant districts in their collection of the revenue. For the natives pay their head-tax in cash or fox-skins; and as the latter rate was fixed many years ago when the tax was low and the skins had little value, now that they have so greatly appreciated, when the ignorant native deposits his fox-skins, the cunning espravnick pockets them, pays the low tax in cash, and through a copert accomplice disposes of the skins at a high figure.

On January 19th, I started off Nindemann and his interpreter, Bubokoff, for Belun, with orders to hasten forward as quickly as possible, and await my arrival. I dispatched him ahead in order that he might prepare the way for our provision train, in charge of Bartlett, accompanied by the Cossack sergeant, Kolinkin. Then, as I have elsewhere stated, it was necessary to travel in sections owing to the limited number of transport animals on the road. On the 23d Bartlett got under way; and now I had only to pay the bills, sign the governor's papers, and be off myself. But the governor insisted that I should wait a few days for the provision train to advance and the horses and deer to return. I then vowed that I would never again find fault with Spanish or Portuguese holidays ; for surely after my experience with the *praesnik* I am satisfied that the Russians surpass every nation on the face of the globe in the number of their religious festivals, and ingenuity in devising excuses for avoiding work and getting drunk.

CHAPTER XXI.

NORTH AGAIN.

FINALLY, on January 27, 1882, having signed every paper to the satisfaction of the governor, and joined him in an excellent dinner, I departed from Yakutsk for Belun, in company with Captain Grönbeck. The weather was exceedingly cold, but we made rapid progress, and overtook both Nindemann and Bartlett at Verkeransk. Nindemann and Bubokoff, however, were ready for a start, and set out again on the evening of my arrival, February 4th.

Of course I met all my old friends again, and was glad to note their pleasure at my return. Kasharofski had been relieved of duty by his assistant, Epatchieff, who was now espravnick of the district, Kasharofski being ordered to the Kolyma at Nijni Kolymsk, to succeed the espravnick of that district, who had been recalled by General Tschernaieff for rascality of some kind. Kasharofski did not consider it much of a promotion, for his new station was farther out on the frontier, where food was poor and scarce; but then the general had said he loved and trusted him, and consequently could send him a long way off, while it was necessary to keep the rogues nearer home. He seemed greatly disappointed that I had not brought him any lemons, but they were not procurable in Yakutsk.

I was forcibly impressed, on the journey, with the remarkable differences in temperature. To the southward of the mountain range absolute stillness reigned, and the snow-fall was constant and heavy. The trees were so overburdened with their white weight, softly and quietly heaped upon them, that many had broken down completely and obstructed the wild roads through the forest. Our *yamshicks* were furnished with hatchets, and would stop to cut and drag the trees from our path. Crossing the mountain divide, our march was long and toilsome, and it was with great difficulty the deer could haul the empty sleighs up the steep incline.

When we had finally gained the top of the divide, I at once felt a change in the atmosphere. Whereas to the southward everything was as calm as the quiet of death, in front of us a gale was already blowing; and instead of trees bowed down and breaking with their burden of snow, to the northward of the mountain range not a single flake appeared on the shrubbery or woodland. We seemed to have passed into another climate; behind us all was white, before us, green; for the wind forever swaying the trees kept them clear of snow; and then again, the snow-fall to the northward is much less than to the southward, since the clouds are mostly milked out ere they can cross the mountains. Arrived at Kangerack station we were met by a Yakut *golivar* (head man) with fresh deer, who hurried us on to Verkeransk. He had been dispatched to our assistance by the espravnick, whom General Tschernaieff had ordered to give my party the right of way, and also to accompany me on the search.

To be sure I saw my exiled friends again, and used their bath-house. They were all in high spirits, making pemmican, and preparations generally for their projected escape. I allowed Bartlett and his transport a few days'

start on the road for Belun, awaiting the return of his
first deer-teams. Just before leaving Verkeransk, Epatchieff gave an entertainment at his house, at which all
the *élite*, and others as well were present, including the
priest of the village, his wife and children, and my old
acquaintances, Leon and Doctor Buali. There was great
feasting, and when all was over and we were ready to set
out, the priest prayed, chanted, and sang, and everybody
went through certain ceremonies, after which the indefatigable priest kissed his friends good-by. Then with
a large following we started for Belun; but a little way
out on the road a halt was called, and more farewells,
hand-shakings, and brandy-drinking indulged in; until,
finally, when the performance was becoming decidedly
monotonous to me, we forsook our friends and shot off
into the darkness. My party now consisted, besides myself, of Captain Grönbeck, Epatchieff, and his Cossack;
and we left Verkeransk on the night of February 10th.

The snow was very deep, the mountain roads rugged
beyond description, and the wind fierce and howling. It
blew as though it would never calm. Our reindeer were
poor and weak, and our progress over the first distance
of two hundred and twenty versts consequently slow, for
we were obliged to stop at every *povarnia* to feed and
rest the teams. The next station was two hundred and
ninety versts distant, but we succeeded in procuring fine
large deer, as wild as the landscape. They plunged and
chafed and wallowed about in the deep snow. A doe
hitched next to a buck will labor quietly along, while
her great fat companion worries and frets himself to
death within several hours; and the father and son of
a rich Yakut family at this station who have large herds
of deer, and do all the traffic on this part of the road, told
me that they had killed a great many of their animals in
forwarding Nindemann and Bartlett. I never experienced such stormy weather, and owing to the heavy snow

and high winds the mountain roads could not be kept broken or open.

We overtook Bartlett about one hundred versts from Belun. His teams had lost their way during the night, and had been snowed in; four of their deer had died in harness; they, themselves, had almost perished; and we came upon them in a *povarnia*, where they were repairing damages and resting their deer before returning for a portion of the transport which they had left in the mountains. We exchanged our vigorous deer for the best of their exhausted ones, and then pushed on; leaving them to recuperate over night and follow us the next day. The remainder of our journey was beset with difficulties, but we covered the hundred versts in eighteen hours, arriving at Belun on the evening of February 17th, Bartlett and Kolinkin making their appearance three hours after, and a part of their train a little later; though half of it yet remained in the mountains. Nindemann killed thirteen deer on the road, Bartlett eight, and the rest of his transport still to be heard from; so it was no wonder that the Yakut, as the espravnick told me, was complaining bitterly of his loss of deer, for since his *yamshicks* had done the driving, I could not be held responsible.

I found Nindemann and Bubokoff contented and drinking tea in real Russian style in our old quarters, the *Balogan Americanski*, where they had engaged a native and wife, as cook and wood and water carrier. They had made the journey from Yakutsk in twenty-two days; Bartlett in twenty-four, and myself in twenty-one. Bieshoff said the weather at Belun had been stormy all winter; the gale, in fact, had not ceased blowing for twelve hours since November; and, truly, our severe travel to the southward was a sorry forerunner of what was in store for us further north. Indeed, if the weather continued so unpropitious, I did not see what I could do

until late spring; but storm or no storm I was resolved
to set forth as soon as I could arrange for transportation.
Then, too, I must secure the services of Yakuts, and also
procure a fish supply for men and dogs.

I soon perceived how fortunate I was in having the
espravnick with me, for the speculative coperts, we
learned, had bought up all the fish caught on the Delta,
holding them all undelivered at the fisheries, and their
value had so appreciated that the fish I could have purchased the previous fall for three copecks were now
quoted at seven copecks. For this reason I decided to
go across the mountains two hundred and eighty versts
to Jamaveloch and bargain for our fish supplies; Epatchieff promising to break all contracts between the natives and coperts, save written ones, or such as had been
fulfilled by payment or delivery. Before leaving, however, Epatchieff attended a public vendue of the effects
of a Yakut who had died intestate, and selecting the
best seventeen from a lot of deer, had them knocked
down to me. I bought them for food on the hoof, and
had them afterwards driven to Cass Carta, our northwest
depot of supplies — no trivial undertaking in itself, as
Bieshoff told me that owing to the tempestuous weather
no one had made the journey for three months. Yet I
was eager to reach the ground as soon as possible and
begin the search, in order that I might be able to look
for Chipp and party during the summer.

I contracted with a certain Ivan Patnoggin and wife
to accompany me in the capacities of cook and scullion.
They had a baby which they ardently desired to take
along, but I solaced them for its temporary loss by paying two roubles per month to its grandmother for her
care of the little waif, and providing her with five pounds
of butter and forty pounds of flour as its food. I agreed
to pay Ivan and wife for their services fifteen roubles per
month, and, beside transportation to and from Belun,

gave them one pound of tobacco as a gratuity. Tomat Constantine I reengaged at a salary of twenty-five roubles and one pound of tobacco per month, and stipulated that he should have two fish a day for food. I also perfected arrangements for the transportation of our provisions to Cass Carta at the usual rate of three copecks per five *pood* per verst, storing six months' supplies at Belun until the next fall, should I be obliged to remain; and likewise arranged that the balance of my supplies, *en route* from Verkeransk, should be delivered at Jamaveloch.

With all these affairs satisfactorily settled, I departed February 22d for Jamaveloch, accompanied by Captain Grönbeck and Epatchieff; the party I had put in charge of Bartlett with orders to leave Belun on the 27th, under the guidance of Tomat Constantine, and hasten without delay to Cass Carta, conducting the provision train and seventeen head of deer. I afterwards learned that it would not be possible to drive the deer by way of Bulcour and Mat Vay, as there was no deer-moss in the whole of that district, and the animals would surely perish without it; so I directed Bartlett to proceed to Buruloch, and thence to Ku Mark Surt, where he would find dog-teams sent over by me from Jamaveloch. Here, too, he would part company with his deer, which would be driven across the mountains to the northward, while he pursued the bed of the river, via Bulcour, to Mat Vay, where I would have a sufficient store of fish awaiting him to carry the party through to Cass Carta, the northwest depot of supplies, from which I proposed to start my search parties. I also proposed to, and did, locate provisions at Mat Vay as our eastern depot, and, on my final search toward the mouth of the Jana River, to make Jamaveloch our base of supplies.

We made a fairly rapid journey to Buruloch, the deer station, and induced the starosti to accompany us to Jamaveloch. I endeavored to secure the services of a

native to drive our deer, when they arrived, to Cass
Carta, and was greatly surprised to have the interpreter
and espravnick inform me that there was but one man
who knew the road, and he was far too old to undertake
the journey at that season of the year. But I had learned
to my sorrow that there were more rogues than saints
in Siberia, and so insisted that there must certainly be
some one else acquainted with the way. Still they assured me that he was the only man, and so upon my
request the espravnick sent for him. He was, of course,
compelled to come, brought from some distant quarter of
the village, leaning on the shoulder of a young girl, and
otherwise supported by a great staff taller than himself.
He was blind and half-naked, only a few deer-skin tatters
clinging to his decrepit body, which here and there was
entirely exposed to the weather. He tottered into our
presence saying " Drastie, drastie," and at sight of him
I was ashamed of myself, but Epatchieff interrogated
him. How long was it since he had driven across the
mountains ? "About *twenty years*," he said. Did he
know of any one other than himself who could find the
way ? He did not, except his two sons, and they were
both dead. No one used the deer-road now; all the travelers went the other way, with dogs. So I left word for
Bartlett to hire a man at Buruloch or Ku Mark Surt to
drive his deer after him to Cass Carta ; and if he thought
there was any likelihood of his provisions giving out on
the journey, to kill for food what deer he could carry
with him. Bartlett received my note, but said it was so
badly written he could not read it.

There being no dogs at Buruloch, we started in deer-
teams along the river bed toward Jamaveloch, entering,
to the eastward of Ku Mark Surt, a ravine, or the dry
bed of a mountain stream, in which we wended our way.
It was a hundred and thirty versts to a *povarnia* in the
mountains, from which to Tamoose it was a hundred and

19

seventy versts more. Though I had been up at three
o'clock, and ready to start at four in the morning, yet we
had not succeeded in getting off until six, which, after
all, was an early hour for this section of the country.
The storm raged in a terrible manner, and the *yamshicks*
had no desire to venture forth and cope with it. We
would find no fresh deer on the road; and, should we be
storm-bound, not only should we suffer greatly ourselves
for want of food, but I should also be delayed in forwarding relief to Bartlett, who, under his orders, would move
promptly. After crossing the mountains, we had before
us a *tundra* passage of one hundred and thirty versts;
and, upon second thought, the natives decided to abandon
their intention of visiting the *povarnia*, and lay instead a
course across the *tundra* that night. Darkness came on
with a furious tempest of snow, and, according to custom, the *yamshicks* lost their way. We slept on the sleds,
while the deer, made fast, rested and browsed, and the
yamshicks, doubled up in their reindeer-coats, sat down
with their backs to the wind, and let it blow.

We were lost, but not so badly but that we could find
our way again at daylight. This we did, and traversing
the *tundra* reached by night-fall the shore of the Bay of
Bukoff; but the natives did not dare to cross the bay,
and so we skirted around it; and finally, at ten o'clock,
after many turnings and abortive movements, during
which the wild gale dashed clouds of snow in our faces
and half-buried the teams, we arrived, nigh frozen to
death, at a hut in Tamoose; having traveled that day
one hundred and ninety versts, and gone without food or
water since our three o'clock breakfast the morning before at Buruloch. And there was some satisfaction in
having kept these people active whether they wanted to
move or not, and in seeing them eat scraps of raw frozen
beef or deer meat from the provision sacks stored in our
sleds.

At Tamoose I met again our exile friends, and found that the Russian and Yakut coperts are as well acquainted with the philosophy of "corners" in fish as our Chicago merchants are with "corners" in grain; for Kusma told me that they had bought up all the fish in Tamoose. I also received the more agreeable intelligence that I could go to Oceansk from this point by journeying directly across the bay and forward for three hundred and fifty versts. We passed the night at Tamoose, and early the next morning set out for Jamaveloch. Immediately upon my arrival there, I arranged with the natives to send two dog-teams to Ku Mark Surt for the transportation of Bartlett and party to Cass Carta. These I loaded with one hundred and fifty fish, the natives carrying an extra supply for themselves and dogs; and sent orders to Bartlett directing him to hold the natives and teams until they had advanced him as far as Cass Carta. I also instructed him in my note that at Mat Vay he would find a depot of fish; for as soon as I had started off four teams to Ku Mark Surt, I collected four more and dispatched them to Mat Vay, with instructions not to fail in getting through, but to make a depot of fish at that point, and then hurry on to Ku Mark Surt and assist in the transportation of Bartlett's party. There was a brief lull in the storm, and during its continuance I succeeded in starting the natives, who were very loath to go. They delayed a long while in tomfoolery before their icons, and in kissing their friends, so that they had barely set forth when the wind arose again, and I felt sorry to see them depart. Still this was our contract, and Bartlett would depend on me for a fish supply if he should be unable to take his deer with him, which was doubtful indeed.

But the natives had scarcely started when they came back and declared that the wind was too strong, and in very truth it was impossible to face it. They all corrobo-

rated the story of Bieshoff and the people at Belun, that
no team had ventured to cross the mountains for three
months ; but they promised to begin their journey again
as soon as the weather would permit.

I paid for the fish at the usual rates, or at a slight premium, and Epatchieff seized all that the natives had for
sale or had bargained to sell to the speculative coperts,
whose ring was thus broken, and who were consequently
very savage.   They threatened the natives with condign
punishment, vowing that they would never thenceforth
sell them any salt, tea, or tobacco, and advised them
spitefully to procure these luxuries from the Americans
— " they were so much better."   Luckily I had an abundance of tea, tobacco, and other articles, with which I
could pay the natives in lieu of cash ; and as I only added
the first price and the cost of transportation together, they
received almost double the traders' allowance from me in
payment for their fish.   And yet they were not benefited,
for I soon became aware of the presence at Jamaveloch of
an organized band of thieves, who gambled night and day
with the natives for their articles in trade, and, in fact,
conducted their business as systematically as any gam-
bling-house in the world.

Noticing a number of these well-dressed, sharp-looking
knaves in the village, I at first inquired of Epatchieff
who they were.   He laughed, and answered, " Coperts ; "
and then dealing an imaginary pack of cards around a
table, he made a sweep with his hands, as though gathering in a heap of money, which, in fancy, he forthwith
stuffed in his pockets.   These thieves, bringing a quantity of tea, tobacco, and small money, come down to the
Delta and live among the natives, from whom they purchase anything they can, paying cash, which they immediately win back again ; so that, at the end of the winter
season, when the coperts are ready to decamp, they take
away with them everything that the natives possessed

the fall before, leaving their victims ragged and starving, but, strange to say, more anxious than ever to be fleeced. I have seen them sell their deer-skins, clothing, copper kettles — their little all — to the gamblers, who would pay over the money, sit presently down, and in less than half an hour win it back again. The kettles and other heavy articles which they could not conveniently carry, they would sell back to the natives at a high rate, or take a lien on their next summer's hunt or catch of fish.

And yet the Yakuts seemed to like it. They gambled away their fish before my eyes; and then, ranging their wives and children in a row, would show me their empty kettles, and push their stomachs in to indicate that they were empty, too; and this while the gambler sat placidly beside his spoils in the same apartment. On one such occasion I asked the native, who had just lost one hundred fish, how many he wanted for supper, and, upon his answering "Ten," I took that number from the ill-gotten stack in front of the gambler, and gave them to him. "All right," said the copert, "but you must pay me seventy copecks for the fish" — that is, one hundred and thirty-three and a third per cent. more than their market value. The native stood anxiously by to see if I would be dunce enough to buy back for himself and family a supper which he had recklessly squandered away; but I coolly drew forth from our hamper enough fish for my party and cooked them at his fire, thinking it might prove of value to him to be sent supperless to bed. But I doubt if it did, since the gambler gave him five fish on the promise of ten from the next catch; and so it continues; the old and young of both sexes gamble whenever an opportunity presents itself, and I believe this to be the direct cause of most of the misery and starvation which haunts the Delta.

When I landed the previous fall at Jamaveloch, the balogans stood about eight or ten feet above the ground;

now they were but slight undulations on the surface of the snow, and a column of smoke by day and a fountain of sparks by night alone indicated the exact location of each hut. The sleds ran evenly over the roofs, and the dogs halted at the chimneys to sniff the good things cooking below in the dinner-pots, so completely was the village snowed under. While awaiting a lull in the storm, I assembled all the Bukoff folks and paid off my old scores for fish and geese. To Kusma I also gave the reward I had promised him for carrying my message to Belun; but as he had not been able to supply the deer-teams and clothing for the transportation of my party, I only paid him three hundred of the five hundred roubles I had agreed to give him, and handed the balance to Bieshoff to cover his expenses in the performance of that service. Kusma, in addition, received a paper prepared by the espravnick, which gave him a proper title to the whale-boat, with the reserved privilege, however, to myself or any American party to make free use of it in searching along the coast during the summer for the people of the second cutter.

Epatchieff had notified all the natives to present their claims, and some of them tried to double the amount of provisions they had furnished us; but I had a tally-sheet which Danenhower had kept, and upon my leaving Jamaveloch I had enjoined upon him the necessity of preserving a correct record of all the stores we received. This he did, and handed me the list at Yakutsk.

It will be remembered that Nicolai Chagra, the starosti, had acted meanly in foisting upon us the smallest fish he had, whereas the other natives, notably one Androuski, had always been liberal in their allowances. And I had determined to be revenged on Nicolai if the chance offered; and it did. The payments were made in his hut, where we were quartered, and the starosti produced his tallies, which agreed with my memoranda;

but when I asked him if his fish had been large or small,
he winced a little and said "Medium." I then told him
to procure a sample for the espiavnick to see and judge,
and he brought in a medium-sized *mucksoon*, which was
far too bulky a representative of the kind he had given
us ; and I finally selected one myself which I believed to
be a fair sample. He looked very much confused and
discomfited, and while his neighbors gathered around,
Epatchieff rebuked him, saying he deserved no pay, and,
if I chose, he should not receive any. He was promised
fine, imprisonment, the knout, and kindred attentions
should he treat strangers as badly again, and if I would
simply say the word he should be punished then. Nicolai
had no defense to make, other than that fish was scarce,
we were a large party, etc.; but I counted his score and
paid over the exact amount due him, whereupon he retired amid the jeers and laughter of his neighbors.

Mrs. Nicolai was then led to the front, and presented
with needles and thread, and enough calico for a new
gown. She was also assured that she was a good woman,
and had done well by the strangers. Androuski came
forward with his bill, which tallied exactly with my account, and as he had regularly supplied us with large
fish, and usually thrown in several extra ones, all the natives were agog to see what would come to Androuski.
I doubled his score and paid him off, while the rest
shouted their approval. And so I settled with all ; wherever I found an inclination to cheat, I told the culprit of
it, and in one glaring instance, where a man put in a
claim for more fish than he had supplied, I deducted the
overcharge from the correct amount, and paid him the
balance. Old Spiridon, the pirate of the Delta, Mr.
Danenhower's much-trusted pilot, fared very badly. His
claim was entirely rejected by Epatchieff, who likewise
threatened him with a variety of penalties for his ill-
treatment of us. He was very penitent, yet, I could see,

a rascal at heart, and when some of the natives taunted him his looks plainly showed that he would never forget or forgive them. To complete his humiliation, Epatchieff deposed him from office, and elevated V́asilli Kool Gar to the high station of starosti in his village.

Several of the Jamaveloch ladies who had fixed our fire-place, plastered our chimney, and had done a number of kind services for us, such as repairing our boots, mittens, clothing, etc., received each a present. To Mrs. Androuski I gave calico for a gown, with thimble, needles, and thread to make it up ; nor did I forget Iniguin's sweetheart, the one whom he had called his " good little old woman," but gave her some small articles, saying through Epatchieff that the American Tunguse had sent them to her — an unexpected remembrance which filled her with delight. And thus did we mete out punishment and reward among the villagers.

CHAPTER XXII.

STORM-BOUND.

Arctic Weather. — Pedestrian Difficulties. — Lost in the Village. — Outstripping the Typhoon. — Continuance of the Same Old Gale. — A Yakut Solution of a Financial Problem. — Off for Arii. — Chul-Boy-Hoy — *Goliwar* Compass. — Turkanach. — An Afflicted Family. — Ordono. — Qu Vina. — At Cass Carta. — Our Palatial Quarters. — In Distress. — Timely Relief. — Together at Last. — The Art of Broiling Steaks. — A Reminiscence. — A Twenty-Pound Drink. — Yakut and Tchuchee Filthiness.

To economize space I will now transcribe from my journal.

February 26th. — Blowing harder than ever, and I do not see how I can manage to get away in such weather. The natives cannot endure it, and, indeed, refuse to try; so it would be folly for me to venture forth, for the present, at least.

Vasilli Kool Gar and Nicolai Chagra returned to-day baffled in an attempt to reach Ku Maik Surt. They are terribly frozen in face, hands, and feet. A young Yakut also arrived from Arii, more dead than alive, having been lost in the storm for two days. He seems dazed, and sits crying and swaying to and fro in a corner of the hut, without strength or wit to tell his story. A yamshick has come into the village seeking relief for a young Russian copert who lost his way on Borkhia Bay to the east of Jamaveloch, and is now in a povarnia to the southeast of us. He and his yamshick ate all their provisions and afterwards their dogs. Then he cached his stores in the ice and snow of the bay, and reaching a hut, dispatched his yamshick on foot to the village for succor. The latter is unable to stir about, but knows the name of the hut (Ka-ra-oo-aloch) where his master is; and a huge, tall, wild-looking Tunguse, whom I have engaged as dog-driver on my eastern search across the Bay of Borkhia and on the mouth of the Jana River, has gone to his rescue, no one else caring to go. The coperts in this region carry very little food, in order that they may transport the more goods for traffic, and many of them are consequently weather-bound and lost. I bought one thousand fish from Kusma, and about nine o'clock in the evening it so cleared that in less than an hour I managed to start the dog-teams toward Ku Mark Surt, with one hundred and fifty fish for Bartlett.

February 27th. — The storm rages more furiously, if possible, than yesterday. I hope the teams which set out last night succeeded in crossing the mountain range; but from present appearances it is doubtful to me if I shall be able to begin operations for a month to come

I have contracted for the hire of three teams of fifteen dogs each from Bukoff, harness and sleds included, for which I pay fifteen roubles per month and feed the dogs. Each yamshick receives twenty roubles per month and feed, which comprises fish, tea, and tobacco. Teams and drivers are to go wherever I direct them, and to haul our provisions when not otherwise engaged; but many of the fish I shall have to transport at road rates — of one hundred to a load, and three copecks per verst. I completed my fish purchases to-day, having bargained and paid, as is the custom here, for 5,150 for our central station at Mat Vay. I will haul away the fish as I need them; leaving 3,000 however, at Jamaveloch, for use at this end of the line.

I find that a number of the natives abandon Bukoff on account of the floods, though many of them live here

throughout the year. Still it blows a living gale. No
one willingly forsakes the shelter of his hut, and those
unfortunates who are forced to expose themselves to the
pitiless weather must either cling to some support or sit
down. There is no compromise. An old native started
from our hut to reach another not more than one hundred yards distant. So blinding was the snow, and so
fierce the wind, which lifted and whirled him around,
that he lost his head, and consequently his way. I do
not know how the alarm was given : but in a few seconds
all the men were getting into their boots and furs, the
women assisting and urging them to make haste. With
old Nicolai at their head they set forth on the search,
and I followed to watch their actions. Noting the direction of the wind and their present location, they all sat
squarely down on the snow, and then crawled away before the wind, shouting vigorously for the lost one ; and
they found him but a little distance off under the lee of
a store-house, crying aloud for help.

I have never seen such tempestuous weather as this,
either in the Arctic or elsewhere. I am anxious to see
it blow itself out, and give me a chance to get under
way. The winds are, and have been all winter, mostly
from south by west to south southwest, and at times from
south to southwest, all the heavy gales proceeding from
that quarter, — while occasionally the wind is variable,
and blows from all points of the compass. And yet how
still it was to the southward of the mountain divide !

When our sleds struck the trunk of a tree we were
buried beneath an avalanche of snow. Here the *tundra*
and high table-lands are swept clean by the gales, and
the valleys and gulches are gorged with snow. This
delay is intolerable ; for I am eager and impatient to
survey the territory where my comrades are.

February 28th. — An exile ventured forth last evening to pay a visit, and lost his way. He managed to

wander to the windward of the village and of Nicolai's
hut. Some of the people heard his cries, and after a
half hour's hunt found him, where he had dug and
crawled into a hole in the snow, which banked up at his
back and kept him warm. Besides, he had on his long
deer-skin *cooly-tang* or *parky*, and would doubtless have
survived the night and been rescued. This was old Simeon Alexoff, a Russian exile, and his adventure frightened him badly.

Towards morning the gale calmed, and the natives
busied themselves in cleaning up their outside surroundings and hauling wood and ice for another siege of
weather. I took advantage of the lull to dispatch two
dog-teams with two hundred fish to Mat Vay or Cass
Carta, as opportunity offered, and thence to Bartlett's
assistance at Ku Mark Suit. An exile arrived here to-
day from Upper Belun. He has been five days on the
road, and had the wind behind him all the way. He
says that La Kentie Shamoola has gone to Ku Mark Surt,
and may carry Bartlett through, or come over to Bukoff
for orders from Epatchieff.

The weather to-day and this evening is the best we
have had for weeks. The natives have lost nearly a
month's fishing, and are hard pressed for wood and
water, as they have an antipathy against using snow;
and although they have plenty of fish for the present, yet
a large part of their catch must be sold for cash to procure tea, salt, and tobacco, and to pay their taxes. I find
that the cost of transporting fish is greater here than
their price. I tried in vain to buy them delivered at
Mat Vay, but it was too abstruse a calculation for the
Yakut brain; they have always sold their fish at Bukoff,
and separately engaged to haul them at three copecks
per verst; that is all they know about it, and all they
will do about it. I heard to-day that there is a famine
at Oceansk and the Omalai, and a great scarcity of food

at Upper or North Belun. Epatchieff advises me to buy
up all the fish I may need, as he thinks after April 1st I
shall not be able to buy any at all. Were I sure of the
salt and dried beef from Verkeransk, I could do with less
fish, but I am not, and cannot rely upon an uncertainty
like this, when I shall have twelve or more persons to
feed, beside three dog-teams for the search, and probably
five other teams for hauling fish, and various purposes;
and the transportation of food by these Yakuts is such
a slow process that I may have to attend to it myself.
The native cannot induce himself to go in a hurry and
keep going — they all "go as they please," halting at
huts, povarnias, and villages, as the spirit moves them.
Yet if I can succeed in separating them from their huts
and women, I will drive them and their dogs too.

During the month of May it will be perilous to work
on any part of the Delta. The inhabitants here say
that at times when an ice-dam breaks a vertical wall and
flood of water will rush down the river for miles at the
speed of a race-horse, carrying everything before it, until
it vanishes in the many outlets and seeks the sea. This
is repeated again and again, until the southern flood-
waters have swept away the northern ice-gorges, and the
river is free to flow.

The weather changed once more. It was snowing silently and slowly, but by eight o'clock in the evening
the wind blew mightily as ever.

March 1st, 1882. — At early daylight there came another calm; but it was very evanescent, and by eleven
A. M. the storm had resumed its sway. The wind seizes
sleds or any other exposed object and hurls them across
the bay; so I do not wonder, as I used to, why the natives stake fast their empty sleds to the ground.

This morning, while the weather was still fine, two natives, who have come here from Belun to fish, went out
to haul their nets. Shortly after their departure the

wind arose, and at three P. M., when they had failed to
put in an appearance, the villagers set forth to look them
up.   These poor devils have a hard time of it; starvation
on the one hand, and danger of death from cold or exposure on the other.   I have now been here four days,
and during that time four men have been rescued in this
village alone.   Another Russian copert, Sennikoff, was
"jacksoned" by the storm for thirty-five days at Turkanach.   He and his two drivers had food enough to see
them through, but the dogs had nothing; so they killed a
dog a day for the other animals to eat, and finally started
a driver with a team of six dogs to Bukoff Moose, seventy
versts distant, where relief was procured.

The wind still blows at such a rate that neither man
nor dog can face it, but must needs crawl into any available hole for shelter.   I stepped outside to-day simply
to experiment, and see if it were possible to stand up or
hold to the hut.   I could actually discern nothing for the
blinding fury of the storm, for the wild rushing air was
opaque with snow and fine particles of ice.   I lost my
grip on the door-jamb, and with difficulty crawled back on
my hands and knees to the top of the snow-steps, down
which I took a header and rolled into the hut.   The natives will not allow any of us to go out alone, but insist
upon sending one of their number to keep us company.
I have seen a typhoon blowing in Japan, when the ane-
mometers on three ships registered ninety-nine, one hundred and one, and one hundred and three miles per hour
respectively; when weak buildings were demolished, vessels at anchor dragged along, and *jin-rick-shas* turned
over like willow baskets, — yet I was not carried off my
feet, nor was the typhoon in its most furious mood a circumstance to this irresistible boreal blast.   Thunder and
lightning are entirely unknown in the Arctic Ocean.
Towards the pole the aurora is the only form in which
the presence of electricity in the atmosphere is displayed;

and the question arises, Why the aurora, instead of the discharges of light, attended by thunder-claps, seen at the equator?

To bring about the usual atmospheric phenomena heat must be applied or extracted. Perhaps, then, the want of heat in the polar regions may account for the absence of thunder and lightning, — or can it be that the immense blanket or non-conductor of ice and snow prevents the discharge of the electric current? So that, if a certain degree of heat were introduced, the aurora would burst forth into vivid flashes?

March 2d. — An aggravated continuance of the same old gale. The snow has closed up the weather-door of our hut for twenty-four hours, and no one has had the temerity to expose his head to the outer air. I now despair of doing aught but abiding a change of weather. One blessing has been vouchsafed to us — there are no children in the hut; though Madam Nicolai persists in lavishing her motherly affection upon her son Abonasshi, a full-grown young man, borne by her to her first husband. She takes him on her lap, and hugs and loves him, and wipes his nose, and cares for him as though he were four years old. Abonasshi is a good boy, and assists his mother in carrying wood, ice, etc. Upon madam devolves the duty of training the young dogs, and during four or five hours of each day she has half a dozen of puppies marshaled in front of her, each one with a thong around his neck attached to a stick from twelve to sixteen inches long, which, in turn, is lashed to the edge of the low-down berth or bench. The effect of this arrangement is that the dog is kept constantly pushing forward; for if he attempts backsliding, the stick is thrust into his neck and reminds him of the folly of such a course. The house-dog is usually employed in cleansing the children; and I had rather sleep in a snow-bank than in a hut full of small progeny.

La Kentie Shamoola should have been here days ago to carry us up to Cass Carta, but he is doubtless storm-bound at Ku Mark Surt, or a povarnia, *en route.* I have dogs enough for our transportation to Cass Carta or Mat Vay, but there is no man in the village acquainted with the road ; and then, too, it may be necessary for us to carry a large supply of fish for a long siege of bad weather, and I must wait here for more dog-teams and guides. Our venison and bread have given out, but with plenty of fish and salt we are much better off than when here five months ago. The natives, who live for months with-out a taste of bread, begged hard for some, and I was a little lavish with it, considerably more so than I would have been had I known the length of our detention here. Kusma besought me for just a little for his supper, and I gave him enough for a dozen suppers.

Yapheme Copaloff, the " Red Fiend," is quartered at Nicolai's hut, living in a corner and sleeping on the floor. Upon our arrival he immediately installed himself as major-domo for Epatchieff, Captain Grönbeck, and my-self. He implored me to take him into my service as a general hand, which I at length consented to do at a salary of fifteen roubles per month. He attends to all our wants, cooking our fish and tea, and making himself very useful indeed. [So I took him with me to the northwest, then back again to Bukoff Moose, and finally, with General Tschernaieff's permission, to Yakutsk, where he ran headlong into trouble, and consequently into the calaboose.] I am worried most about Bartlett and com-pany, for they are on the border of a perfectly barren region ; albeit there is a good-sized village about seventy versts to the west of them, where Tomat Constantine may procure provisions.

March 3d.— The dog-teams, which I started towards Ku Mark Surt, returned to-day, having failed in their at-tempt to reach that point. They succeeded in arriving

at the first povarnia, forty versts distant from Jamaveloch, but on the road to Tas Arii they became lost, and
had a terrible experience. Four of their dogs died, some
were cut adrift, and others were brought back in a helpless condition on the sleds. Spiridon's leader, a young
and well-broken dog, valued at seventy-five roubles, is
among the lost, and the old fellow is inconsolable. The
natives are in a pitiable plight, frozen terribly in face,
hands, and feet. They lost their way both going and
coming, and I shall scarcely be able to persuade them to
dare the weather again until it has fairly settled. Two
of the teams have not yet returned, but those that have
report them as safely progressing. By this miscarriage,
Bartlett and company are detained at Ku Mart Surt, and
we are involved in more expense and delay.

I now fear that the teams I sent to Mat Vay will also
return defeated. It is the devil's own job to secure
transportation here, — the much-vaunted dog and deer-
teams being very insufficient; yet I cannot expect better
means of conveyance in such a wild country. The deer-
teams which brought us to Jamaveloch left to-day for
Ku Mark Surt. They could not carry any fish for me,
but promised to give Bartlett sixty of their own, which
will be returned to them from the one hundred and fifty
which I am and have been trying to get through to him.
I dare say that some of the cabinet Arctic travelers and
critics who have been within three thousand miles of this
place will wonder why the blank fool did n't *do* something —if only they had been there — Ah ! — Yes, I warrant if they were the nucleus of their trousers would be
nigh touching the ground ! One hundred fish, it will be
remembered, constitutes a load, as fish for the dogs and
drivers going and coming must also be carried; though
this burden is partially relieved by deposits made in the
snow along the road.

The day was comparatively fine, but the wind cut like
a knife.

March 4th. — Our few hours of good weather came and went like a flash of the sun from behind a cloud. It is blowing again with, if anything, augmented velocity. I have walked three hundred yards to windward, while with the Jeannette, to read the instruments, and without experiencing any serious inconvenience ; but I cannot even stand up against this ferocious tornado, which surpasses anything in the way of weather I have ever seen. One of the returned natives called upon me to-day, and told a pitiful story of his wanderings and sufferings. The poor fellow's face and hands evidence how cruelly Jack Frost can bite. His cheek-bones are two raw spots as large as silver dollars, and his nose resembles a pickled beet. He does not wish to carry more fish until spring time, and I don't blame him. One of the dogs turned loose found his way back to the village last night, and died. This swells the loss of the natives considerably, for besides the pain and misery they underwent, it was all for nothing, since they must deliver the fish according to contract or receive no pay. We are now down to our last loaves, and boiled fish is our daily fare — very excellent when one can get no other.

I witnessed to-day a division of the profits of a team of eleven dogs among three natives, who owned respectively three, five, and six of the dogs. But the one with the smallest number of animals in the team had contributed his services as driver, while the owner of five had supplied the dog-feed. Here, then, was a financial problem of no little intricacy to the natives, and this is how they solved it. First, by placing as many small sticks in the centre of the table as there were dogs (eleven), to which the driver added three more for himself. They then drew from the money-pot copecks as per sticks, and when all the earnings were exhausted each settled with the man who had furnished the dog-feed for his quota, according to the composition of the team, three, five, and six.

March 5th. — The wind has veered around to the eastward, doubtless a change for the better. Vasilli Kool Gar came over from Arii, bringing with him three men from Long Island, near the Alanek, and a guide for the Omalai. The former will act as drivers and guides to the Alanek should I require their services, and the latter, a Tunguse, I can find here whenever I want him.

The two teams I started off with fish for Mat Vay have evidently pushed through, since they have not returned to Arii; and so I can hope that they are now on the road to Ku Mark Surt. The other teams are still here, but if the wind holds from the east and on their backs, they will make a second attempt. We are miserably located here — the Yakuts and my party of three quartered in a hut twenty feet square. The sights! and the odors!! . . . Still we are sheltered, which is a blessed comfort just now; and have tea and boiled fish every day, which altogether is quite enough.

If the weather continues as it is, we must face the music ourselves. I have five dog-teams ready for our transportation to Cass Carta, but besides the fish for the dogs I can only carry fifty for the use of my party. The wind is slowly calming down, and the natives are about to feed their dogs, with the intention of starting for Ku Mark Surt in the morning. This is the chief cause of delays; the natives refrain from feeding their dogs until the weather clears, and then twelve hours afterward they set forth, *providing* the weather has not changed. But in the mean time they have lost twelve good hours, while the dogs are digesting their food, and if the storm comes on again there is a depletion of the Yakut fish pile, a repletion of the dogs, and a senseless delay.

March 6th. — Light easterly breeze this morning, with snow. The drivers proposed to go as far as the balogan of Spiridon at Arii, ten versts to the northward; and finally getting under way by 11.20 A. M., we made the journey, the day growing brighter but colder.

The yamshicks insist upon tarrying here over night to feed the dogs, and then make an early start to-morrow for Chul-Boy-Hoy, seventy versts distant, which they hope to reach by daylight. There are a number of natives at Arii, from North Belun, who have come, in answer to a summons from the espravnick, to assist me in hauling the fish, etc. They report a famine at North Belun; and their frost-bitten bodies testify to the terrible suffering they underwent on their journey; all having been caught by the storm away from povarnias.

We are stopping at the hut of old Spiridon; and strolling about the village I saw the wife and grandchildren of Vasilli Kool Gar, to whom I gave some tea. There were large quantities of fish stored here, but all have been sold to the coperts. I suffered severely in my feet to-day, during our short run of an hour, owing to my damp socks; for I had worn my deer-skin stockings in the house several hours before starting, and as a consequence they so froze on the journey that when we arrived here my feet were blistered.

March 7th. — My party, which consists of Epatchieff, Captain Grönbeck, Yapheme, myself, and five dog-drivers, set out this morning about seven o'clock from Arii. The drivers said it was seventy versts to Chul-Boy-Hoy, and laying a northwest course we reached that place, a collection of three tumble-down huts, about two o'clock in the afternoon.

When we had been four or five hours on the road I asked Vasilli, in order to impress the course on my mind and afterwards mark it on the chart, how far and in what direction was Barkin. The distance, he said, was forty versts; and, to indicate the direction, he laid his dog-stake upon the snow for me to set and read the compass. The natives have a wonderful sense of locality, and in sunlight, moonlight, or the darkest night, seem able to exercise it equally well. They only lose their way when the snow is swirling in clouds or columns. I have many times, simply to test their ability, asked them to show me the *sever zaputh* (northwest), or some other point of the compass, and they would as often indicate the required direction with their staves, which, first balancing until the iron point was fixed to their satisfaction, they would finally place on the snow, and my compass invariably proved their calculations to be correct.

Old Vasilli delights in using this faculty of his. He has learned the word compass, and tapping his head he laughingly told me — "*Golwar* compass" (head compass). And when I inquired of him the direction of Barkin, he pointed his stake east by north and said, *byral* (sea). So I now have the prominent points on the Delta located as definitely as possible on my chart, and can approximately designate the situation of all the villages and huts, and the course of my journey. To be sure there are some inaccuracies; for upon asking different yamshicks how far it was to certain places, they have answered, if our dogs were good, "Fifty versts;" if bad, "Seventy," or even "Ninety versts." Yet I have learned to measure distances by time and conditions and marked them accordingly.

March 8th. — Clear and cold, when we renewed our journey at nine A. M.

To the southward of Chul-Boy-Hoy runs a small range of detached hills, not unlike the foot-hills of the southern mountain range. They are from five to ten miles back of the great bay, or *gooba*, and though the weather was clear, yet a dense haze, which the Siberians call "the frozen air," enveloped and obscured them.

We passed a cold, smoky, and miserable night in the povarnia. It is vastly more disagreeable to sleep in a hut full of holes than in a snow-bank or on a sled; for the cold winds are forced through the chinks like blasts from the muzzle of a bellows. We traveled all day due west, and towards two P. M. approached Turkanach, where we expected to spend the night. As we drew near, I was surprised to see smoke curling up from one of the three huts forming the village. Our dogs as usual dashed forward in a perfect pandemonium of yells, but no one came out from the hut to greet us. Then Capiocan, one of our yamshicks, crawled inside on his hands and knees, but reappeared in an instant, looking wofully frightened, and muttering something about " pomree " and " propaldi," from which I inferred that there were dead or dying people within. While he was explaining matters to Epatchieff, I observed issuing from the hole in the hut a miserable object half clad in an old deer-skin blouse. It was wailing and groaning dismally, and for a minute I could not determine whether it was a man or a woman. Face and hands were swollen and covered with frostsores, and the stooped and limping figure, leaning on a long staff, bowing, crossing, and beating its breast, at length cried out between its sobs and moans, " Drastie, drastie, drastie ! "

It was a man, we found, and the espravnick, quieting him down, inquired into the cause of his lamentations. Thereupon he invited us into the hut, with the assurance that there was no corpse within, for among the natives no person is permitted to enter the huts of the dead save the kin of the deceased, and even they are quarantined thereafter for thirty days from the rest of the community. Crawling into the hut, we found it occupied by six natives, — a grandfather and grandmother, their married son and his wife, and two children; a young girl of fourteen or fifteen, and a baby several years old. Their cries were heart-rending, for they were all crippled from frost-bite. The grandfather, who was partly blind, sat in a corner swaying back and forth; while his aged spouse, barely able to lift her head, held the baby near the fire, and chafed its almost lifeless body. The mother, her head covered with a deer-skin coat, sat on one of the bed-places and shrieked out her agony ; and the young girl, with her arms around the neck of her mother, wept convulsively.

Presently we listened to the sad story of the afflicted father, who told us that a famine was raging in the western section of the Delta, that he had heard there was plenty of fish at Bukoff Moose, and that, acting upon this rumor, his father and mother, his wife, himself, and their five children, had attempted to *walk* from Long Island, at the western discharge of the Lena, to Bukoff, — a weak team of five dogs carrying their household goods. They had marched through the furious storm for eight days and nights, repeatedly losing their way; three of their children had died of cold and hunger, and were buried close by in a snow-bank; after which they had managed at length to crawl into this hut, where they had been for several days absolutely without any food, having lived for more than a week previous on the rawhide and untanned portions of their clothing and outfit. They were far too enfeebled to hunt or gather wood, so they had torn down and burned the inside of the hut. And here Capiocan was off like a flash, dumping our freight from his sled, and shortly reappearing with a load of drift-wood. Meanwhile our tea-kettle was boiling for their benefit, and keeping scarcely enough fish to see my party through to Cass Carta, I had the rest buried in the bank near the hut for the use of the wretched family. I also gave them a cake of tea, and told them to stay where they were until my teams, returning to Bukoff for fish, could carry them to Arii, whither they were bound. Of course they were almost overcome with joy. And I find that the lives of these poor people are only a succession of such distresses and rescues, as they journey from place to place seeking the bare necessaries of existence, which they do not always find.

When we had thus somewhat alleviated their misery, we were obliged to push on to the next povarnia, there being a great number in this district, since in summer it is thickly populated. As we ran along to-day, old Vasilli pointed northward and said, " Borkhia ! " not meaning the great cape which bears that name, but Little Borkhia, the point at which I chanced upon the three natives in their canoes. Vasilli told the espravnick all about our meeting, and how he had piloted us around to Jamaveloch. And here I am coasting over the same course which I was dissuaded from following last fall. Yet whether I would then have pulled through — perhaps have met my comrades of the first cutter — *quien sabe ?*

We hauled up at Ordono at three P. M., and will stop here over night. Vasilli says Mat Vay is fifty versts to the southwest, or west southwest, and Qu Vina is fifty versts to the northwest; so I will next proceed to the latter place, as it is only a short distance from Cass Carta. The question now arises, — Where is Bartlett and the provision train ? The natives told me to-day that the two teams which I dispatched to Mat Vay with fish halted on account of the weather at every available point along the road, and ate half of Bartlett's fish before they reached Mat Vay, and then were forced to take the other half with them on their journey to Ku Mark Surt. This is dolorous news, indeed ; for if Bartlett is delayed in transporting the provisions, my party will be short of food. We have nothing to eat but tea and fish, and of these only a couple of days' supplies.

March 9th. — A good night's rest in the povarnia of Ordono. Last night we were only cold, the night before we were frozen. Making an early start, we ran a northwest course, passing about four versts to the north of Qu Vina, which was plainly in sight, for the natives wished to reach Cass Carta, and I had no desire to check their laudable ambition. So here we are quartered in two povarnias, miserable holes, and two palatkas, which will answer very well as magazines for fish, etc.

The day has been clear and cold, with a light breeze blowing from the southeast. When we were about two miles northwest from Ordono, we passed a high island which is sometimes mistaken for Stolboi. Our course lay so far to the north that we did not see Stolboi, and then, too, the snow was drifting so heavily that the southerly mountain range was hid from view.

March 10th. — At Cass Carta. Our hut is palatial, particularly in its dimensions—ten feet square, and four feet high. It has no chimney and no door. We put a deer-skin over the smoke-hole, and will make a door to-morrow. The smoke from the fire in the centre of our residence is blinding. Our faces burn, our feet freeze. We are miserable, believe me.

Here is an inventory of our larder: ten fish, no tea, sugar, salt, or bread. I dispatched one dog-team to North Belun for La Kentie Shamoola, and all the dogs in the village; and Epatchieff issued an order calling upon the natives throughout the Delta to send all of their dogs here. I expect two hundred fish to-morrow, if the teams follow me up as I directed; and if we are to stay in this hut I must devise a chimney of some kind, or else smother to death.

March 11th. — We are out of fish — out of all food, and have absolutely nothing to eat. If our provisions do not arrive in good time, I will send Yapheme, the "Red Fiend," to Kigolak or North Belun for help. We have no dogs left, so he will have to walk — not a very great distance, providing he does not lose his way.

March 12th. — Clear, with a strong cold breeze blowing from the south. I was bent upon sending Yapheme to North Belun this morning; so I got him ready, loaned him a compass, and instructed him in its use. He has been to North Belun, but went over the road in company with natives. By road I do not mean a trodden path, but the unbeaten track between two places; and Yapheme may or may not be able to recognize the landmarks. He did not seem very anxious to go, but then what were we to do?

Before starting him off we all ascended to the top of the hut, and eagerly scanned the great waste of snow to the south for a sign of succor. We looked and listened in vain, however; and then turned our eyes towards North Belun. While gazing in that direction I fancied I saw a crow fly over a bank or ridge of snow, and disappear in a ravine. I informed my companions of this, and together we all looked intently at the ravine and waited for the crow to soar upwards. Suddenly a dark object, like a boa, wriggled out of the hollow and crept towards us. It was a dog-team — so we shouted simultaneously; and watched until finally we could hear the yelp of the dogs. Relief at last; for no matter about the quantity, if any, of provisions they might bring us. We would have the means at least to procure food; so we crawled inside of our hut and warmed ourselves; and presently went out again to see how close the teams had approached.

Then, much to our surprise and delight, we detected the baying of dogs far to the eastward, and in a little while caught sight of the provision train wending its way across the snow; now in full view, now swiftly disappearing only to mount and show itself against the whitened hillocks.

The teams from North Belun arrived first. There are five, driven by my old friends, La Kentie Shamoola, Geordi Nicolai, Starry Nicolai, Young Kerick, and Starry Kerick. The last-named was in possession of a saddle of venison which I at once bought, and ordered part of it to be cooked for ourselves and coming party. In an hour

or so Bartlett and the provision train, with Nindemann,
Kolinkin, Bubokoff, Mr. and Mrs. Patnoggin, and five
dog-drivers, reached us, all more or less scarified on face
and hands by the cold, but nevertheless jolly and hungry.
As the deer could not be driven along because of the
dogs. Bartlett killed as many as he could carry ; and Vasilli Kool Gar and son having arrived a short time ago
with two loads of fish, we have now an abundance of
food. I have also one hundred and twenty-five yelping
dogs staked all around me, and twenty people to feed.
Geordi Nicolai and La Kentie Shamoola I have engaged
as dog-drivers, and have three good teams selected for
use : but with one of these I will have to send Yapheme
and Tomat Constantine to Ku Mart Suit to bring forward the balance of our deer. Bartlett was compelled to
leave the bread behind, and I have already dispatched
teams for it. The fish I sent to Mat Vay were all eaten
en route by the natives and dogs, so I will send for more
for our use while prosecuting the search, and for the
teams journeying between here and Belun. Bartlett
lost one hundred and twenty pounds of loaf sugar, but
hopes it will be recovered, as he at once set the party,
from whose sled it was missed, to finding it, under penalty of punishment ; for this is a trick which the yamshicks do not hesitate to practice on the unwary. Kolinkin likewise lost his clothing.

Bartlett, obedient to his orders, started on time, and
against the combined protests of the natives, the starosti,
and the Cossack commandant. It was storming terribly, he says, when they set out, and the train was long
and heavy, and the deer barely able to drag themselves
through the deep snow. By the usual misadventures in
such traveling, Bartlett forged ahead of the train instead of keeping to the rear, and meanwhile the teams
last in line got to plunging and parting their halters,
bolted off to one side, dashed up a bank, and took to the

woods, upsetting the sleds and spilling the provisions as they ran. As soon as the mishap was discovered, a team was sent in pursuit of the runaways, which were found resting on the snow, tethered fast to the sled that had caught in the thick timber. Gathering, then, the rest of their teams together, they sought shelter at the new abode of Kusma, near Ajaket, and next day starting on again, they continued their journey without further interruption to Ku Mark Surt, where they awaited the coming of my long-delayed dog-teams. They passed more than one night in the snow, Mrs. Patnoggin burrowing a bed in nature's white fleece along with the rest. She is bright and merry, and can serve our simple bill of fare quite nicely, for it only consists of fish, boiled or fried, and venison, boiled or broiled — the broiling being done on the bare coals.

I have brought with me a lot of butter and tallow, mixed in equal quantities and called "Verkeransk butter," or *jzere*, with which we fry our fish, and butter our dry broiled steaks; for reindeer is a miserable meat, coarse, black, dry, and tough, and requires in cooking the aid of additional fat. I have initiated Epatchieff into the luxury of a broiled steak put on his plate hot from the coals, and properly salted, peppered, and buttered; and he says he shall devote all of his leisure when he returns home to cooking beefsteaks.

And now I recall a day at Verkeransk when Kasharofski informed me that he would have beefsteaks for dinner, served hot as Englishmen liked them. I expected, of course, a great treat; but imagine my surprise when a wrought-iron pan (the same in which the meat was cooked) about a foot in diameter and about three quarters of an inch deep, saucer-shaped, and carried on a peculiarly-cut stick, by means of which the pan could be readily converted into a dish, was placed in the centre of the table. It was full to running over with fiery hot

and spluttering butter and tallow. And the steaks, —
ye gods of the art cuisine! They were little three-quar-
ter inch cubes of beef, browned like doughnuts, and I
need scarcely say that I was disappointed and my appetite repulsed. I partook of something else, greatly to
Kasharofski's astonishment, while Leon, the exile, who
was present, explained to him the points of difference
between an English and a Siberian steak. As the dinner
progressed I saw both Kasharofski and Leon supping the
molten grease from the pan with table-spoons, and I remarked to them that I thought it extraordinary that they
could do such a thing, at least without sickening. They
laughed at this, and said it kept them warm and fortified
them against the cold weather; and Kasharofski then
told me of the great fondness of the Yakut for hot butter. asserting that one man could drink half a pood of it,
or twenty Russian pounds (about eighteen pounds, eight
ounces, avoirdupois). As I seemed incredulous, and, indeed, plainly expressed my disbelief of this statement,
he sent his Cossack in search of a Yakut, and ordered
half a pood of butter to be melted for our experiment.

. When the native appeared, Kasharofski informed him
of the golden opportunity that was open to him, and then,
after a preliminary drink of *vodki*, handed him a stone
jar containing the butter. A broad grin of satisfaction
lit up the native's face, which immediately after was hidden within the stone jar, and he guzzled away as though
he were swilling buttermilk. A second pull and he
owned the whole half pood; the jar was empty. Kasharofski then inquired if he wanted more. No, he did not,
at least of butter, but he would relish another drink of
vodki. This he got, and thereupon bowed himself out
of the room.

I could not help expressing my astonishment at the
man's capacity, whereupon Kasharofski overwhelmed me
with a statement which Leon confirmed, and the truth

of which I have since proved beyond all doubt. It was this : that if the Yakut was a good and loving spouse he would go directly home, and eject the contents of his stomach into a vessel of water, which would then be placed out of doors to cool and collect, and from the rich, floating vomit his wife and children would afterwards enjoy a hearty meal. The lucky possessor of a stomachful of *vodki* may in a benevolent mood similarly dispose of a part of his repletion, minus the water ; and away to the eastward, among the Tchuchees, families are oftentimes regaled, even to inebriation, with the natural fluid discharge from the bodies of fortunate tipplers. Among these same people it is a well-known custom to use the urine of both parties to a marriage as a libation in the ceremony ; and likewise between confederates and allies to pledge each other and swear eternal friendship. It is also a useful article in their household economy, being preserved in a special vessel and employed as a soap or lye for cleansing bodies and clothing, and curing or tanning skins. Saving the natives themselves, it is their most disgusting institution ; and if any Christian missionary be earnestly seeking a fresh field to labor in, I can assure him that no soil is more desperately in need of cultivation than the Tchuchee country.

These reminiscences of Verkeransk and St. Lawrence Bay have made me forget for the nonce our now thickly populated village of Cass Carta. So I shall return to my journal, transcribing its daily record, and paraphrasing as I proceed.

CHAPTER XXIII.

FINDING THE BODIES.

Getting Affairs in Shape. — My Map of the Delta. — Searching for Ericksen's Hut. — Revelations. — Contending with the Storm. — The Yakut Fashion of Lighting Fires. — A Miserable Night. — Which Cape? — A Clue. — The *Myack*. — Found. — De Long's Ice-Journal and its Sad Entries — Positions of the Bodies. — De Long's Pistol. — A False Report. — Dr. Ambler. — Appearance of the Dead. — "Dwee Pomree."

March 14th. — Sent the "Red Fiend" and Tomat Constantine to Ku Mark Surt for our reindeer. I secured one good team of dogs to-day for fifteen roubles per month and their feed, — hiring the driver at the same terms. As soon as I can procure two more teams of equal excellence, and fish to feed them, I shall proceed to Usterda, accompanied by Nindemann, and pick up the trail. When Nindemann reached Cass Carta the other day, he at once declared that I was too far to the westward, for although I was on the river along which De Long and party took up their march, yet the place where he and Noros separated from their comrades is away to the eastward. This is an unfathomable mystery to me.

March 15th. — I received four good dogs, to-day, from North Belun, and will put them on rations until I can "complete the set." Three teams arrived from Bukoff this afternoon with three hundred fish, which I stored, paying their road money. I have now opened communications with all parts of the Delta, but have received no tidings as yet of the bread teams. I am in constant receipt of fresh information from the natives regarding the surrounding territory, huts, islands, etc.; and have a queer map conjointly designed by old Vasilli Kool Gar, La Kentie Shamoola, and Geordi Nicolai. The names of the islands and huts I have written down as the natives pronounced them; and I find that in the centre of the archipelago there is a section of country about which the natives know absolutely nothing. Of course, it is more than likely that some one can tell me who built the hut in which Ericksen died, but I have not yet been able to discover that person. The weather just now is calm and glorious, and I trust it will remain so indefinitely.

March 16th. — Clear and cold. I have provisioned two teams of twelve and thirteen dogs for six days, to start Grönbeck and myself, and Nindemann and Kolinkin on the search. I have not enough dogs as yet to dispatch Bartlett from the southward to the northward, as my intention is, so that the three parties may meet and spread again; but instead of waiting for him, Nindemann and myself will proceed to Usterda, and, crossing the river where De Long did, will follow his trail to the southward. We have one hundred and twenty fish beside other provisions, and consequently our teams are heavily laden, but the river-bed along which our course lies is hard and smooth. The huts to the eastward of Usterda, and in its immediate vicinity, are called Macha, Mesja, and Bulchoi Mesja; these being the names of the islands on which the huts are located. I have an idea that one of them is Ericksen's hut.

We set out about nine A. M., and before noon came up with the little old hut on the west bank of the river which I visited last fall, and first supposed was the scene of Ericksen's death. I am now told that its name is Do-boi-dak. As we approached Macha, Nindemann recognized the place at a glance, and identified Usterda, one mile further to the northward, as the point at which they crossed the river. We returned by the west bank as De Long had done, and when about a mile .south of Macha I learned for the first time that De Long had crossed over to the east side of the river. and had not followed the west bank as his record declared he would.

It is now plain why I failed to find the party last fall. Guided by the record and my conversations with Nindemann and Noros, I searched the west bank of the river all the way up to Mat Vay, and so lost the trail. When Nindemann to-day indicated the point of land at which they crossed over to the eastward. I took a good survey of the river, and immediately the reason of such a move was made clear to me. The Lena here takes a great bend to the westward. De Long wished to go south. His chart, and mine likewise, showed a branch of the river running to the westward, and to the southward of Mat Vay, —so there is where he imagined he was; and this is why he supposed himself to be at or near Tit Arii (Tree Island), or Tas Arii (Stone Island), which is close by.

We searched around the bluff, but it was snowed under, and too deeply for us to excavate. We discovered evidences, however, of the party's presence, and then, after following along the dry bed of a small stream in a futile hunt for the hut, turned round at length, and have come back to Macha for the night.

March 17th. — A northwest wind; the snow falling, and every indication of a coming storm. Off by nine A. M. I crossed the river at the place pointed out by Nindemann, and found a small stream flowing south. Following this about five miles we crossed it again to an effluent branch running southeast, which we pursued for perhaps ten miles, finally arriving at a hut of which the natives had told us. But it was not the one we were in

21

search of. I then returned to the main river, and, making a fresh start, followed it south to the point where it takes the long westerly bend. Nindemann here recognized the place at which the party camped the first night after leaving Macha; and says he thinks they marched about fifteen miles that day. We traversed the bed of the stream until it ran out. and was lost in the sand-spits and *tundra*. This was as Nindemann predicted it would be.

Continuing then our southeast course across the low *tundra*, we expected to meet a large river running south, with a high western bank; but reaching it about twelve versts from our starting-point we were surprised to find the eastern bank very high and the western bank low — the very reverse of what we were looking for. The natives call this river the Oshee Macha.

By this time it was blowing so furiously that the dogs would not face the cutting wind; and as night was near at hand we bethought ourselves of shelter. The whole face of the country is changed, and Nindemann can recognize nothing; so I must depend entirely upon the natives to guide me to the various huts in the vicinity until I chance upon the one in which Ericksen died, and then I can follow south the west bank of the river until I come up with the lost party.

We started for Sister Ganak, thirty versts distant, and the wind had now grown into a gale, and we could not see ten yards ahead of us. On the way we halted at a hut called Chogen, which I visited last fall, and at which I was now very much tempted to camp; but since it was my intention to leave part of our goods at Sister Ganak to lighten the sleds, we kept on, losing our way and wandering about in the storm for more than an hour. At last we found some fox-traps belonging to La Kentie Shamoola, and from these he started off at once, and in a little while brought us safely to Sister Ganak.

The hut is so rickety that we are robbed of the heat of our fire, which cooked our fish, however, and we have plenty of hot tea.

March 18*th.*— It was too stormy this morning to make a start, and as it is but a short run to Cass Carta, I decided to return there for an additional supply of fish, leaving fifty at Sister Ganak for any possible emergency. Since I have now extracted all the information I can get from the natives, I will leave our two interpreters at Cass Carta and ease the dogs of so much weight; for as soon as the weather will permit I shall return to the Oshee Macha, and follow it down as far as Mat Vay, or until I find Ericksen's hut.

We reached Cass Carta about three P. M., and found the camp quiet and flourishing, for the bread sleds arrived on the 16th, bringing nine bags of bread and one bag of flour. The fish-sleds have not yet returned, for the last arrivals, which brought two hundred fish for Mat Vay and consumed ninety on the road, were sixteen days in coming, and will probably require four days on the journey to Bukoff. No fresh dogs as yet. Bartlett is anxious to go on the next search, but I cannot send him without dogs, though as soon as practicable I shall start him north from Mat Vay to meet Nindemann and myself on our search south from Usterda.

March 19*th.* —Though still fitful, the weather is improving. A south southwest wind, and the sun straggling through snow-clouds. I am arranging the tents, and getting things into shape for another start and trial to the eastward. I must first find Ericksen's hut, and so shorten in the distance to be searched, and shed light upon my labors. When Bartlett joins us, we can separate and spread over the country in quest of the hut, which is the certain key to the problem. Nindemann does not know from which of the many rivers they issued into the bay or *gooba*, but does remember that the island of Stolboi bore about south of them during the whole of the march. If I fail to discover the hut from Usterda, Bartlett will have a chance of finding it from the southward; and after I have searched as far south as Mat Vay, I will then investigate every branch of the river running north from the bay.

Towards evening the sled which carried Tomat and Yapheme to Ku Mark Surt returned. The driver brings word that Epatchieff is weather-bound at Mat Vay, where he has been for three days awaiting an opportunity to reach Bulcour; for the wind at this point rushes out of the river-gorge like water from a fire-hose, cutting and sweeping everything from its path. He is on his way back to Verkeransk, having faithfully secured for me the support and coöperation of the natives. Our deer, under the guidance of Tomat and Yapheme, left Ku Mark Surt the day after the dog-team, and are due here to-day.

Four teams have arrived from Bukoff with four hundred fish; so I now have plenty of dogs and fish to equip my three search parties.

March 20th. — A clear day, with a pleasant breeze blowing from the south southwest.

We made an early start; Bartlett steering for Mat Vay with instructions to follow the main river or one of its large branches north of Stolboi. He has a team of sixteen dogs, a tent, six days' provisions, and Geordi Nicolai as yamshick. Nindemann and myself similarly equipped, with La Kentie Shamoola and young Kerick to drive us, set forth on a straight course for Bulchoi Mesja. Arrived there, Nindemann confirmed his previous recognition of the locality, but was totally bewildered and uncertain as to the direction pursued by the party south from that point. So we ran off southeast until he thought we were making too much easting, when we veered to the southwest to a point he vaguely remembered. Then south by east, then east and west, following a large stream to the southward, until the dogs began to weaken, when we halted and erected our tent under the lee of a hill.

There was very little drift-wood in the vicinity, but we were too tired and cold to care much whether our supper was hot or not. Still the warm tea and raw frozen fish found great favor in our eyes. The tent was too small to allow of our building a fire in it, so, notwithstanding the high wind, the natives dug a hole outside in the snow, wherein they soon had our scant drift-wood ablaze, and our tea-kettle boiling.

The Yakut mode of building camp-fires is as follows · The pot or kettle is hung on a tree branch of sufficient length and strength to project from the snowbank in which it is thrust, over a hole excavated in the snow beneath the kettle, and at such a distance from the bank that the heat will not melt the snow from the butt of the limb. To start the fire, a dry piece of wood is procured from the high river banks, many sticks being cut with the axe and rejected, until one entirely free from moisture and fit for kindling is found ; which is then carefully split and kept dry. The best of the drift-wood is next selected and also split up and chopped into proper lengths. Thus far, so good ; but the natives are ignorant of matches, and with only their flint and steel it would seem a difficult matter to start a fire, since they have no rags, either cotton or flax, or any highly inflammable material like sulphur sticks. But here is where the Yakut and Tunguse ingenuity asserts itself.

The buds of the arctic willow are forever trying to peep from beneath their blanket of snow, and within these buds is a light flossy substance in the nature of thistle - down. Whenever he can, the native gathers a handful of these, and robs them of their down, which he then moistens slightly and mixes with ground charcoal, prepared by cooling a lighted piece of birch wood in the ashes of his hearth. The dampened floss thoroughly rolled through the charcoal is next covered up and dried before the fire on the same board whereon it was compounded and the charcoal powdered. It is now an excellent tinder, igniting quickly into a hot and durable point of fire. But in addition to it, some light match-stuff is necessary, and to supply this need, a bundle of fine soft sticks, about thirty inches long, is always kept drying over the fire-place. Before the native sets out on a journey, or, indeed, as often as the material is required, the old women of the house take down several of these sticks, and carefully shape them into sword blades. They then rest their knives in beveled notches cut in the flat sides of small pieces of wood, about three eighths of an inch broad, one eighth of an inch thick, and one inch and a half long, and the operation proper begins. Along the wooden sword, which is held against the shoulder like a violin, the knife in its gauge is drawn continuously and rapidly, and at each draught a thin coiling shaving drops to the floor or in the lap of the operator. A bag full of these fine curls — which, when matted together, very much resemble the Amérìcan manufactured material known to upholsterers as "excelsior" — is always ready for the traveling native, preserved dry in the huts beneath the sleeping-skins, and carried in a fish-skin bag on the journey.

So now, with the materials at hand, we will start a fire : The native takes from his skin pouch a bunch of the "excelsior" about the size of a robin's nest, rolls it into a ball, punches a hole in it, and then lays it carefully on the snow. Next, taking a pinch of tinder from the bag which always hangs at his hip, he places it on his flint, and with a quick sharp stroke ignites and incloses it in the centre of his nest of shavings, which he then lifts up, holding it lightly with his fingers spread

apart for the passage of air, and whirls rapidly around
his head at arm's length. At first a faint, pleasant odor
of burning birch steals upon the air, then a light streak
of smoke follows the revolving arm, and when the heat
within his hand notifies the native that a proper degree
of ignition has been attained, he suddenly ceases his
gyrations, tears open the smoking nest, and with a quick
puff blows it into flame. Then depositing the blazing
ball on the snow he soon piles his fagots over and around
it, and in a very few seconds his fire is in full blast.

I have watched this operation a hundred times, and
never saw it fail. When I tender matches (*spitchkies*)
to the natives they invariably refuse them, because the
shavings so lighted burn inwardly and give off but little
heat, whereas by the Yakut treatment they are almost
instantly a glowing mass, never missing fire. And so it
was to-night.

We turned, at length, into our sleeping-bags, cold and
tired, La Kentie and Kerick sleeping with us in the tent,
but nearer to the "flies." We had no oil-cloth to sleep
upon, but nevertheless were soon warm and comfortable,
for the snow is soft and dry, and forms a much better
bed than hard ice, or even the harder boards in the Yakut hut. When camping out in winter time, this is a
point to remember and observe, but in summer, when the
snow is wet, it should be avoided.

We had barely composed ourselves to the sleep we
sorely needed, when the wind began to pipe and the
clouds to drift swiftly across the sky. The natives said
" pagoda, bar, bar," and before midnight the snow had
sifted through the tent and into our sleeping-bags, where
it melted, and then our wet clothing froze fast to our bodies, and we could not move. So we endured our misery
until six o'clock this morning (the 21st), when I drove
the yamshicks out to make some tea. They succeeded
in starting a fire, but the snow soon smothered and extin-

guished it. The natives then sliced some raw fish which
they and Nindemann ate, for the weather had stolen my
appetite, but at seven o'clock I caught sight of the sun
through a rift in the clouds, and determined to get under way. It was my wish to reach Mat Vay on this line
of search; but as neither the dogs nor the natives could
face the fierce wind, I stood to the northward of west
and, ran for Qu Vina, where we arrived about eleven
o'clock.

It is a leaky, wretched hut; but we were glad to accept of its shelter, for it enabled us at least to prepare
our breakfast of hot tea and boiled fish. Towards noon
the gale abated, and we were making ready to set out
for Mat Vay, when seven teams arrived on their way
from Bukoff to Cass Carta, having put in at Qu Vina to
escape the storm. Young Kerick found this kind of
service too severe for him; he would *propaldi* (break
down, or die), he said; so I discharged him and took Capiocan in his place, who seems twice as plucky. Two of
the loads of fish I turned back to Mat Vay, where we
arrived to-night. Bartlett left here this morning on his
search to the-northward. He was lucky to have been
under cover last night, but is catching it now, for the
gale roars outside. Still he has a tent, in which with all
its discomforts I am inclined to believe he has as pleasant quarters to-night as have we in this rickety old hut,
devoid of chimney. For the smoke is blinding, and it is
horrible to lie on our backs with mitten-covered fingers
over our eyes, or as a recreation to lie on our stomachs
with our faces on our hands.

March 22d. — It blew violently all night, and has continued to blow all day. The hut is nearly filled with
snow, those of us lying to the westward being half buried
in it. We have stayed in doors all day, almost blinded
by the smoke, and forced to sit or lie down; so night
and sleep are very welcome to us.

March 23d. — At early morning the weather was still squally, but as day advanced it cleared. I will now make another attempt from the southward, and if I can only find the high promontory from which Nindemann sighted Mat Vay, there is no doubt of my ability to follow the trail as far as Ericksen's hut.

The sun came out in course of time, and although the snow still drifted before the wind, I could yet discern the points of land making out into the bay. Our eyes are still weak from the effects of smoke, and the sunlight tortures them. The problem that now puzzled me was, — Which of the round dozen points of land before us is the one that Nindemann turned when he reached the bay or *gooba?* Cold, hungry, without compass, and with orders "to keep the west bank aboard," he only knew that he had journeyed south and a long way from the eastward — but how far? So with nothing to guide me, I decided to start at the northwest and follow along from point to point until I found *the point.* Nindemann was anxious to go east, skipping many of the headlands, but this I would not do for fear of missing the particular one I wanted. Then again, as De Long had said he would follow in the track of Nindemann and Noros, on which point was it that he had camped and died?

So I visited from cape to cape, taking a good survey of each river, until finally we came to a large rough stream, the Kagoastock, where the land ran far out into the bay. Nindemann was still uncertain, and sat on his sled gazing dumbly at the Stolboi which had been a landmark for himself and Noros on their march to the southward, and which now showed nearly to the south of us. Meanwhile I had ascended to the high ground of the point, and stumbled upon a fire-bed, perhaps six feet in diameter, with many foot-prints frozen in around it, for the winds had fortunately kept the promontory clear of snow.

"Here they are," I shouted, and Nindemann, closely followed by the natives was soon at my side. It looked like a signal-fire, the logs were so large, and when I asked our drivers if the Yakuts had built it, they confidently replied, —

"Soak; Yakut agoime malinki, malinki " (no; Yakut fire little, little).

I had not yet found the bodies, but had certainly fixed the trail; for I now reasoned that the party had rounded this point and I would discover them somewhere to the westward. Still I was desirous of securing the record and other relics at Ericksen's hut, and so set out at once to explore the banks of the river. Nindemann had told me that one of the prominent landmarks along this stream was an old flat-boat which lay stranded on the shore of the river, and in which he and Noros had camped a couple of days after they parted from De Long; and now in his anxiety to find it he started off ahead of me, with the dogs of both teams in full cry. I always kept a sharp lookout for strange objects, having directed the others to do likewise, and presently, as Nindemann suddenly sighting the flat-boat drove at full speed towards it, I espied a black thing sticking out of the snow, about three hundred yards to the southward of the boat, and at once rolled off my sled, whereupon the yamshick, having seen me perform this feat before, drew up his team and joined me. I hastened to the black object which attracted my attention, and found it to be the points of four sticks held together at the top by a small piece of lashing stuff, and across the forks of the sticks was hung by its strap a Remington rifle, the muzzle of which peeped about eight inches above the snow. In my eagerness to reach it I fell forward on the sticks, severely cutting and bruising my face. Pulling the rifle from the snow, I cleared the barrel and instantly identified it as Alexia's. There was no record in the barrel as I hoped there

FINDING THE BODIES OF DE LONG AND HIS COMPANIONS

would be; so I sent my driver, La Kentie, for Ninde-mann, surmising that De Long, unable to carry his books and papers further, had cached them here and erected this *myack* as a landmark. The fire-bed, too, that I had just found on the promontory confirmed me in this opin-ion; so as soon as Nindemann came up I set the two natives at work digging out the snow. It was a tedious operation, and in a few minutes Nindemann said he would take a look to the northward. I then climbed to the top of the bank, intending to obtain a round of com-pass bearings for Stolboi, Mat Vay and other points in order to locate the place, as I hoped to make Mat Vay for the night. La Kentie accompanied me, carrying the compass, and as we walked along I noticed some old clothing, mittens, etc., lying on the high ground above the river. Nearing the spot where the fire had been built, I observed something dark in the snow, and upon going towards it was rewarded by the discovery of the party's tea - kettle, a cylindrical copper vessel blackened by many fires.

"Kack, chinick!" (What, the kettle!) exclaimed I to La Kentie, and so saying advanced to pick it up, when suddenly I caught sight of three objects at my very feet; and one of these, the one I was about to step over — *was the hand and arm* of a body raised out of the snow. La Kentie gave one look, and dropping the compass started back in terror, crossing himself.

I identified De Long at a glance by his coat. He lay on his right side, with his right hand under his cheek, his head pointing north, and his face turned to the west. His feet were drawn slightly up as though he were sleep-ing; his left arm was raised with the elbow bent, and his hand, thus horizontally lifted, was bare. About four feet back of him, or toward the east, I found his small note-book or ice-journal, where he had tossed it with his left hand, which looked as though it had never recovered from the act, but had frozen as I found it, upraised.

Turning, then, to the last entry in the journal, I read : —

" *Oct.* 30*th,* *Sunday.* — Boyd and Görtz died during night. Mr. Collins dying."

The other two objects in the snow proved to be the bodies of Dr. Ambler and Ah Sam, the Chinese cook. A few small articles lay scattered around, and these I gathered together and put in the kettle. Besides the journal I also found a medicine case, and a tin cylinder, three inches in diameter and almost four feet long, which contained the drawings and charts of the cruise. Dispatching La Kentie in quest of Nindemann, I occupied myself until he arrived in perusing the sad record, beginning at the final date and reading backward. I learned from it that, after Ericksen, the next man to die was Alexia, and that he had been buried from the flat-boat in the ice of the river. I therefore supposed that the whole party must be lying within an area, north and south, of not more than five hundred yards. After leaving the flat-boat they had advanced about three hundred yards, but the southerly gales were too fierce for them to face; so they had camped where the *myack* was, and there all but three had died. The journal relates how the remaining members of the starving band were so weak that they could not carry Lee and Kaack — the first two who succumbed after Alexia — out on the bed of the river, so they "carried them around the corner out of sight," and " Then," says De Long, " my eye closed up." (Nindemann tells me that during the march the captain suffered severely with his eyes, and when he left him he was almost blind, which explains this passage in the journal.)

One after another died until only three were left, and then De Long perceived that unless the books and papers and the bodies of his comrades were removed from the low bed of the river, the spring floods would sweep them all out to sea. So the surviving three had tried to carry

the records to the high ground for safety, together with
a cake of river ice for water, the kettle, a hatchet, and
a piece of their tent-cloth, but their little remaining
strength was not even equal to the task of lifting the
cases of records up the steep bank, so they sank down
from the effort, after securing the chart-case and other
small articles, leaving the records to their fate. At the
root of a large drift tree that had lodged on the bank
some twenty-five or thirty feet above the river, they
built a fire and brewed some willow tea; and the kettle
when I found it was one quarter full of ice and willow
shoots. The tent-cloth they set up to the southward
of them to protect their fire, but the winter winds had
blown it down, and it now partly covered Ah Sam, who
lay flat upon his back, with his feet towards the fire and
his hands crossed upon his breast; a position in which
the last two survivors had evidently placed him. De
Long had crawled off to the northward and about ten
feet from Ah Sam, while Doctor Ambler was stretched
out between, — his feet nearly touching the latter, and
his head resting on a line with De Long's knees. He
lay almost prone on his face, with his right arm extended under him, and his left hand raised to his mouth.
In the agony of death he had bitten deep into the flesh
between his thumb and forefinger, and around his head
the snow was stained with blood. None of the three
had boots or mittens on, their legs and feet being covered
with strips of woolen blanket and pieces of the tent-
cloth, bound around to the knees with bits of rope and
the waist-belts of their comrades. Ah Sam had on a
pair of red knit San Francisco socks, the heels and toes
of which were entirely worn away.

When Nindemann joined me I showed him the three
bodies as yet undisturbed, and the articles I had gathered
together, including the journal, from which De Long had
torn away three quarters of a page; but as the opposite

one on which the last entry had been made was not filled
out, it was plain that no record was missing. I then told
Nindemann to thoroughly search the bodies, directing
him to cut the clothing in the vicinity of the pockets,
and all of the many small things he found, I tied up
in separate .packages and marked, so that no scrap of
paper or article of any kind might be lost. I did not
then take an inventory of these things, because of the
intense cold. In all the pockets were scraps of old seal-
skin clothing, boots, and trousers, which had been crisped
in the fire, some of it with the hair on the hide. De
Long's pistol was missing. I knew he had one, and that
he had carried it from the time the ship was crushed
until we parted company. It was originally the property of Mr. Danenhower, who, while we were encamped
on the ice preparing for our long march, had thrown it,
together with some ammunition, into the sea — as he
then supposed. But a thin sheet of ice covered the lead,
which shortly before had been open water, and over this,
instead of sinking, the pistol went skimming. So afterwards, when De Long found himself without a pistol, he
directed one of the men to secure Danenhower's for him;
and now failing to see it on his person I thought no more
at the time than that he had thrown it away because of
its weight. Chipp had given his pistol to Ah Sam, who
clung to it until death.

The three bodies were all frozen fast to the snow, so
fast that it was necessary to pry them loose with a stick
of timber. In turning over Dr. Ambler, I was surprised
to find De Long's pistol in his right hand, and then, observing the blood-stained mouth, beard, and snow, I at
first thought that he had put a violent end to his misery.
A careful examination, however, of the mouth and head
revealed no wound, and, releasing the pistol from its
tenacious death-grasp, I saw that only three of its chambers contained cartridges, which were all *loaded*, and then

knew, of course, that he could not have harmed himself, else one or more of the capsules would be empty.

[I am so particular in noting this fact, because of a painful story which has gone the rounds of the press, to the effect that Dr. Ambler took his own life. This is utterly false. The doctor was ever cheerful and fearless of death, and I know he faced it calmly and manfully as he had done before on the field of battle. He came of a brave family, and if the world might read a single page in his private journal there would be no doubt of his unfaltering courage and fortitude to the bitter end.

I believe him to have been the last of the unfortunate party to perish. When Ah Sam had been stretched out and his hands crossed upon his breast, De Long apparently crawled away and died. Then, solitary and famishing, in that desolate scene of death, Dr. Ambler seems to have taken the pistol from the corpse of De Long, doubtless in the hope that some bird or beast might come to prey upon the bodies and afford him food, — perhaps alone to protect his dead comrades from molestation, — in either case, or both, there he kept his lone watch to the last, on duty, on guard, under arms.]

When the bodies were searched, I rolled them, with the aid of the natives, in a piece of tent-cloth, and then covered them with snow, for I could not as yet haul them to Mat Vay. The faces of the dead were remarkably well-preserved; they had all the appearance of marble, with the blush frozen in their cheeks. Their faces were full, for the process of freezing had slightly puffed them; yet this was not true of their limbs, which were pitifully emaciated, or of their stomachs, which had shrunk into great cavities. Dr. Ambler, ostensibly to ease the gnawing pangs of hunger, had wrapped his little pocket diary in his long woolen muffler, and then thrust this great wad under the waistband of his trousers.

From the reading of the journal I now expected to

find the balance of the party near the *myack*, or where I had sighted the tent-poles. I therefore started the natives to digging, telling them that the *bumagas* and *kinneagas* (papers and books) were there. Exerting themselves then to their utmost, they soon came upon the wood and ashes of the fire-place, when, digging around the base of the cone-shaped pit, they presently exhumed, much to their delight, a tin drinking-pot, some old scraps of clothing, a woolen mitten, and two tin cases of books and papers.

Suddenly the two men scrambled out of the pit as though the arch-fiend himself was at their heels, gasping, as soon as they could, —

" Pomree, pomree, dwee pomree " (the dead, the dead, two deads).

Dropping into the hole I saw the head of one corpse partly exposed, and the feet of another; and then ordered the natives to continue their labors. They obeyed, and finally disclosed the back and shoulders of a third. It was now dark and the snow was drifting wildly, so I concluded to return to Mat Vay for the night, and send instant word to Cass Carta for the rest of my party to join me here and assist in excavating the bodies.

CHAPTER XXIV.

THE BURIAL.

Bringing in the Dead. — Writing under Difficulties. — Selecting a Burial Ground. — "Around the Corner." — The Finding of Lee and Kaack — Monument Hill. — Constructing the Coffin and Cross — Nindemann discovers Ericksen's Hut. — Erecting the Tomb-Cairn. — The Simple Obsequies. — A Superstitious *Soldatski* — A Yakut *Bumaga*.

March 24th. — When we arrived at Mat Vay last night, it was to find Bartlett here. I at once dispatched Capiocan after Grönbeck and the others, and wrote out telegrams to the Secretary of the Navy, the Minister at St. Petersburg, and to General Tschernaieff, of which Grönbeck will make the necessary translation. The smoke in the hut is blinding, and to write I am forced to lie on my stomach with my head towards the fire, and the ink planted in the ashes to keep it from freezing.

- This morning I sent Nindemann and Bartlett to complete the exhumations which I began yesterday. Bartlett was caught out in the storm and beset by it for forty-eight hours. He ran north until he encountered our tracks, and met the natives who were carrying fish to Cass Carta, when he returned to Mat Vay *via* Qu Vina. He had tried to camp, but the wind heaped the snow upon his tent, and broke it down.

This evening Nindemann and Bartlett came in, bringing the bodies of De Long, Ambler and Ah Sam. We wrapped them in the tent-cloth, and covered them, close

22

by the hut, with snow. The other two bodies are those of Gortz and Boyd.

March 25th. — I started the party off again this morning to continue the disinterments, Gronbeck, who has arrived, along with the rest. It is twenty versts across the bay to Pomree Moose (Dead Cape), as the natives have already named the disastrous point, and as it will doubtless be known hereafter among them.

Grönbeck returned at noon with the bodies of Boyd and Görtz ; and towards evening Bartlett and Nindemann followed with Iversen, Collins, and Dressler. They have not yet found the ensign.

March 26th. — I finished the preparation of my dispatches to-day, and Grönbeck having translated them set out for Cass Carta, where he met and forwarded to me Captain Bubokoff, who will act as courier as far as Belun.

Bartlett and Nindemann returned this evening from Pomree Moose with the pistol which Chipp gave to Ah Sam. They have not been able to discover Lee and Kaack, or the ensign. Bubokoff arrived here from Cass Carta about ten P. M., ready for duty; but I am progressing very slowly in my effort to copy the record of the last thirty days from the journal. The smoke is blinding, and my fingers are so blistered and swollen that it is with difficulty I can hold the pen at all — and when I do it is only for an instant at a time, while I write one or two words.

The clothing of the dead is badly burnt or scorched, they lay so close to the fire; and those who perished first were stripped of their rags by the half-naked survivors. When Mr. Collins died some one covered his face with a shirt. Boyd lay almost in the fire, but though his clothes are scorched through his flesh is not burnt. There is not a whole moccasin left among them, or a piece of hide or skin, save the arm and shoulder of a coat found under one of the men, — the rest of which was evidently cut off

the body for food. A strip of a moccasin leg was also discovered on top of the bank — everything else had been
eaten.

March 27th. — Bubokoff started for Belun to-day with
my dispatches. I shall soon complete, thank God! my
task of transcribing the journal, and shall no doubt be
able to leave this horrible smoke-house to-morrow, and
prosecute the search for Lee, Kaack and Alexia. I have
little hope, however, of finding Alexia, who was buried
in the ice, for I can see where the river bed has tumbled
in and run out in several places. But I must find Lee
and Kaack, if I have to excavate the whole bank. It
will be necessary, in burying or caching the dead, to
transport them to a point about five versts to the southward of Mat Vay — the foot-hill of a mountain which
extends into the bay and forms the left bank of the river,
looking north. All other land in the vicinity, as indeed
the entire Delta, will be shortly inundated by the
spring floods. The crest of the hill, or head of the great
whale-back mountain, is nearly four hundred feet above
the level of the sea, and is visible in clear weather for
twenty miles in any direction to the north, northeast, or
northwest. Here I shall build a box with the timber of
the flat-boat, which, with the bodies, we shall have to
haul twenty miles.

I was troubled about our inability to find Kaack and
Lee, and an idea kept running in my head, as I read and
re-read the journal, that they were the two men who had
been carried *around the corner.* But what corner? The
bank ran nearly northeast and southwest, and there were
no corners in it, unless some fissure be meant as such.
The snow had been dug out away to the north of the
myack, but very little to the south of it; and finally it
occurred to me that as all the gales were mentioned as
blowing from the southward, they would naturally set up
their tent-cloth to the southward of the poles, and camp

to the northward of them ; so when Lee and Kaack died, and the remnant of the party were too weak to carry the bodies out up n the ice for burial, they simply took them "around the corner" — *of the tent.*

Convinced of this, I seated myself upon a sled, and followed Bartlett and Nindemann to the scene of their labors, arriving as soon as they. I informed them of my theory, and staked out with my staff a goodly plot to the southward of the tent for them to excavate. I then returned to Mat Vay and finished a sketch of the tomb-cairn to be erected on " Monument Hill." When Nindemann and Bartlett came in they brought with them the remains of Lee and Kaack, having exhumed them where I indicated. They also found the ensign, mahogany medicine box, hatchet, etc., so that I have now secured all the bodies (save those of Ericksen and Alexia), and the records of the expedition, which are packed in tin boxes. The effects of the dead we have carefully done up in separate packages, marked with the owner's name. Nindemann and Bartlett searched in my absence the balance of the bodies, and turned over to me all the articles they found upon them, which I have stored in a messbox.

And now that the search is over, the sad duty remains to us of burying our dead shipmates. The earth is frozen too hard and deep to be excavated, so I shall follow the Yakut custom of surface interment beyond the reach of the floods.

The burial ground is a bold promontory with a perpendicular face overlooking the frozen Polar Sea. The rocky head of the mountain, cold and austere as the Sphinx, frowns upon the spot where the party perished ; and considering its weather-beaten and time-worn aspect, it is altogether fitting that here they should rest. I attained the crest of the promontory by making a detour of several miles to the southward of its majestic front,

and then toiling slowly to the top. Here I laid out by
compass a due north and south line, and one due east
and west, and where they intersected, I planted the
cross which marks the tomb of my comrades. The mountain top is swept almost clean of snow by the fierce
winds which are forever blowing at such altitudes ; and
the massive rocky face is riven and torn. For the snow
melting in summer runs into the crevices, and then in
season Jack Frost, the subtle mining engineer, sets his
machinery in motion, bursting the great rock bed into
myriad fragments, so that its surface, though flat as a
table, is broken for a depth of several feet into the sem-
blance of regular masonry work.

With much difficulty I picked and pried out the rock
from the centre of my cross lines, until I had uncovered
a pit about three feet deep and two feet in diameter.
While I was thus employed the rest of the force was
tearing apart the flat-boat and hauling the planks for the
cairn coffin. These planks, seven inches thick, rough-
hewn and fastened to the boat frame by means of trun-
nels, were sawn off, and those about twenty-two inches
wide, which formed the sides of the boat, were selected
for the ends and sides of the coffin, which, when mortised
and tenoned together, was seven feet wide, twenty-two
feet long, and twenty-two inches deep. With the remaining timber I made a cover or lid of planks seven inches
thick, from the centre of which arose the cross, twenty-
five feet high, with cross-arms twelve feet long. This I
shaped from a round spar of spruce thirteen inches at
the base, and tapering to eleven inches in a length of
forty feet, which I found in the bay water-logged and
frozen in, and hauled to Monument Point on two sleds,
drawn by sixty dogs. The upright, which I left round,
but barked, was cut from the butt end of the spar. The
cross-piece was hewn and faced fair to receive the inscription, and it was dressed taperingly away from the centre,

where it was hollowed out to fit in a corresponding notch in the vertical post, the two being fastened by a wooden key after the cross was raised.

The timber was first hauled to Mat Vay, where with axe, saw, and chisel, brought from Yakutsk, the coffin and cross were fitted together, and the inscriptions cut, which include the names of the dead and a brief statement of the time, place, and cause of their death. Grönbeck and myself attended to this part of the work, deeply cutting in the names with chisel and mallet in block letters two and a half by one and a half inches square; the balance of the inscription being comprised in two lines, eight feet long, of letters four inches square. These letters are all regularly formed, spaced, and cut to a depth of a little more than one quarter of an inch.

When everything was ready for the burial, I dispatched Nindemann with a dog-team and Capiocan as driver in search of the hut where Ericksen died, with orders to bring back with him the epitaph board, record, gun, and ammunition, which had been abandoned there. His journey was unsuccessful, and he only found the pan and lid of the fire-pot, which the party had thrown away on its march. The day after his return he set forth once more, and, remaining out over night, came back the following day with the objects of his search. The inscription on the board read : —

"IN MEMORY
H. H. ERICKSEN
OCT. 6, 1881
U. S. S. Jeannette"

De Long's record was as follows : —

"*Friday, October* 7, 1881.

"The undermentioned officers and men of the late U. S. Steamer Jeannette are leaving here this morning to make a forced march to Ku Mark Surt or some other settlement on the Lena River. We reached here on Tuesday, October 4th, with a disabled comrade, H. H. Ericksen (seaman), who died yesterday mórning, and was buried in the river at noon. His death resulted from frost-bite and exhaustion, due to consequent exposure. The rest of us are well, but have no provisions left, — having eaten our last this morning.

> " GEORGE W. DE LONG,
> " *Lieu't. Com'd'g, et al.*"

Meanwhile the teams were engaged in the transportation of the coffin and cross to the mountain-top. Arrived there, with the wind blowing half a gale, I found it a greater undertaking than I had imagined to raise this round stick of timber in place. It was impossible to work without mittens in the freezing air; we had no rigging, other than the guys which I improvised from the dog-traces, and a forked stick for a rest. There were but three of us who spoke English, — Grönbeck, Bartlett, and myself, — and the natives could not grasp the situation or my orders; nor did they seem to appreciate the great weight of the spar until they saw it swaying back and forth, when they all ran wildly away from it. But finally, after many narrow escapes, the cross was raised, and slewing it around to face the east I quickly chocked it in place with four large stones. Then, sighting it in a perfectly upright position, I filled in the base with small rocks, pouring over them a bucketful of ice water, which soon froze and cemented them together. The box being tightly mortised and wedged, and the round planks prepared to close in the top, we next covered the bottom with brushwood and some old rags, and on these we laid our poor dead comrades, arranged in the order of their names as inscribed on the cross, with Captain De Long at the southern end and Ah Sam at the northern end of the coffin. They were all stretched out with their heads to the west, and the faces of those with whom it was possible were turned to the east and the rising sun. Nothing was left upon the persons of the dead, save a

large bronze crucifix belonging to Mr. Collins. When
Bartlett and Nindemann, searching the remains, inquired
if they should remove this, I was at first inclined to answer in the affirmative, thinking that his relatives would
doubtless wish to preserve so valuable a souvenir; but,
reflecting for a moment, I decided, and ordered, that as
part of his religion it should be buried with him.

It was a memorable sight. The long train of dog-teams
winding their tortuous way across the *tundra* and ice-
field, and up the steep ascent to the lone mountain-top,
where, in the awful silence and solitude of that vast
waste of Arctic snow, with no requiem but the howling
of the remorseless storm, which cruelly cut our faces, we
tenderly laid our dead comrades to rest, — as I then supposed, forever. There, in sight of the spot where they
fell, the scene of their suffering and heroic endeavor,
where the everlasting snows would be their winding
sheet and the fierce polar blasts which pierced their
poor unclad bodies in life would wail their wild dirge
through all time, — there we buried them, and surely
heroes never found fitter resting-place. We were over-
awed by the very simplicity of the obsequies, the oppressive stillness, the wonderful wilderness of white rolling
endlessly around us; and, more than all, by our sorrow-
ing memories of the dead. No unhallowed lip mumbled
an unmeaning prayer, but only a low "good-by," "sleep
well," broke the silence, as, natives and all, we took our
last look.

Then covering the bodies with bits of canvas and
some other material at hand, we laid the planks across
the box, weighting them down with stones, and night
had fallen. The day following the Yakuts hauled load
after load of round timber from the flats and sands below.
Large logs were rolled in at the sides and ends of the
box, and a pyramidal frame-work erected therefrom, to
lie upon and strengthen it. A ridge-pole was notched

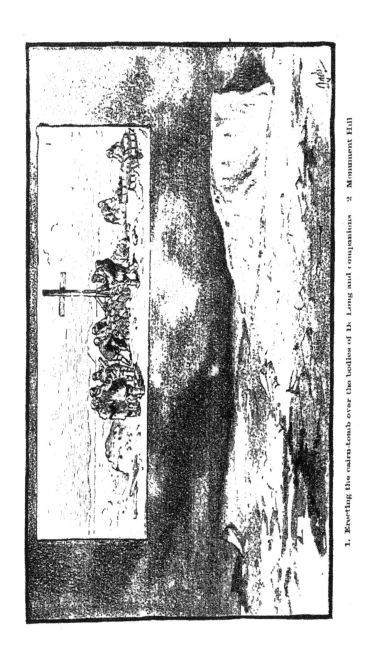

1. Erecting the cairn-tomb over the bodies of Dr. Long and companions   2. Monument Hill

into the sides of the upright post, and diagonals fitted to
brace the cross and support the structure built around it.
The top and sides were then covered with round timber,
resting upon and against the ridge-pole, and completing a
frame about twelve feet wide, thirty feet long, and nine
feet high. Upon this we heaped the huge rocks, some of
more than a hundred weight, which Jack Frost had so
cunningly quarried for us, until the entire cache was
roofed in ; and it was my intention to cover it the following summer with sods from the *tundra*, and to start the
Arctic willow to grow upon it.

By this time Nindemann had returned from his journey to Ericksen's hut, and the only remaining thing to
be done was the elevation of the cross-piece into place ;
which, after several ineffectual attempts, we finally accomplished. Nindemann then drove in the wooden key,
and a cross-key to keep the other from working out, and
— poor fellow ! — in so doing he froze his fingers, nose,
and ears ; for it was bitterly cold, and the wind was
sharper than a two-edged sword. I completed the burial
April 7th, and am altogether satisfied with the work.
The tomb is the largest structure north of Belun, and
the natives hauling fish from Bukoff Moose for my coastwise search for Chipp told me that they saw the *Bulchoi
Crass Americanski* twenty versts away.

Bubokoff and Geordi Nicolai returned from Belun a
few days ago, and I have sent them to Cass Carta to
arrange the provisions for my departure to the Alanek.
Yapheme, our " Red Fiend," tried to impress the Yakuts with his great valor and importance, having been a
soldatski in the Russian army. But Capiocan, who is a
wag, entertained doubts of his courage, and often assured
Yapheme that the *pomree Americanski* lying outside in
the snow would arise some time and enter the hut to pay
their respects to their living *soldatski* friend. At each
sally, Yapheme would look very brave, and say, " Da, da,

(yes, yes); the good sailors could visit the good soldier whenever they thought best." Yet I observed that when we all left camp on any duty, Yapheme did not care to remain behind alone with the dead, but invariably managed to accompany us; and when busy about the hut Capiocan set little traps to frighten him, throwing sticks or causing the deer-skin curtain at the door to sway in and out, greatly to the misery of Yapheme and the amusement of the natives.

On one occasion Capiocan, procuring a gun, wrapped himself in some old tent-cloth, and backed into the hut, calling in sepulchral tones, —

"Drastie, drastie; Yapheme kack Americanski soldatski."

By this time, the nerves of the valiant exile were so unstrung that, seeing the apparition with an American gun, and hearing its unearthly voice, he almost died of terror, and with an unsoldierly shriek plunged across the fire-place and into the farthest corner, upsetting the fish-kettle in his flight. Of course we were all entertained, but poor Yapheme protested that the joke must not be repeated, else *soldatski*-like he would seize the gun and shoot his tormentor.

It was at this juncture, and we had just ceased laughing, when a strange noise was heard without. I saw Yapheme steal a glance at those around him and shuffle away from the door. There was a movement at the back of the hut, where lay the dead, and presently some one ascended the side, and an instant later a face swollen and blackened with frost-bite, smoke, and scabs peered through the hole in the roof, and cried out, "Drastie, drastie!" At the same time our deer-skin curtain was pushed aside, and Bubokoff appeared in the door. This was too much for Yapheme. Dropping on his knees in front of me, he fell to crossing himself as though his life and peace thereafter depended upon the rapidity of his motions. We

all burst into a roar of laughter, and thenceforth the
would-be warrior had no peace on the score of his valor
and affection for the *pomree Americanski.*

I soon had all the effects of the dead, together with
the books, papers, etc., etc., packed in a box ready for
shipment to Yakutsk; for it would be decidedly unsafe
to keep these relics on the Delta during the approaching
season of floods. So, leaving a small supply of fish at
Mat Vay for future emergency, I decided to dispatch
the whole of my force to Cass Carta, as a more central
base of supplies from which to start the three parties on
a final search along the coast for Chipp. The plan I
had in view was to send Bartlett and Nindemann with
four sleds and as many native guides up the river Kagoastock to Cape Barkin, the northeastern point of the
Delta, there to separate, Bartlett following along the
east coast as far south as Jamaveloch, where he would
await my coming. Nindemann was to search the north
coast to the northern mouth of the river proper, and
then, if his provisions were exhausted, to proceed to
North Belun, where I had stored one hundred fish for our
use. If, however, his provisions held out, he was to journey along the coast as far as the river Ketack, and then
follow it south to North Belun, and thence on to Cass
Carta, where he would abide my return. The course I
laid out for myself was along the western discharge of
the river by way of Long Island, searching the coast-line
and visiting the villages until I arrived at the river Alanek. Then turning on my track I would come eastward
as far as the western end of Long Island, at the village
of Turak, and thence along the coast, north and west,
to the river Ketack, down which, *via* North Belun, I would
journey to Cass Carta. From this point it was my intention to repair with the rest of my party to Jamaveloch, and continue the search the same season to the
river Jana.

I had procured guides for all the Delta, except competent ones for the north and east coasts. Simeon Alock
and Vasilli Kool Gar were the two I selected to accompany Bartlett; and as guides for Nindemann I settled
upon Starry Nicolai and Simeon Tomat. So I now impatiently awaited at Mat Vay the coming of the sleds
from Arii; and, as there were some arrangements which
I desired to make at Cass Carta, I, at length, determined
to start immediately, leaving one of the yamshicks behind to inform Vasilli Kool Gar and Simeon Alock of
our whereabouts. I appointed Capiocan to this office,
but he declined it fervently, and we all laughed at him,
for it was evident that he feared the spirits of the dead.
Still he found a way out of the difficulty, saying: —

"Yakut bumaga, Mahor" (Yakut letter, Major).

What about it, I asked? Why, he would write one to
Vasilli Kool Gar and Simeon Alock, directing them to
follow us to Cass Carta. This was all I wanted, and so
gave orders to harness the teams; and when everything
was ready for a start, Capiocan proceeded to write his
Yakut letter. He made four tracks alongside of each
other to represent our four sleds, and, driving a forked
stick in the snow, rested a long pole in it pointing towards Cass Carta. He then drove in another large stick,
inclined in the same direction and supported by a smaller
stick, and called it *Mahor*, meaning myself. Shorter
sticks, one for each man in the party, were arranged in
like manner, Capiocan explaining that they represented
us in the act of walking; and between the tracks he set
up some more in two rows, as the *savaccas* (dogs), and
his letter was written.

The ride to Cass Carta was a cold one, but we arrived
betimes, and I at once busied myself in fitting out the
three search parties, sending one hundred fish to the west
and one hundred to the north for the use of myself or
Nindemann. On the morrow, sure enough, in came

Vasilli Kool Gar, and Simeon and his sons, with four dog-teams and a supply of fish. They had stopped at Mat Vay, and Vasilli's mouth stretched into a monstrous laugh as he said, " Yakut bumaga." But Capiocan was delighted that his letter had been so readily understood, and told me that I wrote with ink and paper, but he used sticks.

I learned from Simeon that it was he who built the hut in which Ericksen died ; but as the country round about was very poor in game, he had never completed, but abandoned it. He and Vasilli also assured me that the huts at Barkin had not been inhabited for many years, and that they had not visited them for two summers. They were both well acquainted with the coast, but knew little, they said, about the interior of the Archipelago, as the Russians term the islands of the Lena Delta. " Paddy " Achin accompanied Vasilli, so I was now well supplied with yamshicks for my journey to the west, " Paddy " being quite at home in these parts, and Geordi Nicolai being the son of the late *golivar* of Jaolak, a western settlement. And so, on April 10th, I started Bartlett and Nindemann towards Barkin.

CHAPTER XXV.

SEARCHING FOR CHIPP.

April 10*th.* — This morning I received a letter from
Pay Clerk Gilder of the Rodgers, who says in it that
he is a correspondent of the "New York Herald." The
message was brought to me by "Pat" Malloi, the rosy-
cheeked son of Simeon Alock. He came through from
Bukoff Moose with a team of eight young dogs in four
days without a single halt for rest or food. So I gave
little "Pat" some tobacco as an extra reward for his
fidelity. He says that the starosti furnished Gilder with
a team of fourteen dogs to transport him to Cass Carta,
where in all probability he will arrive to-morrow evening.

I have therefore postponed my departure for the Alanek, and sent Grönbeck with a good dog-team to Mat
Vay to meet Mr. Gilder and bring him forward. This
knocks a hole in my plans, as I have only three hundred
and ninety-four fish left in my store-house, and each day's
detention here costs me fifty fish for the dogs alone. I
may have to haul more fish from Bukoff.

The traveling season will be over in twenty days, so I

determined to disencumber myself of all supernumeraries
and dispatch Bubokoff and Kolinkin to Yakutsk. For
we may be detained on the Delta all summer, and then,
too, it is necessary that I advance the records and relics
to a place of safety, and these two *protégés* of General
Tschernaieff will be responsible for them. The box in
which I have packed all the articles is quite strong, being
dovetailed, covered with raw hide, and sealed. I gave
Bubokoff and Kolinkin both written orders and instructions, and will send by them explanatory letters to General Tschernaieff.

I must procure more fish for the dogs. I had hoped
to finish my work within the sledding season, but now
fear that I shall be "jacksoned" all summer. I wonder
how long I must wait to see Mr. Gilder? and for what?

April 11*th.* — Bright sunshine, but the wind is rising
and the snow is drifting. It seemed cruel to start off
Bubokoff and Kolinkin, but if I had allowed them they
would have stayed forever. I gathered together the
teams of eleven dogs each and got them under way by
noon; but I expect they will halt at Qu Vina until the
weather clears.

The hut is filled with smoke which so blinds me that
I cannot open my eyes; and altogether I am very much
disgusted with the cause of my delay.

April 12*th.* — A calm and beautiful morning, just the
time to travel; and here I sit in idleness — waiting for
what or whom? Grönbeck returned this evening *without*
Mr. Gilder; so I have lost three days for naught, and,
besides, may have to keep Gilder all summer.

April 18*th.* — I finally succeeded this morning in setting
out for Kigolak and the west, with two teams of thirteen
and fifteen dogs. Neither team is satisfactory; the larger
one being fagged out and foot-sore, and the smaller one
composed of scrubs and pups. I have worn out all the
dogs on the Delta, and the natives are complaining of the

destruction. My yamshicks are Geordi Nicolai and "Paddy" Achin.

We reached Kigolak and halted for tea, while I saw Tomat Constantine in reference to sending our deer to Cass Carta. This is their feeding-ground, and I sent word to La Kentie Shamoola to carry one hundred fish to the northward for the use of my teams on our return. The fish will be cached where "Paddy" can find them.

Leaving Kigolak I journeyed west about ten versts to a Tunguse village of three or four miserable little huts barely visible above the snow. The natives were watching for wild deer, and had a gun set with a lock-string to shoot at them. They told me that two of my tame deer had starved to death, but I believe they were shot for food, as the people look very hungry. This place is called Sava. One small hut contained twenty wretched natives.

Running west another ten versts, we came upon a single Yakut hut, known as Sabas Kokoo, where a man, his wife, and little boy were fishing in a small lake and river which empty into the western branch of the Lena. Here we had tea, boiling a fish for ourselves and scantily feeding the dogs; for we shall travel all night, or until we reach Turak.

Arrived at Turak about four A. M., and turned in.

April 14th. — Turak is the wreck of a once prosperous village. There are two or three good huts left standing among the ruins of many others; a little church; and a large cemetery of perhaps two hundred graves. An old woman and a lad were the only souls we found in the village; all the other inhabitants are off fishing. Geordi Nicolai pointed to the tall cross in the centre of the graveyard, and said: "Yakuts pomree manorga" (many dead Yakuts). Taking a brief sleep, and some tea and fish, I gave the old woman a little *tabac* (tobacco), and set out for Jaolak. A cold keen wind from the west

blew through the mountain gorges, and swept us sideways. The dogs exhibit signs of weariness, and are
dropping out of harness, for yesterday's work was apparently too much for them. We are now within the
hunting-ground of the Kericks, father and son.

The old woman at Turak had heard nothing of the
loss of our boats and men, and listened in awe, with
mouth ajar, as Geordi Nicolai told her all about us. We
have turned five of our dogs adrift, and others are bleeding badly at the feet. I have worked them too constantly
during the past twenty days, but it can't be helped. I
must go on, and dogs are the only means of conveyance.
Poor dogs! Poor natives! Poor all around! We crawled
into Jaolak at a snail's pace long after dark, cold, hungry,
and miserable.

I find that sleeping in the huts is not conducive to
one's comfort, since a suit of clothing is supposed to last
the wearer without washing a whole season ; and the
"little foxes" inside of one's shirt keep one forever
scratching. But it is an unspeakable pleasure to be
able to hang out the shirt over night and freeze the little
pests to death. In the morning all that is necessary to
be done is to beat the infested garment, turned inside
out, against the door-post, and off they will drop by the
score, frozen white with the accumulated rime. It is
rather rough on the "little foxes," but a glorious relief
to the owner of the shirt.

I noticed, while approaching the end of Long Island,
that the river was wide and had discharged great quantities of broken ice. The water in the bay may be shoal,
but I am sure that if even a boat of considerable draught
found the end of Long Island fair, and followed it around
on the south side, she would discover the river to be
easily navigable. For it is there narrow and its rapid
discharge of water and ice leads me to believe that it is
also deep. This is the way into the Lena from the west-

23

ward, but from here towards the river Alanek, that is, westward from the west end of Long Island, it is a succession of sand-bars and shoals, with deep pools of water.

Jaólak is sixty or seventy versts west of Turak, and is located close to a coast-range of mountains on a river which follows along the base of the foot-hills. It is hidden from the northward by a *tundra* island twenty or thirty feet high and eight or ten versts long. Four huts compose the village, and they are inhabited by a large number of half-starved natives. In former years there were many people here, but they were carried off by a disease, said to have been contracted from eating in times of famine the entrails of a certain fish, which the natives still point out and shun. Doubtless this particular fish had eaten some poisonous matter, for the natives say that only its intestines are noxious to health. Here at the western end of the Delta the villages are all in a state of rapid decay, and there are certainly more dead than living Yakuts. Nor is the fishing so good as it used to be, and there is not a year without its famine.

April 15th. — We were up in good time this morning, and I had a frozen shirt to don next my skin. The weather is clear overhead, but it is blowing half a gale from the westward, and drifting the snow in our very teeth.

Ninety versts to Alanek and fifty to Chanker, a deserted village which at one time contained about two hundred inhabitants. Every soul has departed, and nothing remains but the graves, balogans, and yaurtas, and of these many have fallen in.

We ran along the coast and across the bay, moving slowly against the wind which carried us to leeward. For the ice was glass-like, and the dogs could scarcely keep on their feet. This part of the bay is incessantly swept by gales blowing from the mountains, and our yamshicks could talk about nothing else but the *poorga pagoda* on the *gooba;* assuring us that many of their people had been blown from the ice into the sea. And, indeed, the frantic wind did rush upon us in perfect *willawas.*

About ten versts east of Chanker we halted at a solitary hut in a nook of the mountain spur, and made the usual inquiries after Chipp and party; but the occupants had heard nothing of men or boat. As we approached Chanker we passed many deserted huts, and coming upon the graveyards the yamshicks reverentially lifted their hoods. There was one prominent grave marked by a tall cross and surrounded by an ornamented wooden railing. The natives explained that a Russian officer was buried there, and added in a casual way, "Cushat soak; pomree" (nothing to eat; die). This is the most desolate place I have ever seen.

From Chanker we crossed the peninsula to the river Alanek, meeting with numerous sled-tracks and fox-traps. Journeying northward on the bed of the river we shortly encountered three Tunguse tents, pitched on the shore. The people, wild and wretched looking, half-clad and starved, were all out fishing through holes in the ice. The Alanek at this point, perhaps thirty versts from its mouth, and as far as I could see to the northward, is a noble river, from one mile to a mile and a half wide, sharply shut in between two magnificent mountain ranges, and apparently free of sand-spits and shoals, and consequently navigable. All the way to the sea its banks are dotted with little villages of two or more huts, and the natives are very poor. At one place where we drew up and made tea, the people had absolutely nothing to trade, and were fierce beggars for a little salt and tobacco, seeming more than thankful to secure the grounds from our tea-pot. They looked enviously at my two yamshicks, who had dry bread to eat, and I observed that

both "Paddy" and Geordi Nicolai were generous enough
to give nearly all of it, as well as their frozen fish, away
to their hungry kinsmen, knowing they would get more
at supper-time.

Ten versts further on we hauled up at the village of
Alanek, which comprises three inhabited huts, the *débris*
of many others, and a number of square store-houses.
The village is located at the mouth of the river, on a
sand-spit under the bluffs of the western bank, where a
small stream makes into the westward and southward.
About three versts distant from Alanek we passed a
large unoccupied balogan on the eastern bank of the
river.

It was a little after dark when we arrived, cold and
hungry, and just in time to escape a howling gale.
Starry Geordi, the starosti, gave us a cordial welcome.
Every bone in my body ached.

April 16*th.* Many inhabitants of the surrounding
country were here last night to meet the stranger, but
none had seen or heard of the missing second cutter or
her people. Geordi Nicolai and " Paddy " Achin related the circumstances of our landing and the subsequent burial of De Long and party, and the natives, listening in open-mouthed wonder, frequently crossed themselves.

The morning was bright and still, with the promise of
a fine day. So I determined to make an effort to discover the graves of Lieutenant Pronchishcheff of the
Russian navy, his wife, and a Cossack force, all of whom
died of cold, hunger, or scurvy at the mouth of the
Alanek, while engaged in making a survey of the Lena
Delta, albeit they had come prepared to winter. I read
of this in arctic literature years ago, but had an idea
that it occurred at the Delta, and in one of my conversations with General Tschernaieff, I made mention of the
fact, which led him to tell me the story of a young Rus-

sian officer of engineers, one of the social lights of St. Petersburg, who, because of some indiscretion, was sent out to the Delta, ostensibly to search for the grave of Pronchishcheff but really as an exile. He journeyed to the Lena Delta and returned to St. Petersburg, but without having found the burial-place as directed; and then in a fit of desperation, perhaps at a sentence of prolonged exile owing to the failure of his mission, he blew out his brains.   Thus the general recounted the melancholy fate of the young officer, and requested that I look for the graves of Pronchishcheff and party while at the mouth of the Alanek.

Geordi Nicolai, with whom I have talked a great deal about *bulchoi crass Ruski, starry starry crass,* and *pomree Ruski,* said he knew where there was a number of old graves with Russian crosses, the remains of old Russian huts and balogans, and that the legends of the country told of a poor white lady dying there, and being buried in the same grave with the *Ruski* commandant. Geordi is a kinsman of Starry Geordi, the starosti of Alanek, and when they had conversed together some little time old Geordi agreed to guide me to the desired spot. So leaving my own teams behind, we started off, accompanied by young Geordi as interpreter, for he and I had learned to understand each other's signs and expressions.

The morning that promised so well had meanwhile become overcast, and the snow was drifting before a strong wind. I had little or no time to spare, however, and so set forth upon the starosti's assurance that the place was not far off.   The gale blew fiercely in our faces until it blew itself out, for it was only a *willa-wa* from the mountains; and with our poor team of a half-dozen starved dogs we crept along to the extreme eastern point of the peninsula or promontory formed by the river Alanek and the Arctic Ocean. We had no difficulty in finding the graves, and my old Yakut guide was full of historic information relative to the fate of the people; acting in his enthusiasm the death of each one, and showing me how the living buried their dead comrades with *tass, tass* (stones, stones). He also knew of the house which served as an observatory for the party, and wherein they evidently had an equatorial or telescope, for Geordi tried to describe a dome-shaped hut, and, elevating his dog-stake to his eye after the manner of a long field-glass, spoke of the *zevesdas* (stars).

The graves are close to the spot where the huts were located, on a miniature plateau under the lee of a large mountain cliff, forming the east bank of the river mouth. The little table is about forty feet above the river, and overlooks the sea to the northwest, and a more beautiful point or one better adapted for observation and security, cannot be found in the Arctic regions. Near by is a hut, at present inhabited, and around it are the ruins of several others; for its position and surroundings not only recommended it to the unfortunate explorer and his party, but likewise to the natives. Not a vestige of the observatory remains; the starosti saying that the ruins are all those of Yakut huts.

There are six well-defined graves, marked by head and foot-stones. One has two logs of wood ranged alongside, and the stones with which it was covered are now imbedded in the almost solid rock. There are no mounds, each grave having been tramped or rather weather-worn flat, and only the stones indicate the different spots where the poor fellows laid each other down to an endless sleep. But there must needs have been one or two or more who had no comrade to perform this last kind office for them, and of these there is no trace or memorial. A large wooden cross still stands over one of the graves, and about five yards to the northwest of it are the remains of another, at the base of which some careless person years ago built a fire with destructive effect, and more recently some other vandal cut off the top with an axe. I inquired of the starosti if all of these were Russian or Yakut graves, and he said, "Yakut soak" (Yakut no), pointing out, at the same time, a dozen or more in the two groups of graves, which were unmistakably *Ruski*.

The cross which is left standing has a cant or lean to the southwest of about thirty degrees. It is seven feet high, and hewn to six by five inches out of a round stick of timber nine inches in diameter at the base. Originally it had three cross-arms, and the highest one, still in place, is fourteen inches long by six inches wide. The other piece was let in about two feet from the top, and, to make the cross look symmetrical, must have been about four feet long. Lower down, and within eighteen inches of the base, is a diagonal notch, in which were inserted the arms of a St. Peter's cross, as the Russians call it. Graves and cross face to the northwest or the west northwest, and look down upon the bay and river mouth, and across the icy Polar sea.

The cross is cracked and time-worn, and the characters were so poorly cut in with a knife that they are now simply distinguishable and no more. I copied them, not only *verbatim*, but in perfect outline, including the crack, which extends almost from top to bottom of the upright post.

The face of the country in its winter garb, spotted over with the many Yakut graves and crosses, has here the appearance of a veritable land of death and desolation. And yet the Russian government taxes these miserable people for the privilege of dying here. A sad commentary, indeed, are these ruins and fat cemeteries on the unprincipled policy of a great Christian country, whose priests or "popes," the very lowest and lewdest order of men, engage in no other missionary or humanizing work than their annual nefarious trips, when they

steal from the starving natives their marriage and baptismal fees, and collect their revenue from the sales of
ornamental brass-work — in the shape of icons — and of
wax-tapers, prepared by the priests' wives or purchased
by the gross from the manufacturers or traders, who are
not permitted to sell them directly to the poor duped
Yakuts. And I thought all this as Geordi Nicolai and
his aged relative, looking sorrowfully at the graves and
deserted huts, muttered over and over again, —

"Yakut pomree manorga."

The weather continued blustry, with a light fall of
snow. I tarried an hour at this so interesting spot, and
finally took a long last look across the broken white
bosom of the great North Sea, on which many many
years ago the ill-starred dead at my feet doubtless gazed
and dreamed — dreamed as I did then of a bitter past and
an inscrutable future — dreamed as does the whole world.
Brave Pronchishcheff; heroic and self-sacrificing wife
and Cossack comrades, martyrs all to science and duty!

Immediately upon my return to Alanek, I ordered out
the teams, and followed by the blessings of the poor natives, to whom I gave a little salt and tobacco, started
off, taking a short cut across the country to the northeast
through the mountain gorges, and presently gaining the
shore of the ocean, skirted along the coast to the eastward. About ten versts from the mouth of the river, we
passed a hut pitched far out on a spit in the bay, and I
wondered that the floods or rising ice had not swept it
from its apparently precarious position. But the yamshicks said "Soak, byral" (no, the sea), meaning that
when the floods reached this point they melted into the
ocean; and yet I imagine that in a northerly gale the sea
would roll over and submerge the spit were it not for the
shoals making a long way off shore.

Five versts further on we came up with four natives
fishing through holes in the ice of the ocean. Their

catch was very meagre, and they came running to us and had a long talk with the yamshicks, who gave them my story and a smoke, but they knew nothing of the missing boat. We ran along the rocky coast all day, and I noticed great quantities of drift-wood on the beach, and many fox-traps, and the sled-track of a hunter who had been visiting them. The coast is high and precipitous, with here and there a promontory; and at times I espied deserted huts on the numerous shoals or sand-spits. Without precisely knowing how deep the bay may be, still, from the confined bed of the Alanek, I believe it would be a good and navigable river to aim for in boats, in case of shipwreck on this part of the Siberian coast.

We drew up at a well-built and comfortable povarnia, forty versts from Jaolak, and made tea and rested our dogs. The poor things can barely crawl, and such as gave out entirely to-day, we turned loose and left by the roadside. They would follow after for a little distance, and then, unable to go further, would sit on end and howl dismally. For they seem to realize their fate when cut adrift, and too weak to reach a settlement — which is either to die of starvation or be devoured by wolves.

After tea we forsook the povarnia and toiled on wearily and drearily, reaching Jaolak at midnight, hungry, cold, and stiff.

April 17th. — I turned out this morning to find it blowing a half gale. This is a peculiarity of the weather up here; during fall, winter, and spring, there is always a whole or a half gale sweeping the snow.

I witnessed a strange scene last night between a youthful Yakut bride and her spouse, from whom she had stolen away and sought refuge in our hut, the home of her parents. He came in pursuit of her, but she would not go with him, and her parents would not interfere, for it is a custom among the Yakuts that when a bride returns to her first home the husband loses possession of her, and the

mother may sell her daughter to a new admirer. So in
this case the old lady was in pocket and the best of
spirits, and the bride was for sale, notwithstanding the
angry protests of the young groom; but as there was no
purchaser in my party we set forth on our journey, leaving all three engaged in a fierce wrangle on Yakut marital
rights, which the aggrieved husband was about to enforce
with a huge club shaped like an unshod dog-stake. The
law of divorce that obtains in this region is very primitive
indeed. If man and wife cannot agree, save to separate,
they simply do so, and marry again at will, or rather, in
the wife's case, at her mother's will.

The dogs are so feeble that I fear they cannot sustain
the long northern journey, though now the wind is behind us or on our left cheek, and acts as an aid. One of
the dogs dropped down in his harness this morning before
we had gone five versts, and his fellows bit and shook
him ferociously, but he was too weak to show fight. One
of the yamshicks took off the harness and threw him
aside, when he made a painful effort to stand up and
follow us, but failed and fell over. This was near to
Jaolak, and if he revives he may perhaps return there.

We followed the coast until dark from the river Turak,
one of the western discharges of the Lena. There are
numerous fox-traps along the way, and several shelter
huts for the trappers. These traps are owned by Boba-
rouski Gavirillo and Geordi Nicolai, and we slept in their
hunting-lodge, known as Koobalak, ninety versts distant
from Jaolak. This coast is full of large bays and headlands, indicating the presence of rivers, but in reality
there is no river north of Turak. It is a bad coast for
boats without proper knowledge of the land, for its appearance is very deceptive.

April 18th. — We made an early start from Koobalak,
and continued to follow the coast-line. Geordi says that
he and his partner visit their traps about twice a moon,

sometimes oftener; so there is no possibility of Chipp's people having landed here without their knowledge.

We ran along the high ground back of the bluffs all day, occasionally passing a trapper's lodge, but meeting with no permanently inhabited huts. At one of the lodges we halted and found the fish which I had directed La Kentie Shamoola to deposit for our use. This hut is fifty versts from Koobalak, and fifty from Buruloch. We made tea, gave the dogs a little feed and rest, and were off again, arriving at Buruloch after midnight. "Paddy" declares that I never sleep, and will kill off all the dogs on the Delta: "Spee soak; pomree bar" (no sleep; will die). There are three huts here, two of which, the owners having died, are in ruins. This was formerly an excellent reindeer coast, now the hunters laconically lament, "Malinki, malinki olane" (little, little deer).

April 19th. — Lowering weather. We skirted the coast to a small stream west of North Belun, down which we ran to the large deserted village of Tara Janga, where there are abundant remains of store-houses and huts and a multitude of graves, which tell their melancholy tale without need of my yamshicks' explanation: "Tashoo pomree; Yakuts pomree" (all dead; Yakuts dead).

Traversing the chain of small lakes and rivers, we passed through North Belun, and on down the river Ketack to Boikhia, where we halted as guests of La Kentie Shamoola. His hut is the largest and cleanest on the archipelago, and he has a fine wife and three children (*barinchucks*). I learn that Nindemann was here four days ago on his way south.

Geordi Nicolai wants a vacation to sleep; so I will discharge him and one team for the present, since I have now virtually finished the search on this part of the coast, unless Nindemann or Bartlett has found some trace of Chipp.

April 20th. — We set out in a violent snow-storm for

Kigolak, and heard *en route* that the deer which I ordered
to be driven to Cass Carta died on the road. The natives
hitched them to sleds and tried to coax them along; but
they had fared so badly during the winter and were so
weak that they soon broke down, — for the reindeer is a
very tender animal and easily killed. I engaged a dog-
team to carry Ivan Patnoggin and wife to Bulcour, as I
will have no further use for them after we break camp.

I arrived here, at Cass Carta, in the afternoon, and
found Nindemann and Grönbeck looking healthy and
happy. Nindemann saw nothing that would evidence
the landing of Chipp and party. On his journey around
the north coast, he came upon the first cutter lying off
shore abreast the flag-staff or pole which marked the
cache. It is fully four versts off shore, he says, and entirely snowed under and frozen fast, inside and outside.

His guides were competent, and he had no difficulty in
making the journey. At Barkin he found one good hut
and a store-house, and five versts to the northward a *palatka* (peaked hut). There were many fox-traps set
along the cape and coast, and he calculates that the huts
are located about thirty versts southwest from Barkin.
His report agrees with the observations I took last September while off the coast in the whale-boat, for he says
the point was so low that he could not tell whether he
was on land or sea.

If the people of the second cutter landed on the coast
traversed by Nindemann and myself, they would certainly have left some trace of their presence visible to the
most careless observer. De Long and party, for example,
when wood was scarce or wet, burned the fox-traps in
the vicinity of their camp. And then, too, the natives
have traveled back and forth over both courses without
seeing a sign of either boat or people. About thirty
versts southwest of Barkin is the large river Duropean,
full of massive broken ice, — the same as, or a branch of,

the rivers Kagoastock and Barchuck, which Nindemann fully and fruitlessly searched; and I now consider that the whole coast-line, from the Alanek to Bukoff(ski) Cape *via* Barkin, has been thoroughly gone over.

I have received tidings from Bartlett. He completed his journey, guided by the trappers Simeon Alock and Vasilli Kool Gar, from Barkin to Bukoff, inland and along shore, meeting with the same result as Nindemann and myself.

April 21st. — A stormy, miserable day. I paid off all the natives, and made preparations to abandon Cass Carta forever. Tomat Constantine told me that he left word for a dog-team to come here to-day; but, as usual, he lied, so I have sent for one to carry us all to Bukoff.

April 22d. — It was too late to make a start when the teams arrived this morning, as I wish to reach Chul-Boy-Hoy without stopping. I shall require four teams of fifteen dogs each, two of which I have, and the other two I shall hire, paying road money.

April 23d. — Out at three A. M., and off by six A. M., in a light eastern breeze. At eight o'clock the clouds began to gather, the wind increased, and by ten it was blowing a full gale.

We crossed nine wide rivers between Cass Carta and the Kagoastock, and further on crossed the Duropean, at the entrance to the mouth of which poor De Long and party perished. Here we ran into the heavy broken ice of the *gooba*, or eastern main branch of the Lena; and the storm had grown so fierce that the yamshicks lost their way and floundered wildly around on the rough bed of the river. We laid a southeast course by compass, not knowing our position, save that Ordono lay somewhere to the southward of us; and when we at last reached the land our dogs fell over and howled, so blinding was the gale. The yamshicks searched vainly about on hands and knees for a path or sled-track, until finally I ordered the tent to be set up, and with the dogs coiled

around it and the sleds on end, we all crawled inside and
shivered from two o'clock in the afternoon to two the
next morning, — Nindemann freezing all the fingers of
his left hand during the night.

April 24*th.* — When the weather cleared, we opened
our eyes and found ourselves but half a mile away from a
povarnia, whither we went, made tea, and dried our clothing, and thence set out for Turkanach. Halting there for
a few minutes, we ran on to Chul-Boy-Hoy, where we encountered the young man whom I found at the same place
one month ago with his starving and bereaved family. He
is now engaged in the construction of three little coffins
for the burial of his children, and I watched his labors
with interest. He has three solid pieces of timber, sufficiently long, broad, and deep, when dug out, to receive
the bodies. They are similar in shape to the Egyptian
mummy cases seen in museums, elliptical in cross-section,
broad at head and shoulders, and tapering toward the
feet, with both ends neatly rounded off. With wooden
wedges he split the logs longitudinally, matching and
"fairing" both pieces with pins. The top and bottom
of these two sections were next hollowed out, and the
corpse inclosed within, the case being then bound around
with three plaited birch hoops, in the manner of a long
taper cask. These coffins are disposed of in various
ways by the natives, — elevated on trestle-work, placed
in crevices of rock, set upon the ground and protected
by little houses built over them, or buried in the earth,
though this is the most difficult kind of burial, since in
digging a grave it is necessary to thaw out the ground
by a succession of fires.

We traveled along all night, the weather steadily improving, until by midnight it was delightful; and at two
o'clock in the morning of the 25th, we drew up at the
hut of old Spiridon in the village of Arii. Here we had
tea, and then resumed our journey, arriving at Jamaveloch about six o'clock.

CHAPTER XXVI.

MY FINAL SEARCH TO THE JANA RIVER

The Rodgers. — Mr. Gilder's Culpable Conduct. — Harber and Schuetze. — Noros and Jackson. — Mr. Larsen — Jackson's Vandalism. — Eating Wood. — Shumomea. — Oceansk — Mammoth Ivory. — Off for Verkeransk. — The Exiles once more. — A Letter from Berry — On Horseback to Yakutsk. — Our Miserable Equipment and Ride to Kangerack. — The Rapid Thaw. — "Jacksoned." — The Experience of Bubokoff. — Kolinkin and Gilder. — Siberian Cattle.

I AT once interrogated Bartlett concerning the whereabouts of Mr. Gilder, the correspondent, and learned that he had departed the day before for Tamoose.

From one of the many letters which Mr. Gilder sent to me, I gleaned that he belonged to the relief ship Rodgers, commanded by Lieutenant Robert M. Berry, U. S. N., and that, after making an extended cruise in the Arctic Ocean, and visiting the islands of Herald and Wrangel, the Rodgers was finally burned at St. Lawrence Bay, south of East Cape, in Eastern Siberia ; that after the destruction of the vessel, Lieutenant Berry ordered Gilder to proceed along the coast to Nijni Kolymsk, on the Kolyma River, and thence to Irkutsk, the terminus of the telegraph line, there to communicate the news of the Rodgers' loss to the Navy Department, and then follow the telegram to the United States as a bearer of dispatches. But upon his arrival at the Kolyma he met my old friend Kasharofoski, the ex-espravnick of Verkeransk, who told him of the Jeannette's fate and of our exploits on the Delta. Gilder, in turn, sent the information to Berry, and then held on his course until he reached Kangerack stancia, where he met the Cossack courier who was posting to Yakutsk with my sealed dispatches to General Tschernaieff and the Navy Department. The Cossack, who had heard the news at Verkeransk, told Gilder of the contents of the sealed packet, which that spirited journalist straightway induced the derelict courier to surrender into his hands, and coolly broke open. He abstracted the desired particulars, and then forwarded the packet to General Tschernaieff, sending, however, in advance to the "Herald" an account, taken from my report, of the finding of the bodies of De Long and comrades. He here turned over to his traveling companion, the ex-espravnick of Kolyma, Lieutenant Berry's dispatches to the Secretary of the Navy, directing him to mail them to the United States, and likewise to forward his telegram to the "Herald." It is needless to state that General Tschernaieff expressed great surprise to me at the very questionable liberties taken by Mr. Gilder, but dropped the subject at length with the remark that he supposed the breaking of a seal was a matter of little or no consequence in a free country like the United States, but in Russia it was a penal and serious offense, and he assured me that the Cossack would not go unpunished for his part in the transaction.

When I reached Tamoose it was to learn that Gilder had gone, so I returned to Jamaveloch and began my preparations for a final search to the mouth of the Jana River; for the sledding season was now over, and I would soon have to leave the country or be detained until fall. I at once started Captain Grönbeck to Belun, with orders to seal up our stores and send a list of them to Epatchieff, espravnick of Verkeransk. I likewise took an account of stock at Bukoff, — bread, salt, dried beef, tea, and tobacco, — which I had stored in bags and sealed: for

I now received word that two American officers had been detailed to assist me in the search, and that they were about to charter the steamer Lena for the purpose, — an expensive piece of folly, to prevent which I redoubled my efforts to reach Yakutsk, since it would be plainly impracticable to navigate the Lena with this vessel of seven feet draught. Luckily the Department had detailed two very sensible young fellows for this special duty, and upon their arrival at the head-waters of the Lena they saw at once that the steamer was utterly unfit for the navigation of the shallow stream, and so stepped out of their charter-party. Lieutenants Harber and Schuetze then built a small schooner and several boats suitable for their work, and finally reached the Delta months after I had completed the search.

When, at last, I was on the eve of starting for the Jana, I received a message from Tamoose informing me that two Americans were there, at the hut of Kusma. Thither I drove on a sled, thinking that I was about to meet the naval officers of whom I had heard, but picture my surprise when, instead, I beheld Noros, who had set out for home in January with Mr. Danenhower. He was accompanied by a Mr. John P. Jackson, correspondent of the "New York Herald," who, journeying to the Delta to "write up" the Jeannette disaster, had met the Danenhower party at Irkutsk, and telegraphed their stories to his journal. He had then secured permission from the Secretary of the Navy to take Noros along with him to the Delta as companion and *aide*, and here they were with all the paraphernalia of Oriental travelers. Noros had shed his deer-skin rags, and was clothed in purple and fine linen, so to speak. Jackson had a Cossack escort and two covered sleds filled with toothsome foods and other good things.

I invited him over to Jamaveloch, where he learned from Bartlett and Nindemann the details of the search,

24

and how and where we buried the dead. And now a
Mr. Larsen, artist and correspondent of the " Illustrated
London News," appeared on the scene. He and Mr.
Jackson had been fellow-travelers as far as Yakutsk, and
now joined company, and wished to visit together the
places of interest on our recent search. Mr. Jackson desired that I would detail either Nindemann or Bartlett
to accompany him ; but, as I had no authority to detach
any of my party for such service, I declined to do so,
greatly to the displeasure of Mr. Jackson, who seemed
to imagine that he had only to order in the name of his
master, and I would obey. The egregious egotism of
this kind of person is amusing in the extreme. At our
first meeting he told me, with a great show of importance, that he would be obliged to me if I would turn
over to him for his perusal and inspection the log-books
and journals of Lieutenant De Long and Mr. Collins ;
that Mr. Bennett had so ordered, etc. ; that if there was
anything I wished to have done, he would be pleased to
forward all my projects, etc. ; or if I wanted any money
he was empowered to draw on Mr. Bennett, etc., etc. In
short, he was prepared to take me in charge and complete in a proper manner the work I had almost finished.

Very much to his astonishment, I was in need of no
assistance, and not at all inclined either to surrender myself into his keeping, or to be captured by force. Had
I supposed it was the intention of this ghoul-like party
to break open the cairn-tomb, I would certainly have accompanied them, and prevented such a desecration. But
I never dreamed that a person born in a Christian land
would so far forget the respect due to our honored dead
as to violate their sacred resting-place for the purpose of
concocting a sensational story, and making sketches, or
out of idle curiosity. Yet this, I afterwards learned, was
done ; and the timbers were sawn off and tumbled down,
and the structure left so weakened that it no longer
served the purpose for which it was intended.

Finally, with everything in readiness, I withdrew my whole force from Jamaveloch to search the bay of Bor-khia, and round the peninsula to Oceansk. Ere leaving I bade good-by to all my old friends, and divided among them what provisions, etc., I had to spare; previously storing, however, all the valuable articles, such as tea and tobacco, — an account of which I left with the espravnick of Verkeransk for the use of any other search party that might be sent to the Delta.

One incident I have almost overlooked — one that well illustrates the extremities to which our poor Tunguse and Yakut friends were reduced by my wholesale purchases of fish. Gabrillo Passhin, one of the natives who had supplied us with food when we first landed at Jamaveloch, and a man of considerable reputation as a deer-hunter (he having repeatedly promised to sell me venison, but as regularly failed to do so), with his wife and children, was on the verge of starvation, and begged of me two hundred fish, promising to pay for them. I agreed to give him that amount, but, although he had called upon me several times, I did not believe that he was actually in want until I was informed, at length, that he was eating *masta* (wood); and upon visiting his hut, sure enough, I found him scraping chips or fine shavings from a log of spruce. These he mixed in a tub with snow and a frozen fish, pulverized, bones and all; and the wretched inmates were filling their pinched stomachs with this mixture, the fish affording them a little sustenance, the wood, distention, and the snow making the mess comparatively palatable.

My party, now consisting of myself, Nindemann, Bartlett, Yapheme, and dog-drivers, got away from Jamaveloch about eight o'clock in the fine morning of April 28th. We halted at the east end of the island of Tarrahue, and ate our supper of tea and raw fish. Then skirting the island we came upon a couple of old huts and a

myack marking several caches of fish, made by some natives.   Vasilli examined the distinctive marks, and stated
whose they were.   We found no signs of Chipp or party;
and so traveling all night arrived at Shumomea, the opposite shore of the Bay of Borkhia, stopping at an octag-
onal povarnia.

Following the coast-line and crossing the shoals, we
came to the little village of Bulcur, at one of the mouths
of the Jana Delta.   Here we had supper, and procured
reindeer to carry us to Oceansk, two hundred and ten
versts distant in a bee-line.   The natives treated us most
kindly, but had seen nothing of Chipp's party.   We then
passed from village to village, — to Maxim Bottono, to
Batter Arack, to Isverska, and thence to Oceansk; where
we arrived at two o'clock on the morning of April 30th,
having completed the coast-wise search for Lieutenant
Chipp and party all the way from the mouth of the
Alanek to the mouth of the Jana, a distance of more
than five hundred miles in a direct line, and more than
one thousand miles by the sinuous coast-line.

Oceansk is quite a large town of three hundred inhabitants, composed of Yakuts, Tunguses, exiles and their
keepers, and quite a number of traders, who buy up the
pelts and the fossil ivory which is found throughout this
section of Siberia.   I saw many thousand pounds of the
mammoth tusks stained black as night by age and the
tanning qualities of the *tundra* peat or bog, in which a
great quantity of the ivory, or mammoth (as the natives
and Russians call it), lies buried.   Some of these tusks
which I measured were nine feet in length along the
curve, and at the large or skull end were thirty inches in
circumference; hollow and elliptical in cross-section.   I
saw one train of thirty sleighs laden with these tusks, all
marked with the owner's name, *en route* for market, and
upon inquiring its destination was told *Keti* (China), the
great ivory-working country of the world.

After a sleep and breakfast, I arranged for as rapid a run to Yakutsk as possible, *via* Verkeransk; for I now felt that my labors at the Delta were completed, and if Messrs. Harber and Schuetze intended to prosecute the search during the coming summer, it would be necessary that I give them the benefit of my experience, and caution them against employing a steamer of too deep a draught.

My journey from Oceansk lay across the edge of the *tundra*, and over the mountain district to Verkeransk. The road does not follow the banks of the Jana, on which both Oceansk and Verkeransk are pitched, though in summer time boats are floated down the stream. We traveled by reindeer teams, but the sledging season was fast drawing to a close; the snow was melting and running off in little streams, and the bare earth was beginning to show on the sunny sides of the hills.

I bade farewell to the shores of the Arctic Ocean on the first day of May, 1882, and left behind me the ragged and faithful Yakuts. Though lousy and dirty, they had done more than a Christian part by me when cast ashore among them. They were not above lying or stealing in their own original way; and yet they gave us of what little they had, and no matter if I did pay them double rates for all I received, I still have not forgotten them, as I trust they will all live to know.

The distance to Verkeransk from Oceansk is about nine hundred versts, and the stations between bear the following euphonious names: Tallowguil, Kool - Gark-Soak, and two or three small ones, such as Belcur, Dwee, and Aimee.

The snow was fast forsaking the roads, making hard travel for the deer, which suffered sorely from the heat. Many herders were already driving their deer to the mountains, and the natives were loath to work them at all, since a large number were with fawn, and indeed it did seem cruel; but then I would have to reach Verkeransk at once or wait until fall.

I arrived at Verkeransk on the evening of May 6th, and received a hearty welcome from espravnick Epatchieff. Again I saw all the exiles. They were in high spirits, and full of talk about making their escape down the Jana River to the Arctic Ocean, and thence along the coast *via* Behring Strait to America. I could not, of course, aid or abet their attempt, but they had still in a double degree my sympathy and good-wishes; for surely they have been treated harshly, outrageously. It is astonishing, the number of young men who are sent into exile for mere participation in students' rows, such fracases as are frequent in our university towns, and for which offense our law locks the youngsters up in a police station over night, and fines and liberates them in the morning. But in Russia the students are the groundwork of the educated classes, who are likely to think in too free a manner; and if they express their liberal opinions too freely, away they go to Siberia. When a Russian is convicted of a felony, he first serves a prison sentence, and is then banished to Siberia out of harm's way, going further to the northeast in proportion to the heinousness of his crime, — the mouth of the Kolyma River being the easternmost penal settlement on the Arctic Ocean.[1]

Upon my arrival at Verkeransk, I received a letter from Lieutenant Berry, written at Kolymsk on the Kolyma River, and dated April 7th. It informed me that he was working his way west and searching along the coast as far as the Jana. Had I known this before leaving Oceansk, I would have waited for him or traveled to the eastward and met him, and thus finished the coastwise search for Chipp from East Cape to the river Alanek. But it was now too late to turn back; and, indeed, the season was so far advanced that it was dangerous to proceed to the southward, for the country was flooded by the rapid thaw, and I had no time to spare. So aban-

[1] Appendix : Letter from M. Leon.

doning my sleds and deer at Verkeransk, I began a weary
horse-back ride to Yakutsk, a distance in winter, by the
short river cuts, of nine hundred and sixty versts, but
now drawn out into twelve hundred versts, or about
eight hundred miles. And the horses' Old "crow-
baits," which had survived the wear and tear of the recent winter's baggage trains. And the saddles!! Cavalry recruits never had such an experience, for they are
generally broken in on that prince of saddles, the "Mc-
Clellan tree." We had the tree, it is true, but it consisted of two curved sticks fastened to two pads which
rested on the back of the horse — *and we rested on the
sticks;* at times filling in the aching void with a bag of
hay. Deer-skin straps and wooden stirrups completed
the harness; and very long and uncomfortable was that
ride.

I had four pack-horses to transport our provisions and
personal effects, and my party consisted, beside myself,
of several guides, Bartlett, Nindemann, and Yapheme, —
now better known to us as the "Red Fiend," because of
his flowing red hair, and turkey red cotton blouse, which
the Russian peasant so much admires; and for his rascality in losing (stealing and selling) our camp equipage,
tobacco, and tea, and in getting drunk whenever an opportunity offered.

Altogether, there were about ten horses in my train,
a rather large number for that section of the country,
and at times we could not be furnished with full relays.
We ourselves were a hard-looking half-dozen of ragamuf-
fins, and mounted on our lean and worn-out, but vicious,
steeds formed as shabby a cavalcade as ever marched
through Siberia. Yet we were jolly withal. The rivers
and creeks were swollen by the torrents of melting snow
and rain; and our weak and miserable horses, without
shoes, slipped and staggered about on the ice, pitching us
over their heads or rolling us in pools of water. Often a

dismounted rider would come marching into camp, leading his Bucephalus after him, and were it not that the
horses carried our extra clothing some of the luckless
ones would have turned theirs adrift, saddle and all.
Thus we trudged on until finally we arrived (May 14th)
at the Kangerack stancia, the mountain divide between
the districts of Verkeransk and Yakutsk. The horses
by this time were barely able to stand, and there was no
relay until we crossed the divide and journeyed far down
into the valley of the River Aldan.

Halting at the Kangerack stancia, we turned the horses
loose to graze on the scant shrubbery which can grow at
these great altitudes. The poor animals scraped away
the snow with their forefeet, like reindeer, and cropped
the dry grass beneath.

We had been seven days on the journey from Oceansk,
during which period none of us had been in a hut to
sleep for more than three hours at a time. Our provisions had given out, and the baggage horses and the "Red
Fiend" were so far in the rear that I had not seen either
for three days; nor would I have been sorry had the
"Fiend" been lost to me forever, he was such a nuisance.
My haste in pushing forward was in order to cross the
Aldan before it broke up, but at Kangerack I met several old deer-drivers and the Cossack whom I had started
to ride post to Yakutsk, who told me that the valley was
flooded for miles in every direction, and that it was not
safe to cross the divide. My meat was exhausted, and I
had purchased a quantity of horse-flesh, or, as the old
Yakut assured me, the flesh of a mare which had been
with milk. It was of a superior quality and of a decidedly superior price. I saw many of the Yakut women
and children hunting around for roots to eat.

After a couple of days' rest I tried to urge the horses
on, but they sank to their girths on the soft wet snow,
and finally lay down, and we were compelled to lead

them back to camp. A fine prospect, indeed, in this mountain gorge, with swollen streams on either side of us, and almost nothing to eat. At one time it looked as though we had only survived the perils of the Arctic Ocean and the Lena Delta to starve to death on the mountains, or be drowned in the rushing torrents. I subsequently learned that my two messengers. Bubokoff and Kolinkin, accompanied by Mr. Gilder of the Rodgers, had been overtaken by the floods in the valley of the Aldan and driven into the tree-tops, where they lived for days, killing and eating one of their horses, whose carcass they moored fast to a tree and hauled up into their perches when they were in need of food. At length the waters subsided and they were released from their lofty captivity, and none too soon, for the odor of their floating larder had become painfully powerful, and their stomachs correspondingly weak.

True to their trust, Bubokoff and Kolinkin hoisted the box containing the precious books and records of the expedition into the top of a high tree, lashing it fast; but the water continuing to rise they became frightened and raised and lashed it still higher, when Kolinkin fell from the tree and was borne away by the current into the branches of another, where he remained without food for several days. Through their negligence they ran the very risk it was my intention to avoid; for I had started them from the Delta so early to insure the removal of the records to a place of safety ere the spring floods set in. But they stopped at the different stancias; idled away a week at Verkeransk; and meanwhile the season crept on, and the floods caught them about ten days before they crossed the Aldan, and their disobedience of orders nearly cost them their lives and the loss of our records, the fruit of so much toil and suffering and death.

Seeing that I should be "jacksoned" for an indefinite time at Kangerack, I dispatched the rascally Cossack to the nearest reindeer station with an order to drive back five deer for food. Our bread had long since given out, and had we not been so fortunate as to meet the two old herders and the Cossack we should have been forced to eat our horses. The Cossacks while traveling in Siberia live on the country, and this fellow, I was told, had come to Kangerack and coolly quartered himself upon the station-keeper until after the floods had subsided. The old Yakut, knowing his man, offered him five roubles to move on. The Cossack pocketed the cash, but remained where he was and ate up all the venison and meal at the station, so that my timely arrival was a windfall to both of them.

On the afternoon of May 16th, an aged Yakut came into camp. He was the driver who had carried Bubokoff and party to the next stancia, ninety versts beyond Verkeransk; and he reported the roads as flooded and impassable. There had been a considerable rain-fall, so that the snow rested on a bed of slush and rendered travel of any kind almost impossible. I purchased his three reindeer for food; and he then followed his tribe into one of the mountain gorges to herd deer for the ensuing winter's work.

May 18th, snow fell a foot deep, whitening all the landscape. Here was a queer sight. The snow lay in some places to a depth of forty or more feet, and occasional avalanches had left certain of the gorges choked for weeks. Yet when the sun had melted all but a few inches of snow, the blue, or whortle-berries, and the wild cranberries, no larger than French peas, could be gathered in abundance; and hardy little plants of the color of dark ivy pushed their shining leaves through the snow and gladdened the hearts of our half-starved horses.

It is remarkable the discrimination these Yakuts make in the care of their cattle as against the poor horse. The cattle are kept housed with the family until spring, and when turned out they are the leanest, hungriest-looking kine to be seen out of Egypt; albeit they are fed on fodder cut during summer with long knives, similar in shape to the Cuban *machette*. And I never beheld such peculiarly constructed cattle. Like most people, I had been only acquainted with the ordinary bovine, furnished with a reverse-curved spine; but the cattle of Siberia have backs like the pig's, on which flourishes a crest of hair, long and matted like the shaggy head of the bison.

CHAPTER XXVII.

THROUGH SIBERIA.

Meeting with Berry and Hunt — In the Valley of the Aldan. — "Dismal John" — Traveling through the Flooded District. — A Tunguse Family *en route*. — Crossing the Aldan. — A Picturesque Scene. — Arrival at Yakutsk. — On board the Pioneer. — Mosquitoes. — Castellated Cliffs. — Passing the Harber Party. — Copert Barges. — The Voyage to Karinsk. — On the Constantine and the Tow-Boats. — In the Tarantass. — "Ivan," the Fictitious Friend. — Irkutsk. — The Story of my Watch.

On the morning of May 21st, Lieutenant Berry and Ensign Hunt, of the lost steamer Rodgers, arrived at Kangerack, accompanied by Mr. John P. Jackson, the correspondent; Mr. Larsen, the artist; a squad of Cossacks; and seaman Noros, who had been traveling over the Lena Delta in the capacity of guide and servant to Mr. Jackson. Lieutenant Berry and Ensign Hunt brought with them a Russian boy, whom they had shipped somewhere along the coast of Kamschatka, and who had since been acting as their interpreter.

I was certainly never so glad to see two white men in my life as I was to see Berry and Hunt. They had journeyed westward nearly two thousand miles along the shore of the Arctic Ocean, from East Cape to the mouth of the Jana, — arriving at Oceansk only two or three days after I had taken my departure. There they first learned of my having found and buried the dead of De Long's party, and also of my inability to discover any trace of poor Chipp. Berry then hastened to overtake me, bringing with him a goodly supply of bread, of which my party was sorely in need; and he thus actually accomplished in part the mission on which he had originally set out: viz., the relief of the Jeannette people. And, I repeat, it was a most pleasurable thing for me to meet two of my own countrymen, and fellow-officers at that, in this remote and cheerless region.

After much hand-shaking and many inquiries about affairs at home and our common friends, we made preparations for another attempt on the morrow to cross the mountain and reach the next stancia. So early in the morning we set forth, and at the end of a hard day's work camped in the valley on the other side. Our party now numbered fifteen, and our horses were just able to drag their legs along. We rode them from povarnia to povarnia, turning them loose at each halt to graze upon the withered leaves and shrubbery. We were greatly entertained on this long journey by one of the late additions to my party. He had acquired a smattering of Russian, and would shout to the natives, —

"Yamshick, yamshick! skulka versta to stancia?"

The yamshick addressed would state the distance as near as he knew it, and then our petulant companion would whine out like a babe, —

"You lie, you — ! You said it was only eighty an hour ago."

And as the saddle-trees cut, so would the greatly aggrieved and only dolorous member in our party growl and anathematize Yakut, horse, saddle, and bridle. Finally a bright idea occurred to him. He saw that in the awarding of horses at the various stancias a few roubles bestowed on yamshick or station-master would secure for him a better outfit. And so for a time he was less miserable; but presently his saddle and stirrups would not fit him; and from the wholesale manner in which he grumbled about his eating, drinking, sleeping, about everything that was around him, the sky and the earth beneath, — I seriously doubt if his halo would fit him should he succeed in edging his way into Paradise, which I must say, however, is, in my opinion, utterly improbable. And what if he were possessed of all Heaven itself, its fancied comforts and glories? Would he not fret and complain and pine for a portion of Hell as a diversion? My language is thus plain and strong simply because in all the miles of my travel or days of my life, I have never encountered such a fault-finder. He quarreled with things that seemed luxurious to some of us, albeit we would have preferred better accommodations had circumstances been different. Yet "Dismal John," as he was soon dubbed, made no such allowance. When his sharp practice of bribing the yamshicks or stancia-keepers was discovered, it of course became the duty of one of our number to mount the wrong horse — any horse, and ride on in advance much to the disgust of " Dismal John."

We camped in the valley west of Kangerack at a little old hut floored with ice and dripping with water from a leaky roof. Some of the party slept outside on a bed made of the leaves and branches of the pine, fir, and hemlock, placed close to a great roaring fire, wherein we roasted some venison, that with tea and sugar and the remainder of the black bread which Berry brought us composed our evening meal. Enjoying a good night's rest, we started betimes for the next stancia, fifty versts away, on the edge of the flooded district. The natives desert the inundated regions, taking to the high lands, and therefore the second stancia would be far ahead, one hundred and twenty versts distant; and in all that tract we would find no people, no horses, no game, nothing to eat. Hence, we halted for forty-eight hours to rest our horses and allow them time to feed on the dry grass, which the dissolving snow was abundantly revealing.

Taking possession of the deserted povarnia, we killed

a deer that we had brought along with us for food, and tethered the others where a little reindeer - moss was showing itself. Just as we arrived at the stancia a Tunguse family was leaving it, mounted, wife, boys, girls, and babies, on the backs, or, rather, shoulders, of reindeer. Two of the smaller children were suspended on either side of a reindeer by means of a strap passing around their bodies and over the back of the animal; two additional thongs adjusting their arms and legs into position. And the little rats did not seem at all ruffled by their situation; and I must here remark the decidedly superior temper and behavior of both Yakut and Tunguse babies over those of enlightened Christendom.

The deluged district, which had been so thickly populated the previous winter, was now the picture of desolation. We passed through quondam villages of ten or fifteen huts, so entirely gutted by the waters that only the ragged uprights were left standing, to be covered afresh with earth for the winter's occupation. The country bordering on the river was overspread with short tussocky grass, with patches of glassy ice between, on which our feeble and unshod horses slipped and floundered, giving us many an ugly fall and ducking. It rained almost incessantly, — a cold sleet-like rain, and we were never dry or comfortable; nor did even the oaths and groans of our dismal companion tend to promote hilarity in our ranks. At the povarnias in the evenings his countenance would light up, providing we had been able during the day to secure a duck for each member of the party, either by purchase or through the sportsmanship of the Cossacks. Then he would bestow a greedy smile upon our cook, Yapheme, who, inclosing the split ducks in a long-handled gridiron, broiled them to a turn with pepper, salt, and a mixture of butter and tallow. Yes, I verily believe that he not only looked, but actually *felt* glad at times, if he received the first pot of tea, the first

duck, if it was a large and tender one, and broiled exactly right, — yet if all these "ifs" were not fulfilled I
am quite certain he was still wretched.

Thus we progressed, for the most part happy and contented, though cold, wet, and lousy, until we arrived on
the 31st day of May at the bank of the river Aldan.
Halting and building a fire, we hailed the ferryman on
the opposite bank, and discharged our guns to attract his
attention, but in vain. Finally we sent a Cossack across
in a canoe which we found moored to the shore; and,
pending his return, we surveyed with interest the sad
havoc which the flood had wrought on the banks of this
great tributary of the Lena. Monstrous blocks of ice as
large as small cottages lay stranded ashore, and the water
had risen in places to a height of forty feet, as indicated
by the drift-wood lodged in trees, thousands of which had
been torn up by the roots or hurled flat by the overrun-
ning ice, and swept out to sea, there to drift on that
long north and west course into the Atlantic Ocean, or
float down on the southerly current and strew the shores
of the east coast of Spitzbergen.

We were ferried across the Aldan River at two o'clock
on the morning of June 1st, in a large flat-bottomed boat
fashioned like a New England fishing dory, but sharper
at the stern and of considerable sheer. It was probably
sixty feet long by ten or twelve feet beam and four feet
deep; open from stem to stern, but furnished with a
raised platform amidships on which the passengers sat;
and it was rowed by eight men, some of our party assisting. The scene was wildly picturesque. Our roughly-
built boat; the savage-looking oarsmen; ourselves, if
anything, more savage in appearance, clad in rags and
the skins of beasts, bearing guns and other trophies of
our Arctic travel; the dark, cold river; the massive ice-
blocks standing in the vast silent woods like so many
white cottages; the utter loneliness of the scene, — I

can never forget it, and though sick at heart and failing in strength, I then rejoiced that I had lived to witness it.

Landed on the other side, we were soon installed in a comfortable hut, and partaking of a good meal of ducks; and here we were first informed of the narrow escape of Bubokoff, Kolinkin, and Gilder. I could not advance all of my party on the morrow, for the horses which were here tendered us had by no means recovered from their constant work during the past winter; so I dispatched a Cossack and Bartlett ahead to prepare the way for our coming, and left a Cossack and another of my men behind to follow us with the baggage.

The journey from the river Aldan to Yakutsk was most tedious and disagreeable, but arrive we did at last, June 7th, and were welcomed by all the officials of the town. Madame Lempert cooked us a capital dinner, and we then repaired to the Balogan Americanski, our old quarters of the preceding winter, where we found Messrs. Gilder and Bartlett.

The day following, accompanied by Lieutenant Berry, I called upon the governor-general, who received me with open arms, calling me his son, and embracing me fervently with tears in his eyes, as though I had been one he loved just arisen from the grave. He complimented me on the success of my search; he was proud, he said, to be associated with such a " son." Returning our call the next day, he invited Lieutenant Berry and myself to dinner, to meet the lieutenant-governor and Lieutenant Irjansk of the Russian Navy, who was then at Yakutsk preparing to set out for the mouth of the Lena River, to establish a meteorological station and make a survey of the Delta.

Lieutenants Harber and Schuetze had not yet departed from Vitim, where they were still engaged in fitting out a small schooner and several light-draught boats for their search of the Delta, they having sensibly abandoned

25

their original idea of chartering the steamboat Lena. I was disappointed at not meeting them in Yakutsk, and so drawing sufficient money to pay off all my debts and the expenses of my party to New York, I decided to proceed at once to Vitim.

The governor-general gave a farewell breakfast at which we were all assembled, and many were our toasts and pleasant speeches and fond hopes for each other's future happiness and prosperity. Accompanied, then, by the governor and hosts of friends we sought the steamer Pioneer, which was to bear us up the river as far as Vitim. All Yakutsk was out to see us off, and about five o'clock in the evening of June 11th we started slowly up the Lena.

The steamer was small, dirty, and hot. Warm weather had now set in, and we knew no rest because of the mosquitoes. The Pioneer made slow progress against the swift current, at times making none at all, or zigzagging her course to avoid the rapids or dodge into eddies, and so crawl up stream. It was a mystery to me how she was managed, for there seemed to be but one set of men who ran her day and night without relief. Our party slept in the two little cabins, one forward and the other abaft the paddle-wheels, and luckily there were no other passengers. We drew water from the river and washed ourselves in a deck-bucket, using our own soap and towels. We had agreed to pay six kopecks per verst for transportation, and two roubles per day for food, but as the unchanging diet of boiled beef and tea soon palled on us, we purchased milk, eggs, and other provisions at the wooding stations, and if the mosquitoes had not interfered we should certainly have spent an amiable and enjoyable time.

One evening, while sitting on the bows of the boat trying to keep cool, we saw ahead of us what seemed to be a great sand-bank, and so we warned the pilot. But what was our surprise as the steamer neared the bank to see it arise and pass over and on either side of us like a column of smoke. It was a bank of mosquitoes. Nothwithstanding our calico head-covers which we tucked under our coats, and which had horse-hair faces or visors attached, the fine midges or black flies managed to get through and into our eyes and nostrils, causing us unspeakable annoyance. To protect our bodies we had buckskin gloves with tie-strings around the wrists. but the miserable little torments got at us nevertheless. There was a ventilation hole in my fur cap through which they bit me on the top of the head. They were everywhere.

The second day on the river, and during the two succeeding days, we passed the most remarkable extent of palisades or castellated cliffs I have ever seen. In places they towered to a height of two thousand or more feet, and for miles and miles the wonderful unbroken rocky façade arose from the river, turret and buttress constantly varying in projections of columnar beauty. The rock had the color and appearance of brown sandstone from the deck of the steamer, but I could not fully distinguish it at such a distance. It was not regular, like the basaltic columns, and I know of nothing that it resembled in nature or architecture but a stupendous front of castle wall.

On the morning of June 15th I was told by the captain of the Pioneer that a steamboat had passed us in the night, having in tow a schooner and two small boats. He thought it might have been the Harber party, whom I had instructed him to stop in order that we might consult together ; but from stupidity or a rascally fear of losing some passage money he had deliberately allowed them to go by. However, it made but little difference, for in several days we arrived at Olekma, where I found a note left for me by Lieutenant Harber, the first communication I had received from him. He therein desired me to return to Yakutsk should we miss each other on the river.

This I did not think at all necessary, since I had already completed to my own satisfaction a fruitless coastwise search for Chipp, and at the only time when it would be possible to find traces of him had he landed. For if any track or vestige had indeed escaped my notice, it had ere this time been swept away by the spring floods. Yet as Lieutenant Berry was about to dispatch Ensign Hunt to join the Harber party, I concluded to send along with him fireman Bartlett, who volunteered to go. I also prepared for the party's guidance a letter of instructions, and a chart of the Delta on which my various tracks were marked. Returning to Yakutsk on horseback, Hunt and Bartlett met Lieutenant Harber, who was riding back in the hope of overtaking me on the Pioneer, and he would doubtless have succeeded had he not encountered Hunt, who gave him my letter and chart.[1]

[1] [Lieutenants Harber and Schuetze pushed forward in their little schooner, the Search, and reached the mouth of the Lena in July, 1882. Their coastwise search for Chipp was barren of results.

In December they received at Yakutsk the order (which had been six months in transmission) to bring back the bodies of De Long and party to the United States, Congress having appropriated $25,000 for that purpose.

Below is a table of the magnificent distances over which the remains, wrapped in felt and placed in metallic caskets, were transported : —

Table of Distances.	Miles.
Mat Vay, by reindeer sleds, to Yakutsk . . .	800
Yakutsk, by horse-sled, to Irkutsk	2,342
Irkutsk, by horse-sled, to Krasnoyarsk	670
Krasnoyarsk, by horse-sled, to Tomsk . . .	867
Tomsk, by horse-sled, to Omsk	582
Omsk, by horse-sled, to Orenburg	1,000
Orenburg, by railroad, to Moscow (about) . . .	900
Moscow, by railroad, to Hamburg	1,390
Hamburg to New York	4,140
Total	12,191

At Hamburg the two officers embarked with their dead on the

I continued on my journey up the Lena; the villages becoming quite numerous, though still small and scattering. The people here carry on a feeble kind of agriculture; own some cattle; a few chickens; cut wood for the steamboats; fish a little, and work as watermen on the river. We now passed great numbers of large barges belonging to wealthy coperts. They are capacious storehouses, forty, some eighty. feet long, built of heavy timber, decked over, and strongly fastened together with wooden tree-nails, the seams being caulked with moss and payed with pitch. They are built on the banks of the river in winter time and launched by the spring floods, being freighted with all sorts of goods suitable for trade; and they float down stream with the current, occasionally flying a sail, and being steered by means of three long sweeps forty or more feet in length, which are rarely used to propel, but only to guide them clear of the shoals. They halt at all the villages along the river, and often congregate at some large settlement and hold a bazaar. These, of course, are gala days to the villagers, who forthwith affect the cheap colognes and gaudy bandanas, considerably to the glee of the flourishing coperts.

We visited several of these barges, our companion and interpreter, Captain Grönbeck, being acquainted with many of the traders, who all received us with distinguished consideration. Some of the barges were nicely furnished, and the merchants were accompanied by their wives, or had a few passengers on board. They all make Yakutsk their terminus, and if they have not entirely disposed of their goods by the time they arrive there, they hold an auction, and also sell their barges for fire-

steamer Frisia, arriving at New York February 20, 1884, after an absence of two years and sixteen days. Everywhere along the whole route, in Asia, Europe, and America, the dead heroes were honored with rich tributes of respect, and the final grand procession and solemn burial in New York on Washington's Birthday of this year (1884) are yet fresh in the memory of the reader. — ED.]

wood or building material, as the timber is all hewn and
easily taken apart. The banks of the river are strewn
with the wrecks of the monster boats, which have either
come to grief or been turned adrift after discharging
their cargoes.

There were many exile Scaup villages along the river.
I visited one inhabited by thirty-three men and three
women, all mutilated and miserable looking, but thrifty
and prosperous; and I am of the opinion that the love of
money is the root of their religious frenzy. They were
anxious to purchase a flour-mill and drive it by means
of a wind-mill, — quite an institution thereabouts. The
Scaups raise great quantities of vegetables, and are the
only industrious farmers on the Lena. Their estates
upon death go into the state treasury, but they always
manage to quietly rid themselves of their property before
they die; so the governor told me.

While paddling along the river I beheld two dead men
floating in the stream, and had seen one earlier in the
day. Mentioning the fact to the captain of our boat, he
said : —

"Yes; we passed two others this morning before you
arose. They are only the men from the mines who come
to attend the bazaar, get drunk, and kill each other; and,
besides, there are many *Judes*, who murder people for
their money. I have seen fifteen corpses floating in the
river at one time." And Captain Grönbeck confirmed
his statement.

These criminals who are sent to the mines are a hopeless collection of cut-throats. Those, I believe, employed
by Alexander Silenikoff are all fed, clothed, and paid for
their labor, and allowed to visit the bazaars and spend
their earnings, which they mostly do in drink, and their
orgies end in death and a watery grave.

We changed steamers at a large village of four or five
thousand inhabitants, called Karinsk, boarding the Con-

stantine, a more powerful and commodious boat, in which our progress up the river became pleasantly perceptible. The Lena's banks were now dotted with villages ten or fifteen versts apart, and at many points within sight of each other. At Omalai we abandoned the Constantine, and continued our journey on the river in tow-boats. These are about forty feet long and ten feet beam, built in the shape of a whale-boat, with sharp ends, but straight, flaring sides, and flat bottoms like a dory. They are steered with a long oar, and drawn by three or five horses attached to a tow-line about fifty yards in length. One or two riders guide the horses, and the charges are three or five kopecks per verst for three or five horses, and a gratuity of ten kopecks to each rider and the steersman. The rate of speed is a lively walk, though the horses occasionally break into a harmless trot. A platform is raised for the passengers; and there is a movable shelter, a little less wide than the beam of the boat, and ten feet long, made of bent poles, covered with canvas and painted with bitumen. It is open at front and back, but in daytime a curtain is hung at the sunny end, and the air drawing through keeps the passengers cool. We traveled about three hundred versts in these boats, at the rate of from sixty to eighty versts a day, buying eggs, milk, and bread at the boat stations, and cooking our tea as we went along.

Our journey of four hundred and fifty versts to Irkutsk was performed in the *tarantass*, a large four-wheeled coach carried on long pole springs and leather straps, in the manner of our old-fashioned carriages, and drawn by three or five horses hitched abreast. We ran along day and night, sleeping in our *tarantass*, or *tallega*, another kind of four-wheeled conveyance covered in against the sun and rain. These vehicles are as heavy as an ordinary omnibus, and are intended to hold two or three passengers, the fare being three kopecks per verst for three horses, irrespective of the number of passengers, one, two, or three; but three kopecks per verst extra for two additional horses, and a giatuity of ten kopecks to the yamshick, the omission of which is sure to rob the horses of all speed.

I found an utter lack of honesty among the station-keepers, who are supposed to forward all travelers at regular rates and in regular order. But as soon as it becomes known that the traveler is in a hurry and willing to pay extra money, so soon is he informed that there are no horses, or that those on hand must be kept for the "posta." But, the station-keeper tells you, he has a friend, Ivan, from whom you can hire horses for double, treble, or five times the usual rates. I allowed myself, at times, to pay these outrageous prices, in order to keep our dolorous companion in the rear, and I managed to do so until we reached a station twenty versts from Irkutsk, where I secured the only post-horses in the village; but while breakfasting, "Dismal John" arrived in delirious haste, and, paying the premium for "Ivan's" horses, managed to outstrip me in the race to Irkutsk.

We all repaired to the Hotel Decco, and I at once telegraphed the Secretary of the Navy of the arrival of myself and party with all the records of the expedition, and requested permission to return home. The answer to my telegram was thus orthographically mutilated:—

"WASHINGTON, *July 8th.*

"May return home with pary.

"CHANDER, *Secretary.*"

I then called upon the lieutenant-governor, Pedoshenko, and other officials, receiving a warm welcome from all.

Irkutsk is the grand emporium of the far Northeast; a city of about 25,000 inhabitants, and well built of brick and wood. A large section of it was destroyed by fire in 1878, and the residents having none of the public spirit displayed in our cities, where such damages are almost instantly made good, the ruins still remain undisturbed. It was a noteworthy sight to see the caravans coming in from the distant south, laden with the tea and other produce of China and Tartary. Many Chinamen were assembled here, all active business men; but not in the laundry line, for which Irkutsk has little patronage. Exiles, too, of every grade are abundant, from the murderer to the unlucky prince or political offender.

Lieutenant Berry was possessed of a fine gold chronometer watch which needed repairing, and he was recommended to a watchmaker of the same name as the celebrated Danish chronometer-maker, Jürgensen. We visited his shop together, and, after Berry had displayed his handsome chronometer, I for amusement drew forth my old time-piece, which for more than twenty years had been measuring off the minutes of my life all over the globe. The old man smiled at sight of it, but undertook to put it in good order, and so I left it with him. And here I will redeem a promise I have somewhere made of telling the story of my watch's vicissitudes during the cruise of the Jeannette.

On the day when the Jeannette sank, her bows were thrown upward, the ice ceased for a spell its fierce intent to crush her; and as the sun was shining brightly, De Long requested me to make a photograph of the doomed ship. So I set up the camera, using my watch to time the plate, and hence when the Jeannette went down I had the watch upon my person, otherwise it would have been lost with the rest of my valuables. While in the dark room developing the plate, it may be recalled that the ice again began its fatal ramming, and the word was passed for all hands to abandon the ship. I left the plate unfinished to attend to other and more urgent duties, and while on the ice handed my old watch to Walter Lee. I was about to throw it away, but he said, " Give it to me, Chief, and I will carry it. If we ever get back to the United States I will return it to you."

So we started on the long march across the frozen sea. Lee was not very sure-footed; he had been shot through both hips during our civil war, and now kept tumbling into the water—and such water!—with almost intentional regularity. Of course the old single-cased watch came in for its share of the wettings, and at each one Lee would calmly empty it of the salt sea wave. And still it continued to keep time; albeit the rusting of its iron and steel parts soon streaked and stained the golden face, rim, and back, and made of it a mirth-provoking thing.

Once while the entire party — men, dogs, boats, sleds, and equipment — were crossing an open lead on a great ice-raft, the rope on which we were hauling (it being fastened at both sides of the lead) parted with considerable force, and one end struck Lee a smart blow on the ribs, stretching him out, and at the same time smashing the crystal of the watch. That evening Lee came to me and reported the damage, saying that both hands had also been detached or broken. I laughed and told him to throw the watch away, it was not worth carrying longer. But no, he said, it was a great pleasure to the men to be told the time of day; and so Sweetman, the carpenter, made a pair of wooden cases for it, like a pair of clam-shells, and Lee with his sheath-knife cut a tin hand and drove it down on the hour spindle, and all was well again. By this arrangement, when the tin hand pointed at twelve it was either noon or midnight; when a quarter of the way between twelve and one, it was a quarter past twelve; half way, half-past, and so on; for a minute hand is a frivolous luxury in the Arctic Circle.

And so the old watch ticked on through many a ducking, since it seemed that Lee was amphibious, going overboard as if "to the manner born." But at length the time arrived when we were all told off to the boats, Lee seeking his fate in the first cutter, and as there was no watch in the whale-boat I was glad to come in possession of my own again, and placed it under the care of Mr. Danenhower, for it would have been very inconvenient for me to attend to it while holding the sheet and sailing the boat with my cracked and swollen hands. One day I noticed Danenhower winding the watch several times. I inquired the trouble, and he said he could not understand it; that he might keep turning the key all day without fully winding it up, and yet the watch ticked on. In short, the mainspring was not broken, but had partially slipped on its spindle, still retaining enough power to propel the works for about four hours. So it was wound every third hour until we reached Jamaveloch, when we hung it up in the hut for our common benefit, and then some one must needs step upon and mash the wooden cases.

When I arrived at Verkeransk one of the political exiles, the " Little Blacksmith," soldered a brass plate in the crystal rim, and as he had no watch-hand the old tin one of Lee's make continued to do duty, and the old watch still ran on. In Yakutsk Bartlett came upon a watchmaker who tampered with it, and was pleased to ask if I wished a second-hand put on. I did not; hours and minutes subdivided Siberian time quite finely enough for me; but on our journey to the Delta I discovered that the faithful old ticker did not tick as well as it did before the exile had it to repair, and upon opening the case I found that one of the jewels was gone. The little rogue had stolen the stone and replaced it with a piece of brass on which the friction was so great that thenceforth I had to ease or compress one of the screws in order to regulate the running of the watch.

And now at Irkutsk the old gentleman watchmaker informed me that the second-hand movement had been purloined by his fellow-laborer in Yakutsk, who doubt-less had as good use for it as had the young exile for my jewel, and I at once understood why he was so particular in inquiring if I would have the second-hand renewed. However, I paid Mr. Júrgensen nine roubles for the kindly interest he displayed in its welfare; but although it had been able to withstand the rigors of Arctic travel, its undermined strength succumbed to the treatment of the Siberian artisan, and I despaired of its future use-fulness.

Finally, arriving at Philadelphia, I laid it away as a relic, but a certain sympathetic friend decided that it should be cleaned and put in order once again. Now it is fair to look upon; the rust stains have departed its poor old face; and as I write these words in the ward-room of the steamer Thetis, it is at sea once more, bound on another Arctic voyage, and within its case I have just discovered the inscription, " *Tobias*, No. 121305; Liver-pool." What, I would like to know, has been the fate of 121304? or of 121306? And I wonder if old Tobias himself has worn as well as his watch; and I sincerely trust, at least, that his inside works have not been so ruthlessly doctored or deranged.

CHAPTER XXVIII.

HOMEWARD BOUND.

Governor Anutchin. — On to Tomsk. — A Current Ferry-Boat. — Agricultural Settlements. — Rascally Stancia-Keepers and Yamshicks. — Their Methods. — Exiles in Droves. — At Tomsk. — The "Hotel Million." — Attentive Mr. Hildenberger. — On the River Obi. — Tobolsk. — Tuamen. — Floating Jails. — Ekaterinborg. — Perm. — Nijni Novgorod and the Great Bazaar. — Moscow. — St. Petersburg. — Our Reception. — A Day at Peterhof. — Home.

GENERAL ANUTCHIN, Governor-General of Irkutsk, was absent on a visit to St. Petersburg, but he was daily expected to return, and when he did, a general holiday was proclaimed, the entire populace turned out to welcome him, and there was a fine display of fireworks in the evening. Lieutenant Berry and myself called upon him, and also paid our respects to his wife and daughter. The following day one of his aids visited us, and arranged for a dinner at the gubernatorial mansion, at which there were present Berry, Jackson, Larsen, Gilder, and myself. The governor and his daughter spoke excellent English, the latter being dressed at dinner in the national costume — a short-waisted and short-skirted gown of white linen, embroidered and inserted with blue and red; her head crowned with a gilt tiara, and her hair flowing loosely down her back. She was very beautiful, and looked every inch a queen.

As soon as possible, I prevailed upon the governor to assist me in securing a quick and safe passage through his territory towards Russia. He provided me with an open letter and a doubly-stamped road passport that directed all of his minions to give me the right of way on the post-road, saving only the mails, which was a loophole quite large enough to permit any rascally road-master to withhold his horses for a valuable consideration.

Lieutenant Berry and myself, traveling together, purchased a tarantass for our own use, and hired those at the stancias for the accommodation of the two men who still remained with me. I also hired a tallega to convey our baggage and the two boxes of relics, and purchased mattresses and leather pillows for the bottoms of our wagons; for we slept in them, riding night and day, and only halting at the stations long enough to change our horses and make an occasional meal of tea, milk, boiled eggs, or such other simple food as we could buy.

Here at Irkutsk I paid off and parted from my faithful interpreter and companion, Captain Joachim Grönbeck, who accepted an appointment under Alexander Silenikoff to explore the rapids of the Yenisei River from Irkutsk to the Arctic Ocean; thence to find and blaze a road from the Yenisei to Archangel; and thence to proceed to Sweden, his native land.

On July 14th, wishing our numerous pleasant acquaintances in Irkutsk farewell, we set forth on a 1,500 verst (1,000 mile) wagon ride to Tomsk. The first river to the westward is the Yenisei, which we crossed on a current ferry-boat, capable of holding six teams and one hundred passengers, moored up-stream by an anchor and grass hawser. This hawser was five hundred yards in length, and its weight was carried on a series of eight or ten small flat-boats or scows. A large square frame was raised about the centre of the boat, extending its whole beam, and perhaps one third of its length, and the hawser traversed the forward part of this frame-work, which was greased to facilitate slipping. With everything in order for a start, a man on the upper platform

above the passengers' heads began walking with a huge
tiller, and the boat moved slowly out into the stream.
Presently the hawser slipped over the forward end of the
frame-work, and then the boat shot rapidly across the
river, and it became necessary to exercise considerable
care and attention, as we approached the opposite shore,
to prevent a too sudden landing alongside of the temporary pier and staging. We were thus ferried over many
rivers of all sizes, between Irkutsk and Tomsk.

The country through which we passed was exceedingly
beautiful, — rolling, well-watered, and wooded. Splendid
crops of rye and some wheat and oats gave the landscape a cultivated look, and everywhere there was an
abundance of cattle and horses. The villages along the
road were not far apart, and contained as many as one
thousand inhabitants. They are all agricultural settlements conducted upon a partial commune system, under
the governing rules of which each member must take a
section of the soil, till it, and pay its taxes, and no person
can hold the same tract two years in succession without
paying for the privilege. It is very strange, at first, to
see a hundred acres of land planted in fifty long distinct
strips of rye, oats, or wheat, by as many husbandmen,
each one of whom the following season or year may rent
and plant a different strip. The cattle are grazed in a
body or herd by an attendant, who keeps them away
from the crops and watches the gate to the village.
Sheep are plentiful, but swine by no means as numerous
as I had expected they would be in a country whose people are so fond of pork.

The distance between relays is from sixteen to thirty
versts, and the speed of fresh horses is about ten versts
an hour. But then the changing operation at the stations consumes anywhere from forty minutes to two
hours, according to the temper and trickishness of the
station-master and grooms, for they are beyond doubt

the meanest lot of beggars and knaves on the face of the
earth. The station-master can never change bills for the
traveler, and so he secures the odd kopecks; but he did
not rob me of mine, for I armed myself with a bagful of
fractional silver and copper currency, greatly to his disgust. If you are in a hurry, he has no horses; but, as I
have stated before, the dishonest fellow can always induce "Ivan," his fictitious friend, to supply you with
plenty if you are fool enough to rent them at two or
three prices. Then, too, the drivers openly pass the word
along from station to station : —

"These people pay twenty kopecks gratuity to their
yamshicks if they are driven rapidly." Or, — "Fifteen
kopecks are all they pay; they are in no hurry. Their
wagon was greased at four o'clock this morning; make
them grease it again and get your money. They are
Americans, and don't drink tea or want the *samovar;*
but if you make good time you will get twenty kopecks!"

Consequently a ten-kopeck gratuity means a walk for
the horses; and five kopecks, a succession of long rests
while the yamshicks smoke and curse.

We passed many exiles, men, women, and children, in
companies of from two to five hundred, marching wearily towards far Siberia. They were mostly of the criminal classes, all their heads being half or clean-shaven.
A majority of the men were in chains, and many were
linked together. Not a few of the women marched
among the men as prisoners, while the rest trudged on
into voluntary exile, holding the hands of their husbands,
brothers, lovers, or children. Many of the sick, aged,
and young were in wagons, but all the others toiled
along the dusty road like droves of cattle, under the
vigilance of a guard of from ten to a dozen Cossacks,
mounted or on foot, and in charge of an officer usually
taking his ease in a carriage. These were distressing

sights. Once we met a family of Jews, husband, wife, and two children, in a wagon, with a soldier, his gun and bayonet fixed, riding alongside ; and we stopped to change horses with the party, as we were then about midway between stations. The father, a bright, intelligent fellow, addressed us in German, and said he had been wealthy and was exiled to the Yenisei country simply because he was a Jew. His eyes brightened with delight when he heard that we were Americans, and the next instant clouded with regret at the bitter consciousness of his captivity. Four thousand of his townsmen, he said, had emigrated to America, and then pointing to his wife and two pretty children, the tears rolled down his cheeks as he faltered out, — " Siberie." Poor fellow, that word has all the import of a hell to many, many more than him. We should have reached Tomsk before noon, but at next to the last station from it there were no horses, and I paid double rates (five roubles for fifteen versts) in the hope of procuring horses at regular rates at the last station ; for there is generally a good supply of them near the large settlements. But the wretches of yam-shicks sent word ahead that we were in a hurry, and as a matter of course the station-master announced a dearth of horses. Then his accomplices asked ten roubles for twenty-nine versts, or an advance of about 3.6 times upon the usual rates. I offered double fare (8 r. 20 k.), but they would not accept, feeling confident that I must finally accede to their exorbitant demands. So I sat down, risking the chance of losing our steamer, which was advertised to sail the following day; and I calmly waited until the post-horses came in and had their hour's rest. The yamshicks who had refused to carry us for eight roubles and twenty kopecks then suggested that they would accommodate us for six roubles, but I was deaf to their blarney, for I had beaten them at last, though at a loss of three hours.

26

At Tomsk we patronized the "Hotel Million Siberie," which internally was the most horrible building I have ever been in. Its corridors were long and dark, and its square cell-like doors so low that I had to stoop to enter them, and on the outside their appearance was rendered the more forbidding by large black iron padlocks and hasps. The proprietor wandered along with an enormous bunch of keys, opening the doors and exhibiting his apartments, and at first I actually believed that the yamshick had misunderstood my order and taken us to visit a Siberian prison, instead of conducting us to a hotel. Upstairs, however, the rooms were much better. Each one was furnished with a bedstead, two chairs, and a chest of drawers, but there were no washing arrangements whatever, or mattresses or bedding; for every traveler in Siberia is expected to carry his own pillows and bedclothes. Here we put up, at any rate, dining at the restaurant, which perhaps was even more repulsive than the rooms.

We called at the telegraph station, and found four telegrams awaiting us, two for Berry and two for myself. Shortly afterwards a Mr. Hildenberger, who was in the employ of the telegraph company as English interpreter, called upon us and proffered his services. He had been a prisoner of war at the Crimea, and had been sent to England, where some charitable ladies and gentlemen interested themselves in his behalf, taught him English, and converted him to the Episcopal faith. He then returned to Russia as a missionary, but is more proficient in his natural character of a remorseless rogue. We foolishly gave him our tarantass to sell for us. It cost one hundred and seventy roubles, and we could readily have sold it; but he persuaded us to leave it with him to dispose of, which I have no doubt he did, though we have never since heard of him or it. Mr. Larsen, of the other party, overtook us at Tomsk, and learning into whose hands we had fallen warned us of our danger, but too late ; we had lost our tarantass.

The Governor of Tomsk received us very kindly, extending every civility, and he invited us to visit the University, which is the pride of the place. We also paid our *devoirs* to the mayor of the city, a fat and jolly old merchant, who treated us handsomely, and repeatedly expressed his regrets that we were to leave his town so soon. In the afternoon we were visited by a couple of gentlemen who spoke English: a Mr. Kuhn, of German extraction, and Mr. De Norpe, a mining engineer and geologist in the employ of the state. He was very clever and well-informed, knowing his geology by heart, and had somewhere met Professor Dana, to whom he sent many kindly messages.

Paying our bills at the "Hotel Million" we repaired for supper to the "Hotel Europe," and found the table garnished in true Siberian style with a single beefsteak — simply this and nothing more. Boarding the steamer then about midnight, we sought our bunks amid a frightful din, for the other passengers were just assaulting their evening meal. Next morning, the 27th of July, we paddled away on the river Obi. The steamer was quite roomy for this region, and laden with people journeying to the annual fair at Nijni Novgorod. Their habits, especially at table, were very disgusting, which was the more unfortunate, inasmuch as the cuisine and service were capital, meals being served *à la carte*. All the Russian passengers were plainly out on a protracted lark. They drank and played cards incessantly, and there was quite an array of gamblers on board, who fleeced the excursionists without mercy. We had scarcely gotten under way when I, too, discovered the loss of fifty dollars in small silver money, of which I had doubtless been relieved at the "Hotel Million."

We arrived at the ancient Cossack town of Tobolsk at

midnight of July 31st. It is one of the oldest fortified cities in the empire, and way back in the history of the czars was taken and retaken again and again by Cossack and Tartar. Long ramps and avenues lead up to the antiquated fortifications, which, pitched upon the hill-tops, frown down upon the modern settlement, and look for all the world like old Moorish towers and forts. The town was ablaze with light when we drew up at the landing, and venders of fruits and confections thronged the main thoroughfare leading to the steamer. We enjoyed a carriage ride by moonlight, which enhanced the singular charms of the quaint old place, and I was sorry to leave it so soon.

Continuing up the Obi until its waters shoaled too rapidly for our large steamer, we were transferred to a very small boat, on which each person appeared to understand that no one else had rights that he was bound to respect; and, although first-class cabin passengers, we were hoisted on board and simply directed to take care of ourselves — an admonition that we proceeded to obey, but with indifferent success.

We arrived at Tuamen before midnight of August 2d, and being unable to secure hotel accommodations drove directly to the coach-station, and from there launched forth on a journey of 450 versts to Ekaterinborg, the next town. Here at Tuamen I obtained a closer view of the double-decked barges, great numbers of which we had seen being towed in the wake of steamers by means of long hawsers. They are built in modern shape, with overhanging guards supported by struts like the guards of our side-wheel steamers, and are from two hundred and fifty to three hundred feet long, with two decks and a lower hold; and along two thirds of the barge's length there is an iron cage reaching from the lower to the upper deck-cover, and having the appearance of a great two story tiger's cage. A passage-way surrounds it on the lower deck, but none is necessary above. In these enormous floating jails are transported the thousands of exiles *en route* to Siberia. Each deck, I should suppose, is capable of accommodating from two hundred and fifty to three hundred and fifty persons; or the capacity of each barge is from five to seven hundred. And I saw ten such in use, four of them crowded with prisoners. We met three exiles, one young lady and two young men, who had been released and were homeward bound from far Siberia. One of them spoke English fairly well, but was rather reticent.

At Tuamen we were most agreeably entertained by an American dentist, Dr. Ledyard, of San Francisco, and wife; and received a pleasant call from a Mr. Waldraper, one of three brothers forming a steamboat construction company. He was a young Scotchman plying his Clyde-gained knowledge on the Obi; sharp as a Yankee, and equally full of aggressive energy and ambition. He talked of going east as far as the Yenisei, and had many pertinent questions to ask concerning the navigation of the Lena.

We were three days, of twenty-four hours each, in making the journey from Tuamen to Ekaterinborg, where we arrived on the morning of August 5th, and established ourselves at the "Hotel Europeanski," a very creditable inn considering the country. The city was founded by Catherine of Russia, after whom it is named, and some portions of it, notably the Public Gardens, are quite attractive. We dined with Dr. Ledyard, and then left by rail for Perm, crossing the Ural Mountains into Europe; for here is the beginning or rather terminus of railroading, although there is a gap of several hundred miles in the line between Perm and Nijni Novgorod, which distance is covered by steamboat. At the railroad station, the first thing that claimed my attention, outside of the usual bustle, was another collection of double-decked cages for the transportation of exiles, for it is part of the business of the railroad to furnish a sufficient number of properly constructed cars for this purpose. We felt, indeed, at sound of the familiar toot and snort of the iron horse that we were at last crossing the confines of civilization, for surely no one like the American can appreciate the wonderful benefits conferred upon mankind by the swift "smoke-wagon," that annihilator of distance and prime agent in the manifold glory of our times.

We reached Perm August 9th and boarded the steamer without delay. It was larger, and far more comfortable in every respect, than any we had as yet seen, for, true to the law of progress, everything was materially improving as we journeyed westward. The river-front at Perm was filled with steamers, and the double-decked barges, no longer a novelty to us, were transferring their captives to the double-decked cars. Poor wretches! they looked so like wild animals, back of their iron bars.

We were now on a branch of the mighty river Volga, and in a few days would reach its forks, where is built Nijni Novgorod, the "New City." The site of the ancient town, whose foundation dates back into the thirteenth century, is about eight miles below. There plies on the Volga a fraudulently-called American line of steamboats, with such names as Washington, Wisconsin, etc., for even here America has the well-deserved reputation of possessing the finest steamboats in the world; and I am confident that if a number of our river boats, so superior to all others in speed and comfort, were placed upon the Volga, they would at once absorb all of the trade, for the Russian is fond of good things and in nowise averse to paying for them.

We arrived at Nijni Novgorod about twelve o'clock in the night of August 12th, and went to the Hotel Europe. On the morrow and the day after we visited the wonderful bazaar which has made this place so famous.

Here can be seen the representatives of all peoples; and every marketable article seemed present in profusion. There were furs from the vast districts of Northeastern Siberia, and furs from Northwestern America; rugs from Persia; ostrich feathers from Africa; tea and carved ivory from China; diamonds from Brazil; cutlery from England and Germany, and some very excellent samples from the United States. The bazaar is held on a point of land between the forks of the river, and is joined to the city by a well-built pontoon-bridge. The grounds are constantly crowded, and a perfect babel of tongues prevails. The good-natured merchant, who talks a composite language, calls loudly to the passersby, exhibiting his goods, and if he fails to make himself understood he shouts out " roubles," and states his price either by numerals or his abacus. That is, the price he *asks;* what his selling price might be, Heaven only knows, for I am sure the merchant himself does not; at any rate not before the end of the season, when articles are often sold to the highest bidders.

Among the attractions at Nijni was the American lion-tamer, Colonel Boone, with his cages full of wild beasts, and it was evident that the admiring natives regarded the colonel himself as the greatest lion in his collection. We visited and dined with a Mr. Dunbar, formerly of Pittsburg, Pa., who had come to Nijni and built a very creditable stern-wheel steamboat, modeled upon those in use on the Ohio and Monongahela rivers. She was apparently a favorite with the people, but lacked speed, as essential a quality in the eye of the Russian as in that of the restless American. I have never learned the result of Mr. Dunbar's venture, but I am persuaded that for the navigation of the large Siberian rivers there is needed a suitable number of light-draught stern-wheel boats. In summer the water is low, the river-beds being so wide that the water of necessity must be shallow, albeit the channels are not so capacious that they can control the fury of the spring floods.

It was a very pleasing and instructive time I spent at Nijni Novgorod and the bazaar. Often in boyhood had I gazed at my school atlas, struggling to pronounce it and the other impossible names, and wondered in my dreams if I would ever see Nijni or Moscow. I had now seen the one and in twelve hours would see the other.

Nijni Novgorod is the eastern terminus of this railroad line through the empire of Russia, and when the short section between it and Perm is built there will be an all-rail communication from Ekaterinborg, just within the boundary of Siberia, to the rest of the continental railroads centring in St. Petersburg, Berlin, Paris, Vienna, and Rome; and what a vast combination there will be upon the completion of the proposed English roads to farther India and Afghanistan!

We entered Moscow on the 15th of August, and were met at the depot by the American consul, who gave us a hearty reception and conducted us in carriages to the Hotel Dessaux. Many persons of note called upon us, and our indefatigable consul kindly drove us to different points of interest in the city.

The following day we " did " the Kremlin, the great bells, and, by special permission, the new cathedral, whose interior decoration is superb. Among other curiosities I saw in Moscow was a jet black female native of Demerara, driving around in a gorgeous open barouche, with all the airs of an old-time Russian princess.

On the evening of the 17th, we left Moscow for St. Petersburg, and were greeted at the railroad station by Colonel Wickham Hoffman, United States *chargé d'affaires*, and quite a delegation of Americans. Driving to the Hotel Europe we rested for a while, and then called upon Minister Hunt, at the quarters of the United States Legation. He had received his appointment but a short time before, having previously occupied the office of Secretary of the Navy. Now he had prepared for us a capital cold collation, and all the distinguished American residents of St. Petersburg were present, and truly a heartier welcome was never accorded a band of shipwrecked mariners. The old and young, rich and poor, of our countrymen flocked around, and generously, warmly congratulated us on what they seemed to feel was a resurrection from worse than death.

Driving along the Neva we saw the bridges which span that majestic stream, and the fortress of Peter and Paul (Petropavlosk), the cold citadel which has witnessed a world of misery and crime. From the river we were afforded a splendid view of the city; and at night we visited the great summer garden and saw the populace. The grounds are magnificent, and at certain seasons of the year are thrown open by royal command for the enjoyment of the people. Here, as in Moscow, the military largely predominated; uniforms were everywhere, and a military band played martial music, and the grand march in honor of Skobeleff, the dead and favorite Russian general, which was encored again and again.

The morrow we devoted to sight-seeing at the Hermitage, gazing upon the relics of Peter the Great, so often described by tourists; his staff, tools, arm-chair, and the rod that measured his height; and here, too, are the celebrated marbles, jewels, and beautiful art-galleries. But I most admired the colossal nudes in black marble which support the portico over the entrance to the Hermitage. They are very striking in their massive proportions, and look like so many living giants bearing sturdily up with straightened limbs under the heavy stone

entablature which rests upon their black and brawny
shoulders. A trip to the Cathedral of St. Isaac's, the
grandeur of whose interior surpassed my expectations,
and we then repaired to the palace of the Emperor
Paul, wherein the late Czar Alexander II. ate his last
breakfast; and finally we visited the spot where he was
killed, and the temporary shrine erected over it, which,
we were told, was shortly to be replaced by a chapel.
That evening (August 20th) Minister Hunt and lady
entertained us at dinner.

During the day we had been waited upon by an aid,
a colonel of engineers, who delivered invitations to myself and two seamen of the Jeannette and Lieutenant
Berry of the Rodgers to be presented to their majesties
the Czar and Czarina, at Peterhof, an imperial summer
residence about sixteen miles out of the city. An usher
in civilian clothes next made his appearance, and directed
us how to proceed. Our dress, if not uniform, should be
full evening dress, with white ties; the carriages at St.
Petersburg were to be of our own providing; but a special car on the train leaving at eleven A. M. would be
placed at our disposal, and the royal carriages would
await us at Peterhof.

Promptly at the appointed time we were at the depot,
where our usher met and conducted us to the proper railway coach. Minister Hunt was to be presented at court
the same day, but he was under a different escort. We
were soon whirled to Peterhof, and alighted from the
train in company with a numerous party of officers, diplomats, and court officials, all under separate and appropriate guards. Here we were shown into an open barouche bearing the imperial arms, and attended by coach-
men in gold-laced liveries, with cocked hats, cutaway
coats, and buff waistcoats, and were driven to the royal
gardens and dormitories, where there was a series of with-
drawing rooms with small breakfast rooms attached. We

were first seated in one of these, and presently an officer in uniform followed by an amanuensis entered, and politely saluting us in English opened a running conversation, which continued for a few minutes, when the amanuensis wrote upon large and separate sheets of paper the name, rank, and nativity of each of our party. Both then withdrew.

The next formality was decidedly less irksome. We were conducted into the breakfast room and regaled with a light repast, which included tea, coffee, and wine ; cognac, *pousse-café*, and *liqueurs ;* cigarettes and cigars. In a little while another officer, in uniform and under arms, greeted us and requested that we follow him. Entering a carriage we were driven to the audience chamber and ushered into a large ante-room hung with portraits of the royal family, battle scenes, etc. Here there was a brilliant gathering of officers and functionaries of high rank, generals, admirals, ministers, and diplomats, all in gorgeous uniform, and glittering with stars and decorations, humbly awaiting their turn for a brief chance at the imperial ear. Our conductor proclaimed to the assemblage who we were, and for a moment every eye was curiously riveted upon us, and some of the Russian officials came forward and spoke to us. At the same time our names were announced to the Czar. Minister Hunt had preceded us, and our attendant now said that we would be next received by the Czar in the reception room, after which we would be presented to the Czarina in an ante-room.

As soon, then, as Minister Hunt had departed we were conducted to a door opening into a passage-way which led directly to the audience chamber. This door was thrown open by an usher, who called out our names and vanished. We advanced a few paces two abreast, and as we did so, Czar Alexander III., Emperor of all the Russias, crossed the room with outstretched hands and greeted us, saying in English, —

"Good morning, gentlemen. These are Messrs. Melville and Berry ; which is Mr. Melville?"

I set him aright by introducing the others, and Lieutenant Berry performed the same service for me.

"Préférez vous parler en Français ou en Anglais, monsieur?" the Czar inquired.

I assured him that English being my native tongue I preferred to speak it. At this juncture the Czarina approached and graciously greeted us, appearing quite solicitous about our health, and asking many questions in regard to our sufferings. She carefully and with a kindly show of interest examined my hands and fingers, which still bore the marks of their old sores.

We then began a cross-fire of conversation, each of the royal pair conversing at times with one or another of our party. Finally the Czar expressed his regret that any of us should have come to grief on his territory, however remote ; and "I trust," said he, "that it was the rigor of our climate alone, and not the coldness of heart of any of my people, which caused the death of your comrades."

The Czarina commended our fortitude and courage, which, she said, were peculiarities of the American character ; "But I hope," she remarked, "that you will not again tempt fortune in the frozen North."

Speaking of our own land she observed, with a gentle sorrow in her tone, "I had hoped in my youth to visit America, but now, I fear, it can never be."

There coming a lull at length in our talk, we shook hands a second time, and with mutual farewells and my honest wish that the imperial couple might be blessed with "future peace and happiness," our audience was ended. It had lasted twenty minutes.

We passed out without the usher's aid into the anteroom, where the crowd of visitors was fast swelling. Our official conductor escorted us in a carriage back to the breakfast room, where he delivered us over to the tender mercies of the civilian usher; tender, since there was here spread out for our edification a delicious breakfast, whose tempting viands we had not the least inclination to slight. We were then driven for several hours through the grand gardens full of lakes and artificial cascades pouring, one over a silver, and another over a gold-plated wall of rock, and called the "Silver Falls," and "Golden Falls." There were fountains casting spray in every part of the grounds; curiously and beautifully trained plants; fish in ponds, which came at the sound of a bell to be fed; geese, swans, and other water-fowl — all forming the most magnificent artificial park I have ever seen.

We returned to St. Petersburg by boat, obtaining a fine view of the great work, then in progress, of making a harbor at Cronstadt, whose shipping was barely visible in the distance. Reaching the city towards dusk, we arranged at once to leave for Liverpool *via* Berlin and Paris. At each of these points we rested for a few days, receiving kindly tokens of interest on every hand; and finally setting sail from Liverpool on the steamer Parthia we arrived at New York on the 13th of September, 1882, — three years and six months from the time I left the Atlantic sea-board to join at San Francisco the luckless Jeannette; and one year from the day when our three boats were separated in that fatal gale.

THE GREELY RELIEF EXPEDITION.

CHAPTER I.

NORTHWARD ONCE MORE.

Failure of the Neptune, Proteus, and Yantic to Relieve Greely. —
My Proposal. — The Plan and Fitting Out of the Expedition. —
St. John's. — Disco Island.

[THE celerity with which the Greely Relief Expedition accomplished its noble mission surprised the world. As I have
elsewhere stated, when Chief Engineer Melville set sail from
New York on the Thetis, he was still engaged on the final
chapters of this work, which we imagined would be in print
some time before that cruise was ended. However, he has returned in time to annex the following account of his last Arctic
voyage, which, it is thought, forms an appropriate epilogue to
the tragic tale of the Jeannette. — ED.]

I only propose for myself a brief outline of the object
and results of the Lady Franklin Bay Expedition, for
the simple reason that the brave commander himself,
Lieutenant A. W. Greely, U. S. A., who still survives
in full vigor of mind and body, will be the best historian
of his own and the adventures of his heroic followers.

And I here desire to say parenthetically that there is
no one living competent to criticise Lieutenant Greely's
conduct of the Expedition, beyond affirming that he performed the greatest amount of scientific work possible at
the least expense, and made good his retreat from depot

1. An exploring party   2 A musk-ox.   3. Landing Lieutenant Greely's party from the Proteus at Lady Franklin Bay.

to depot, until he arrived at the point of safety where our government had promised to deposit supplies and have a vessel awaiting to carry him and his band away from the " Land of Desolation." How bountifully the government furnished the means for the execution of its promise, and contrariwise, how strangely, if not criminally, the government's efforts were thwarted by carelessness, incompetency, or inexperience, the reader knows too well to warrant further comment here.

In August of 1881, Lieutenant Greely's command was conveyed to Lady Franklin Bay by the steam-whaler Proteus, for the purpose of establishing one of the two[1] metereological stations to be fitted out and maintained by the United States Government, under the charge of the Signal Service, in accordance with the agreement of the international meteorological congress. America, with commendable zeal, dispatched her observers to the most northern point attainable with safety and readily accessible to support and relief. I assert this deliberately, notwithstanding the sad fate that overtook a portion of Greely's party ; for he executed his duty with all honor, and can in no way be held responsible for the terrible disaster that resulted exclusively from the miscarriage of the promised relief.

The unsuccessful attempt of the Neptune to reach Fort Conger or Lady Franklin Bay in 1882, and the return of that vessel to the United States with all her supplies on board, which should have been cached as near Fort Conger as possible, —particularly at Norman Lockyer Island, the highest point attained by the Neptune, at Cape Albert, Cape Sabine, the death camp of Greely's command, or at Littleton Island, where Greely requested

---

[1] The other American station, under command of Lieutenant Ray, U. S. A., was established at Point Barrow, north of Behring Strait, and the European powers were to station their observers at optional points within the Arctic Circle.

the depot of supplies to be made, — was followed in 1883 by the double failure of the Proteus and the U. S. steamer Yantic to leave provisions at Cape Sabine and Littleton Island.   In other words, these three vessels succeeded in transporting to and *beyond* the point of disaster sufficient supplies of food to last the Greely party two years or more — and yet, singular to say, either sunk this food in the sea, or brought it back to the United States.

But here it may be asked, — Of what avail would provisions deposited at Littleton Island have been to Greely encamped across the Sound, at Cape Sabine?   I can confidently reply, — It would have been the salvation of the whole band.   Not a single man need have starved to death; for if Greely had known that an abundance of food awaited him at Littleton Island, he would certainly have made greater efforts to cross the channel, and in all probability would have met with success, even in the one boat which remained to him.   Or ten men of his party might thus have escaped, and the other fifteen at Cape Sabine could then have subsisted on that amount of food which proved so sadly insufficient for twenty-five.   As it was, however, he might just as well have starved at Cape Sabine as at Littleton Island, where there was a cache of only 240 rations — nine days' supply for twenty-five men!   Still there are reasons why his chances would have been much better at Littleton Island; chiefly because there is game the year round all along the coast from Cape York, scarce, it is true, in winter time, but quite plentiful in early spring.   Moreover, natives are well known to be settled at Life-Boat Cove, Littleton Island (not permanently), Port Foulke, Cape Parry, Saunders Island, North Star Bay, and Cape York, and from them Greely and his party would surely have received aid, and with their superior weapons could have returned valuable assistance in hunting.   Had supplies

been left at Littleton Island by any of the ships mentioned, Lieutenant Garlington and his men could have
remained there after the loss of the Proteus, and, as
Greely directed, have "kept their glasses bearing on
Cape Sabine for his retreating column." Cape Sabine
and Littleton Island were unquestionably the key to the
situation, and had either been properly provisioned there
is no reason why any member of Greely's party should
have perished.

The futile attempt of the Proteus to reach Lady Franklin Bay, her destruction, and the return of Lieutenant
Garlington, were generally regarded as a death-blow to
all hope of forwarding succor to Greely, and the news
flashing across the continent appalled many who fore-
saw the terrible position in which the isolated little band
would then be placed. It was too late in the season
(September 12th) to fit out a new relief ship, yet one
resource remained at that time, namely, to send a vessel
at once from St. John's, N. F., to Cape Athol or Cape
York, which, after landing a rescue party with supplies,
tents, boats, and sledges, could return immediately before the ice began to make too rapidly. This plan I
embodied in a telegram to the Secretary of the Navy,
September 14th; following it up later on with a letter,
wherein I said, —

"Greely, without doubt, is now at Littleton Island, where
he expected to find stores and other means of relief. The
Yantic is at St. John's, N. F. Telegraph orders to Captain
Wilde to put his guns and extra weights on shore; reduce his
officers and crew to a minimum for safety in working the ship;
fill the ship with coal and stores, all she can carry; buy twelve
first-class whale-boats with outfits; put material on board ship
for the manufacture of boat-sleds, and material for clothing,
tents, and sleeping-bags. All these can be made up on the way
to Cape York. If the ship can be got to the northward of
Cape York, there will be no difficulty in communicating with

27

Greely this winter. Arrived at Cape York, or any point to the northward, land the stores and boats with a small party to guard them in addition to the party that is to advance to Greely's relief. If the ship can go beyond Cape York or Cape Athol, there are plenty of harbors to winter in. If not, work her well in shore and take the chances of wintering in the shore-ice, as far from the running pack as possible. . . . If the ship stands the ice during the winter, and the ice moves out in the spring time and carries her to sea, the chances are yet good to drift out into the southerly pack. Should the ship be crushed, the whale-boats are at the command of her people to make good their retreat.

"If landed at Cape York I will undertake to lead a party to Littleton Island to communicate with Greely, and if his men are able to travel, conduct them to the new base of supplies at Cape York, and encourage them to hold on. This is the point to which Buddington and party retreated after the loss of the Polaris, knowing they would sight the whale-ships about June 1st. After this date there is no difficulty or hardship in making the way to the Danish settlements. It is during the fall and winter time that the great risk is run, and now is the time that Greely needs succor and encouragement."

To this I received the following response by wire: —

"WASHINGTON, D. C., *Sept.* 19, 1883.

. . . "Careful consideration is being given to your letter of the seventeenth.

"W. E. CHANDLER,
"*Sec'y of the Navy.*"

But alas! a board to whom the matter was referred adjudged my scheme an impracticable one for several reasons, mainly the lateness of the season, albeit whalemen have been known to cruise as far north as Cape York so late as October 20th. Thus my project for relief was not accepted, though the effort could certainly have been made without difficulty or danger, it being simply a question of seamanship. The ice was then scattered or entirely driven out of the bay and what

little remains at that season of the year hugs the western
shore from Lancaster Strait to Cape Chudleigh, and
along the coast of Labrador. The autumn gales are terrific, it is true, and continuous, as I learned to my cost
in the fall that I cruised with Captain James Greer in
the Tigress, on the Polaris search. Then, however, the
weather was not so severe or the season so far advanced
as when the endeavor was made from Norway to rescue
the Norwegian fishermen who had been cast on the west
coast of Spitzbergen; and in this instance, although the
attempt was not crowned with success because of ice and
storm, yet the effort was not relinquished until December, and the relieving party did not return to Norway
until January.[1]

I can now only express regret that my proposition was
rejected, and I desire to further say that all I learned
during the cruise of the Thetis simply confirmed my
faith in the entire practicability of my plan of relief, —
and then even though I had failed to reach the starving
band, would not humanity have had the satisfaction, at
least, of knowing that everything possible had been done,
and defeat and disaster had only come when rescue was
impossible?

During the winter of 1883–84 the country was aroused
to the necessity of dispatching at the earliest possible
moment a safe fleet of vessels to accomplish what the
Neptune, Proteus, and Yantic had failed to do in 1882–
83. Congress meanwhile quarreled over the appropria-
tion and the manning of the ships, until the season was
so far advanced that it would have been next to impossible to equip them in time to save the few survivors at

---

[1] The poor fishermen perished miserably. They were established
in a comfortable house well stocked with provisions, which prin-
cipally consisted, however, of salt fish; so scurvy set in and they all
died. The attempt at rescue was none the less heroic, and reflects
the highest honor on the hardy Norsemen.

Camp Clay, who, it afterwards appeared, were slowly starving to death while Congress waxed warm with debate. But Secretary Chandler, with characteristic energy, realizing the necessity of immediate action, assumed the responsibility of purchasing for the government the ships Bear and Thetis, two of the best in the Scotch Arctic whaling fleet. These vessels he then had fitted for sea, with every appliance calculated to insure their safety and promote the comfort of the crews. Provisions and clothing were of the most approved quality and design, and the bureaux of the Navy Department vied with each other in their efforts to make the expedition a complete success. How admirably the plans of Secretary Chandler have been executed, and how well the confidence reposed by him and the country in the commanding officer and the *personnel* of the fleet has been justified, it is needless for me to indicate.

English courtesy, and the chivalry peculiar to that great nation, which has ever cherished the spirit of adventure, applauding and rewarding acts of heroism, and stretching forth her strong hand to rescue those who risk their lives in quest of fame or in the interest of science,—a national policy that has done much to develop the wonderful strength of her army and navy, — exhibited itself on this occasion in a very graceful act. The Queen tendered, as a gift to our government for service in the search, the Alert, formerly the flag-ship of Captain George Nares, R. N., in the English Polar Expedition of 1874. This is the strongest wooden ship afloat. She did excellent service throughout the voyage mentioned, resisting tremendous strains and nips, and at one time was cast bodily out on the surface of the ice without suffering material damage. The British Government put her in perfect repair at their dock-yard, so that upon her arrival in New York only a few trivial changes were required, when she was stored with provisions for her own and the

crews of the two advance ships, the Thetis and Bear.
Meanwhile the iron transport steamer, Loch Garry, of
Dundee, Scotland, had been chartered to carry 1,000 tons
of the best Welsh coal as far as Littleton Island, our
government becoming responsible for her damage or loss.

The general plan of the expedition was for the Thetis
and Bear to proceed as rapidly as possible toward Littleton Island, where it was expected a record of the Greely
party, if not the party itself, would be found. Indeed,
it was the opinion of a majority of the officers of the
fleet that we should come upon the party somewhere
between Cape York and Littleton Island, and there, we
subsequently learned, they certainly would have been
had they been able to cross Smith Sound. The Alert
and Loch Garry were to follow in our wake with all
speed consistent with safety, and land a house, coal, and
stores at Littleton Island; the Loch Garry to depart as
soon as she had deposited her cargo of coal, and the Alert
to tarry as late in September as the state of the ice
would permit.

Aware of the house and depot of supplies at Littleton
Island, the Thetis and Bear could push on boldly together, if necessary, as far as Lady Franklin Bay, or
until their progress was checked by an impenetrable bar-
rier of ice, when the advance was to be made by sledge
and boat to Fort Conger, or until definite information of
Greely's command was obtained; this advance party to
be supported from the rear by the combined strength of
the crews of the Thetis and Bear. The board convened
by the War and Navy Departments to adopt ways and
means of relief recommended the above general plan,
and, to my thinking, it was a perfect one. The details,
of course, were left to the discretion of the commander-
in-chief, Commander W. S. Schley, U. S. N., whose efficient performance of his duty needs no praise from the
pen of one of his subordinates. On one thing alone the

Secretary of the Navy insisted, as advised by the board, namely, that the fleet should sail from New York not later than the 1st of May, 1884; and had we departed sixty, aye, ninety days before, we could not have reached Cape York, the turning-point of the North Water, one hour sooner. As it was, the Bear was dispatched on April 20th to enable her to be in Baffin's Bay as early as any of the whale-ships. She came up with the foremost of the whalers at Disco, and thence advanced as far as the Browne Islands, when, observing the ice that made across Melville Bay as solidly as the unbroken continent of America, she returned in company with several of the whale-ships to Upernavik. The Thetis left New York and proceeded to sea on the day and hour appointed, May 1st, at 3.30 P. M.

The following is a list of her officers: Commander W. S. Schley, Commanding; Lieutenant Uriel Sebree, Executive Officer and Navigator; Lieutenants Emory Taunt and S. C. Lemly; Ensigns C. H. Harlow and W. J. Chambers (the latter for duty on board the Loch Garry, as government custodian); Chief Engineer G. W. Melville; Surgeon E. H. Green; and James W. Norman, ice-pilot. The crew, including the steward (Charley Tong Sing, of the Jeannette), cook, and engineer's force, numbered twenty-six men, one of whom, becoming exhausted, was discharged at St. John's, and another man shipped in his place.

Our passage from New York to St. John's was without noteworthy incident. A minor break in the moving part of the engine caused a delay of several hours; but, with a fair wind, the sails were set, and soon under steam and canvas we pushed evenly along, making an average and pleasant passage. The better we became acquainted with our ship the more we liked her staunchness and sea-going qualities; and although our impatience to advance led us at first to wish for a greater display of speed, yet we were afterwards satisfied that she was quite fast enough for safety in handling and economy of fuel.

On the 8th of May we sighted Cape Race and our first berg, the comments on the size and beauty of which from the "tender-foot" members of our mess were quite diverting. During the day we passed many hummocky bits and bergs between Cape Race and Cape Spear, running along in sight of the black rocky coast, with purple heather aglow on its distant hills, and patches of snow in its hollows and ravines like so many flocks of sheep. We exchanged signals with the light-house: "Adieu! God speed!" it answered; and then the snow and ice in the valleys and lochs seemed to struggle with the white cottages for possession of the little garden spots on shore. Apparently all the inhabitants in this region directly or indirectly fish their livelihood out of the sea, and the coast waters are dotted with the brown bark-tanned sails of the myriad little fishing craft. General hilarity and good-fellowship prevailed among our company, and each member was earnest in his endeavor to spin the best and tallest yarns. There was much better material in the officers' mess of the Thetis than there was in that of the Jeannette.

We arrived in the harbor of St. John's early on the morning of the 9th of May, and diligently set to work getting our dogs and a few articles of clothing on board. We were ready for sea on Sunday morning, but a whole gale of wind was blowing outside the harbor, and the vessels inside were dragging their anchors; so we waited until Monday forenoon, when, accompanied by our coal-vessel, the Loch Garry, we steamed off for Disco and what adventure lay before us. The sea continued very heavy for several days, and we made but poor headway. Towards the 16th the weather became fine, the temperature falling to about the freezing point, and producing quite a jovial state of feeling among the ship's company.

When in latitude 60°, well off the coast or in mid-channel, we met with large quantities of drift-ice and numerous bergs, but we kept steadily on our course straight through the loose pack, sighting many seals and walrus. Nothing unusual occurred to disturb or enliven the even progress of our passage ; though, to be sure, there were some in our party to whom the curious and beautiful berg-forms, the seals, northern birds, and all the other phenomena of high latitudes, were a fresh and constant source of delight.

May 22d we came up abreast of Disco Island. A heavy pack extended about eight miles off shore, and we here had presented our first opportunity of testing the ship's ability to bore her way through masses of broken but closely crowded ice. The wind had moved the pack off shore, leaving a lane of open water known as the "land lead" or "land water." We had no great trouble in forcing our way through the ice, but experienced some difficulty in finding the harbor, although directly off from it, because of the puzzling similarity in appearance of the high rocky headlands and promontories. The proper way to approach the harbor is from the westward, close in shore, where the water is deep and there are no sunken dangers, and where the beacon on a low point of land comes into plain view. As we drew near, a native pilot and four men came out to us in a whale-boat and piloted us into the outer harbor, where we made fast to the edge of the ice, the main harbor still being frozen over. The Loch Garry fastened alongside of us, and the work of filling up with coal was at once begun. The officers of the Thetis next called on the governor and inspector, the two Danish officials who look after the interests of the king, to whose private purse the incomes of the Greenland settlements accrue. Inspector Alfred Andersen and Governor Peter Petersen have each a wife from Denmark, both bright pleasant

ladies who have come into voluntary exile with their
husbands in the hope of deriving a pension from their
government that will secure to them a competency in
their old age

Here at Disco I met again my old shipmate Hans
Christian, the Esquimau dog-driver, who served with
Dr. Kane, Dr. Hayes, and Captain Hall, and drifted
about with the memorable floe party, returning with us
to Greenland in the Tigress. He also cruised with Sir
George Nares in the English Expedition of 1874. I
saw his children, now full-grown men and women, but
both Charley Polaris, who was born on board the Polaris,
and his mother, are dead. Hans, however, has another
little Charley, for he took unto himself a new wife, as all
savage or semi-savage men do, since women among them
are only slaves and drudges attending to all the personal
wants of their lords and masters. I am not prepared to
say that she is "as good as she is beautiful," yet truly if
she be no better than she looks, poor Hans is deserving
of universal compassion. Still I have no doubt but that
she is more useful than handsome, for their hut was kept
in comparative good order, and her teeth were worn
almost to the gums from chewing skins for boots and
clothing for the household and the market, as the Esquimaux carry on a small trade with the crews of the
visiting whalers. It was plain, though she had lost one
eye, that the other was a shrewd business one and could
single out a desirable purchaser.

I visited the native school-master, who keeps the vital
statistics. He told me there were 211 souls in the district of Lievely, an increase of one over the census of last
year, the population having remained stationary at exactly
210 for the past ten years. There had been five births
and five deaths during the year, and one arrival from
another district. From his records he gave me the following particulars of population in the North Greenland

settlements: Disco Fiord, 52; Upernavik, which includes Tassusisack, Proven, and the outlying villages, 730; Rittenbenck, 600; Egemende, 900; Jacobshaven, 800; and Amenack, 800.

I strolled all over the old ground that I had trod twelve years before with a party of jolly young companions. The place looked much the same; the huts and hovels were as squalid, the natives every whit as miserable, as they were then, and I cannot see wherein Christianity has ameliorated the condition of these poor people. Most of them speak a little English learned from the passing whalemen. The women make various kinds of small articles for sale; caps, slippers, miniature *kyacks*, tobacco pouches, etc., which are sold on board the ships. They can say " money," " half-pound," " pound," or " two pound," according to the value of the wares, and all day long the natives paddled around the Thetis selling or trading. Their goods are very poorly made, and I should think that if the government official really had their welfare at heart, he would see to it that they bestowed more skill on their work, subject to the inspection of the storekeeper, and would fix a schedule of prices for standard articles.

That evening the inspector had a cooper shop cleaned out, and a couple of native fiddlers furnished the music for a dance, in which the sailors and Esquimaux belles participated. The fun was prolonged far into the morning, no lights being needed, for at this season of the year the sun is continuously above the horizon.

CHAPTER II.

RACING IN THE ARCTIC WATERS.

Upernavik. — The Whalers. — Captain Walker's Story. — The Bear Aground. — Racing. — Nipped. — The Duck Islands. — Off Cape York. — Conical Rock.

ON the morning of the 24th, having coaled our ships, we set sail for Upernavik, attended by the Loch Garry. Heavy pack-ice lay along the shore and extended far to the westward. The inspector and governor declared that it would be impossible for us to get beyond Hare Island, in the mouth of Disco Fiord, the Bear having been baffled in her attempt to do so a week previous. Breasting our way through the broken ice about three feet in thickness, we succeeded in gaining open water, and ran pleasantly on until six P. M., when we came up with a solid front of ice lying even and unbroken to the north of us, and apparently reaching across Baffin's Bay. The weather was thick and lowering, and considerable sea was rolling in on the weather edge of the pack, rendering the situation decidedly unsafe for the Loch Garry, which, being little else than a great iron tank filled with coal, should she receive a nip, or should a sea drive her against the ice-edge, would most certainly and instantly become spoils for Davy Jones's locker. So lying off the pack all night, it was considered expedient next morning to send the Loch Garry back to Disco, there to abide the coming of an easterly gale which would drive the ice off shore and open up a northerly lead for her along the coast.

The Thetis was driven bravely into the pack and headed towards shore, in the hope of finding land water. On we went, bumping and staggering, but making fair headway; at times colliding with a sharp shock against great floe-pieces and utterly demoralizing those of the company who were inattentive to their environments. We forced our way well in towards Hare Island, and after ramming and butting for several hours in an attempt to break through a narrow neck of ice that impinged against the land, finally retired behind the island to await the action of the ice. While lying fast to a small berg, which was more convenient than to anchor, we descried two steamers approaching us through the Waigatt Strait. One proved to be the Loch Garry, and the other the steam-whaler Arctic, the most powerful of all the Arctic steam-fleet, and commanded by Captain Guy, the most intrepid and enterprising young sailor in northern waters. We thought he would certainly stop and speak us, but not so; he went booming along at the rate of about nine knots an hour, barely deigning to notice us, and, plunging wildly into the ice-neck that had brought us to a halt, rolled about from side to side, banging and pushing right through, until in an hour or so he had almost disappeared from view. We then followed as quickly as possible, but the ice setting in shore delayed us for a few minutes. At length, emulating the example of our whaling friend, we dashed boldly into the pack, leaving behind the Loch Garry and the Wolf, another of the whalers. The ice closed again and shut out our "coal-tank," but presently both vessels got through, the Loch Garry leading. She soon caught up with us, distancing the Wolf, but our good fortune was only of short duration, for running several miles into a blind lead we were compelled to retrace our course, and during this time the Wolf had forged far ahead.

Thus the chase continued throughout our passage from

Disco to Upernavik, where we arrived on the morning of May the 29th. The run was an exciting one, and we came in victorious over both the Arctic and Wolf. Here we found the Bear and three steam-whalers, the Triune, Polynia, and Nova Zembla, making now quite a fleet of us, seven vessels in all, with five more to be heard from. The Bear had been as far north as the Browne or Berry Islands, but was obliged to return by the state of the ice. Several whalers were still to the northward of Kingatook awaiting the opening of the pack.

Coaling the Bear from the Loch Garry, we all got under way and pushed on as far as Kingatook, the Loch Garry remaining behind until the arrival of the Alert, when the season would be so far advanced that there could be no difficulty in bringing her forward to Littleton Island. When we halted to procure a lot of seal meat for our dogs, the whalers Aurora, Cornwallis, and Narwhal were in sight. All the whalemen came on board to visit us and said they were going as far north. if necessary, as Littleton Island; for they all knew of the $25,000 reward offered by the United States for the recovery of the Greely party, and proposed making a desperate effort to win it if the ice was loose when they reached Cape York. Captain Guy, in particular, stoutly asserted his intention of securing the reward; so we were now actually entered in an ocean race with the odds very much against us. There were ten or twelve ships in the fleet, the Arctic having the advantage of speed, and all the whalemen the benefit of many years' experience; yet if we could manage to keep company with them into the North Water our chances would then improve.

Captain Walker, for forty years a whaling skipper, boarded the Thetis, and told us that late last fall he had met some Esquimaux near Cape York who lived on the coast to the eastward of the Etah Esquimaux. Among them was an old native well known to the captain, who at one time had presented him with a gun for some service rendered. Exhibiting this he said that he had been hunting reindeer and seal for a party of five white men from Lady Franklin Bay, who were in excellent health, as were likewise all of their companions up in the far North, save the Doctor, who had been shot, whether accidentally or not Captain Walker could not learn. Circumstances were such, the captain explained, that he could not go in quest of the white men, but the native was a thoroughly reliable old fellow, and it was not at all likely that he would lie to him. But we knew that savage tribes the world over are greatly given to exaggeration, and delight in entertaining their listeners with marvelous perversions of the truth.

We lay at Kingatook waiting for the ice to move, until June 1st, when it began to blow a half gale, and the Arctic fouled with the Bear, the Bear fouled with the Thetis, and we were all forced to get under way and stand around to the north of the island. Driving our ship nearly its whole length into the pack, we had just made fast to the edge of the floe, when we observed the entire whaling fleet spin away to the northward, the Arctic leading by several miles. We, of course, pushed off at once in hot pursuit, passing the Triune and Nova Zembla, and continuing on to the Berry Islands, where we espied the Arctic and the other vessels under the land, tied to the floe-edge. This was a good day's work and an exciting race, but we could go no further. Seeing a long lane of water in shore where the other ships lay, we kept on, trying to force our way toward it in company with the Bear. Presently a lead opened near the Bear, and Captain Schley hailed her to proceed. She shot ahead, and we followed at full speed not more than a hundred yards astern, when suddenly we saw her bring up standing on a shoal or rock. She seemed to leap a foot or two directly out of the sea, and then roll over on

either side. A few seconds of great excitement ensued.
It was a perilous position, and we all fully realized what
the consequences might be. "Back her!" "Stop her!"
"Port!" "Starboard!" was shouted; and then,—"Hard
a port," and "Go ahead!" and the Thetis swung swiftly
around in deep water just clear of the stern of the Bear,
which now lay fast and firm, heeled over to starboard.
From the rate at which we had been going, and the manner in which she rocked from side to side, we knew the
shock had been a very severe one, and had perhaps occasioned such damage that she could not proceed further on
the voyage. But the hawsers were quickly in play; she
soon swung around, and much to our delight and relief
settled again in the water. We then both steamed into a
little cove, where a cursory examination revealed no serious damage, although the forefoot and iron clamps had
been carried away and she was leaking badly. But this
amounted to little, since the water could easily be pumped
out and the broken and shattered timbers would soon
swell closer together. Thus a leak of a foot per hour, as
in the case of the Bear, will in a few days contract to a
couple of feet in twenty-four hours.

None of the numerous rocks or shoals along this coast
are marked on any chart; indeed no proper coast-survey
has ever been made. The whalemen—from experience,
that best of all instructors—are tolerably well acquainted
with the dangers; but as far as the mere thing of grounding their ships is concerned, they do not appear to mind it
in the least. Two of the whalers were leaking at the rate
of twenty-four inches per hour, or forty-eight feet in twenty-four hours! but the pumps were kept in constant operation, and the ships continued serenely on their voyage.
One old salt, a sturdy Scot, who had been whaling in
Arctic waters for many years, related an amusing experience. He once struck his vessel, and, in his own language, nearly tore her bottom out. She was leaking

badly, and the crew mutinied, expressing a natural wish to return to Scotland. He prevailed upon them, however, to continue the cruise, which resulted in a very successful catch; but of course as the ship was loaded with oil and bone, she sank deeper and leaked faster. Finally, when a full cargo was secured and the water was making very rapidly in the hold, the plucky old captain headed for home, and mustering his crew declared that —

"He would like to see old Scotland as well as any of them, but if they ever expected to get there and view her bonny blue hills again, they would have to pump and be d——d to them ! "

And this is the way in which the whaling industry is carried on. The best of ships are originally built for the trade, yet in a short time they become dilapidated, but still continue to do service and roll up wealth for their owners. Apropos of the accident to our consort the Bear, we had with us a brawny but not overly bright son of Erin, who toiled with the strength of a giant at the hawser, and was heard to grumble, — " This is a blank of a coast where there's nayther light nor baycon," — as if he expected the shores of this desolate region to be lighted and buoyed like our green section of God's country.

We were held captive by the ice off the Berry Islands for several days. All the whalers were equally inactive, save that a man with a spy-glass was never out of the crow's-nest, which is simply a large cask of about sixty gallons' capacity minus its head, and with a trap-door let in to its lower end, which is fastened to the fore or main royal pole as a look-out station. We all hunted a little for ducks and dovekies, which abound in these waters, and the whalemen took occasional climbs to the highest point of the islands to observe the movement of the ice.

At length, on the morning of the 3d of June, Mr. Nor-

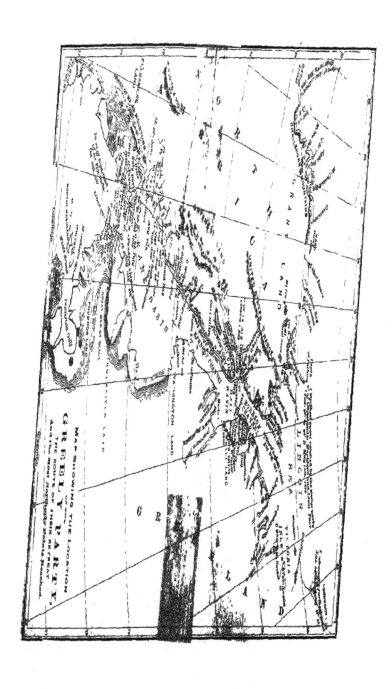

MAP SHOWING THE LOCATION
OF THE
GREELY PARTY,
THE ROUTE OF THEIR RETREAT,
And the Most NORTHERLY Points Reached.

man, the ice-pilot, reported five Esquimaux sleds with *kyacks* and hunting-gear in sight, coming from the westward and making toward one of the whale-ships. Receiving permission to take a whale-boat and volunteer crew, including Mr. Norman, I started across the open lead and then through the soft poshy pack to intercept the hunters, surmising that they might be an advance party from Greely. or natives with information of him. The ice in places was too soft to bear our weight, and as a steamship could scarcely force her way into it, there was no use trying to push our boat through. So we hauled out, and then away we went, breaking through at every pull. First, Mr. Norman sank in up to his arm-pits, then Hicky went in over his head, and Johnson in terror jumped for the boat, calling forth an awful anathema from Harvey; and thus we proceeded about three miles with great fun and labor. Meanwhile the whaler, observing our manœuvre, rammed into the ice toward the approaching Esquimaux and soon had them on board. We clambered up the side of the Narwhal and were kindly received by her master, who extended us the hospitality of his ship and an opportunity to wring out our dripping clothes before returning to the Thetis. As for the natives, they proved to be a hunting-party from Tassusisack, who visited the whalemen to barter some of their articles for bread, tobacco, etc.

Our ships now set sail again and made good progress along the coast in a west northwesterly direction, passing the village of Tassusisack, where several natives in canoes hailed us and pointed out a way into the harbor. But we kept on through a maze of small islands, high, bold, and bare, impressively picturesque in their bleak ice-bound silence. Our artist ensign, Mr. Harlow, here caught a number of excellent views with an instantaneous camera. At last our land lead failed us and we brought up against a solid wall of ice, held immovably

28

in place by the many little islands. We lay at the floe-edge all that night, and next morning the governor, attended by some natives, came on board the ship and urged us to return to Tassusisack, where we could procure oog-joog skins for boot-soles. We consented, at length, keeping the governor on board as pilot; and arriving safely in the little bay where the Tigress had run aground twelve years before, we moored fast to the ice and sought our evening meal. The whaling vessels had been left far behind, but while we ate, our lookout in the crow's-nest reported the whole fleet in sight, skipping along to the westward, with the Arctic, as usual, in the lead. So, regardless of supper, our lines were cast off and we steamed out of the harbor again, vastly to the astonishment of our friend the governor, who solemnly assured us that the ice would not break up for ten days to come. It was greatly a matter of chance, the selection of leads, but we felt very much chagrined that having gained ten or fifteen miles over the whalers, we should now see them fully as far in front of us. Still we pushed forward, and eventually had the satisfaction of passing some of them, and of approaching close to their crack ship, the Arctic. But alas' we chose a lead that seemed fair to look upon, and proceeded six or eight miles in it with a beautiful blue expanse of clear water still stretching beyond. Then taking many short cuts we brushed so perilously close to the bergs that more than once we were obliged to swing the boats in-board to prevent their being swept away from the ship's side. It was a grand race, not only on account of the novelty of the scene and the dangers that spiced our headlong rush, but had not our nation wagered the sum of $25,000 on these whalemen, and was not the honor of our navy at stake? We drove on in a narrow canal-like lead, which cut off at least ten miles of the course the Arctic had taken, and, when within five hundred yards or less of open water, made a

final dash at a small crack or opening in the canal.
With all her power and speed brought into play, the
Thetis rose to the crash like a blooded charger at a high
hurdle, but she was caught fast in the narrow jaws of the
ice as if in a vice, and rising several feet out of the water
she displayed her excellent ability to withstand a nip.
Yet there we lay firmly fixed, unable to advance or
retire. The Bear, astern of us, took our hawsers, both
steel and hemp, but they parted; we hooked two anchors
in the ice and worked the steam capstans, but all to no
purpose; and meanwhile most of the whale-fleet sped
tranquilly through the clear water within a few hundred
yards of us, and halted at the floe-edge not far beyond.
We had done so nobly and almost won; but only to learn
anew, and with painful emphasis, that there is indeed
many a slip 'twixt the cup and the lip.

We now tried to blast the ice in our immediate vicinity. It was about four feet thick, but rafted together in
two or three layers, making a total thickness of from
eight to twelve feet. A ridge was thrown across the
canal under our bows, and we were wedged tightly in
by two projecting points of ice beneath our fore-chains.
Albeit quiet and comfortable, this was by far the most
critical position we were placed in during the entire voyage. Happily for us the ice was still; for, had either
wind or tide set it in motion, the floes would certainly
have passed through the side of the ship like a sharp
knife through an old cheese. Charges of powder and
gun-cotton were exploded under the bows, but with no
effect. Then a single charge was tried on the port side,
as close to the ship as was prudent, since the ice bore
hard against her and the shock would be transmitted
with fearful force; yet this was ineffectual. Finally ten
charges of gun-cotton and gunpowder, five pounds in each
charge, or twenty-five pounds of gun-cotton and twenty-
five of common powder in all, were ranged about ten feet

apart and fifteen feet from the ship in a gentle circle
abaft the swell of the bows. The holes were bored in
the ice with three-and-a-half-inch augers made for the
purpose, and the charges, pushed under the ice and between the layers, were exploded simultaneously by means
of two batteries. The effect, so far as relieving the ship
was concerned, amounted to nothing. The shock was
terrific, but she neither budged nor settled an inch in her
bed, and the engines of both vessels were powerless to
move her. The explosions simply resulted in ten distinct holes, each six or seven feet in diameter. Our ex-
periments at Disco and off Hare Island had been equally
unsatisfactory, and I question if the Thetis could have
carried enough gun-cotton in her coal-bunkers to have
blown her through the ice as far as the coal contained in
the same bunkers could, with judicious handling, have
rammed her. Gunpowder and gun-cotton are useless in
exigencies such as this, and had the ice been in motion,
nipped as we were, all the gun-cotton in the world could
not have saved us. And this, I may say, was likewise
the opinion of our own torpedo officer on board the Jeannette, Lieutenant Chipp, who expressed it as his belief
that that vessel could not have held sufficient powder to
have released her from the bed in which she lay.

We persisted in our efforts until after midnight, and
then turned in, anxious to see what the morrow and the
returning tides had in store for us. The day had been
full of excitement, and now the night was a sleepless
one. Sebree, the executive officer, was out by five A. M.,
and I along with him. A survey of the situation showed
that the ice had backed off and the ship settled down to
her natural water bearings. It was the work of but a
few minutes to get out a couple of kedges instead of the
ice-anchors, which, though easier to handle, were not
sufficiently tenacious. Hawsers were next brought to
the steam-capstan and drawn taut to the point of snap-

ping; another was attached to the bows of the Bear, and we then pulled slowly out of the death-lock as smoothly as an old glove slips from the reader's hand. The ramming process was now continued, and we at last succeeded in forming the series of holes into a branch of the canal extending around the point of obstruction, and in several hours we had again caught up with the leading ship of the whale-fleet.

June 6th. A gloomy and memorable period in the history of the Greely party, for on this day three of their number died.

We had tied up to the edge of the ice, when towards noon the Arctic, ever vigilant and alert, began ramming at a tender part of the floe. When she had advanced about three hundred yards. with the Aurora immediately in her wake, we swung into line closely followed by the Bear, but before we could overtake them the floes came together, and we were once more forced to cut our way through, the Bear and Thetis striking alternate blows. The ice opened, at length, and then away we spun after the Arctic and Aurora, with the remainder of the fleet close behind us. I think that day was the most pleasurably exciting I ever passed in Arctic regions. Ten fine ships steaming along at full speed in a race for the relief of Greely; the ocean white with floe, hummock, and posh; while scattered over the sea of ice, as thickly as stars in the heavens, were countless bergs of all sizes, from the tiny, hummocky bits to the great islands, and of as multiform and fantastic shapes as the clouds floating in the blue vault above us.

Toward evening it grew cold, and rained and snowed; indeed, I should have observed before that we had experienced rain and snow and bad weather ever since we reached the Greenland coast. I went aloft, and could count our ten ships, all in plain view of each other, made fast to the floe, and waiting for a north or northeasterly

gale to drive the ice out of the bay, or so scatter it that
we might fight our way forward. To the westward the
Duck Islands were in sight, and to the eastward the
House's Head. We kept moving, but oh! so slowly; and
there was not a clear space of water visible. Arctic navigation, however, teaches one the necessity of patience,
and of instantaneous displays of energy. Thus, we were
suddenly called upon to exhibit our agility, for all at once
the ice opened and moved before us like a kaleidoscope,
and in an instant the whole fleet was steaming in the
direction of the Duck Islands, where we arrived before
midnight, and again fastened to a floe.

And there we lay about midway across Melville Bay,
still patiently abiding the tardy movements of the pack.
The whalemen who visited us did not expect the ice to
break up within a week, and then only under the influence of a lively gale, which at that season was scarcely
to be looked for. Ducks were not plentiful here, and the
combined fusillade maintained by the various hunting-
parties made them very wild; but we managed to secure
a few of the eider and king varieties, both of which are
fairly palatable. A month later these islands were easily
accessible, and it is singular to me that our yachtsmen
do not cruise to them for a season's shooting. Walrus
and seal were scarce, but at one time we saw a school of
white whales. From observations taken on the south
side of the islands, we found that they were in latitude
74° N., whereas they are erroneously charted about ten
miles to the southward.

On the morning of June 11th we again got under way.
For forty hours a fresh wind had been blowing from the
northeast, and the ice streaming down left plenty of
water to the north and westward. Accompanied by the
fleet, we worked out and around the floe, making but
little headway, in fact not advancing at all on our course;
but then the pack was disintegrating, and that was a

great satisfaction. The ice was slowly going to pieces, but there still remained large cakes of a square mile or more in extent. It was comparatively unbroken or telescoped, and looked as though it might be the previous winter's growth. There were two thicknesses of thirty inches each — quite a small formation ; but the ice we were in was a smooth coherent mass of many miles, seemingly the permanent floe of the coast-line turned loose by the last gale. Perhaps it was a second formation ; if so, we may thus account for the large quantities of ice driven out into the North Atlantic last winter and spring ; for if the floe remained intact, it should certainly, in that latitude and exposed position, have grown from six to eight feet thick.

Slowly we edged our way to the westward, until Wilcox Head appeared in sight, and had the weather been clear we could then have seen the Devil's Thumb, one of the most remarkable landmarks on the coast. But a dense fog or cloud hung across the land, screening from view all the mountain peaks and promontories. It was plain we would be late in crossing the bay, for last year Captain Phillips, of the Narwhal, traversed the North Water in Prince Regent Inlet by the 9th of June, and some of the whale-fleet were off Cape York as early as June 3d. The passage from Upernavik to Cape York has been made in thirty hours, but we felt that we should be lucky indeed if we could accomplish it in three weeks. The $25,000 reward was not the only incentive the whalemen had to reach the North Water as quickly as possible ; for a difference of three days will sometimes determine the success of a catch. As soon as disturbed, the whales become very wary and take to the ice, barely affording the alert whalemen a chance to strike them, or, if struck, they dive under the ice and escape. A majority of the whalers go annually as far north as Dalrymple Rock or Saunders Island, and they intended doing so now if they could not force a passage through the middle pack to Lancaster Sound; but Captains Guy, of the Arctic, and Fairweather, of the Aurora, announced their intention of proceeding directly to Littleton Island in search of Greely. According to the terms of the reward, if they could communicate even one hour ahead of us, the prize would be theirs.

On the 14th we again got under way, the Arctic and Wolf starting a good hour in advance of us. There was an abundance of open water in sight, and we worked steadily along the edge of the shore ice within respectable distance of the two leading whalers. In clear water we slowly overhauled them, and at one time were within a mile of them, but in passing through the leads they seemed to have the advantage of better judgment, and overcame their ice difficulties with much greater ease. We observed from the first that Captain Guy invariably drove his ship just where the ice broke under her bows. We now had a good lead in view, and if luck continued with us would doubtless make Cape York on the morrow. Guy in his powerful vessel still kept in the van, and it looked very much as though he would capture the reward. Beyond the excitement of our race, the days were uneventful. One of our company at this time journeyed forth on the floe to shoot a seal, and a bear following on his tracks was seen and shot by several men from the Arctic. Its skin was purchased by an officer of the Thetis, and a portion of the carcass was given to us for our dogs. Some of the men brought the liver in and fed it to the canines, a number of which, however, refused to partake of the dainty, although a little Esquimaux dog gorged himself upon it, and apparently without injury. Yet, strange to say, "Growler," one of our finest Labrador dogs, ate of the liver and died in convulsions.

On the 15th of June we brought up against the floe almost in sight of Cape York. I fitted an eight-foot Mel-

ville sled for a dash in shore should we be beset close to
the Cape. It was loaded with ten days' provisions for
four men and eight dogs, and weighed in all four hundred and fifty-seven pounds, the weight of the sled being
sixty-six pounds, as against that of a ten-foot McClintock, which tipped the beam at ninety-four   The outfit
consisted of one and a half pounds of pemmican per
man a day; half a pound of bread; one ounce of sugar;
one half ounce of tea; four sleeping-bags; an alcohol
stove; pots and pans; two gallons of alcohol; three gallons of water; one shot-gun and one rifle, with one hundred and fifty rounds of ammunition for each; half a
pound of pemmican per day for each dog; two paddles;
two boat-hooks; one heaving-line and small grapnel; a
rubber blanket (no tent); binocular glasses; a bunch of
rope-yarns, and some small stuff for extra lashings; eye-
goggles, etc.; and some provisions that I knew nothing
of. It was astonishing how far the guesses came from
the actual weight; no one except myself estimated more
than two hundred and fifty pounds for the total load;
and I draw attention to this incident merely to illustrate
how loads grow heavy by the addition of "just one more
article," which "does n't weigh much," in fact, "nothing at all;" but by the accumulation of which a light
flying sled-load becomes an unwieldy burden.

That afternoon we all moved onward, with the Wolf
in the lead, the Arctic and Aurora next in order, and the
Thetis and Bear following. We cut off the Aurora, and
then a race began, resulting at length in our defeat, the
Aurora forging half a mile ahead of us. Meanwhile the
mettlesome Bear had been snorting at our heels, impatient
to exhibit her speed and engage the whalers; so Captain
Schley now signaled her to advance. At the command
she shot past us and gradually crept up on the Aurora,
but it was only a spurt. She failed to overtake her, and
during the attempt we had so increased the speed of the

Thetis by gentle jockeying in the engine-room that the Bear did not exceed an advance of six thousand feet. Thus we raced until the morning of June 16th, when once more the pack closed up in front of us. The ice was moving rapidly and half a gale was blowing, forcing us to shift our anchorage to the edge of the floe. The Arctic was embayed, the first time we had seen Guy in an unfavorable position; but he had still a chance inshore to lead us in the final heat for Cape York, which was now plainly visible.

The pack opened on the 17th, and speedily we were all under way, the Aurora leading by fifteen minutes, and the Thetis, Wolf, and Bear following in the order named; but the Arctic, the pride of the whale-fleet, was soon left far in the rear, ice-locked. Her plucky captain, however, still rammed and butted at his prison walls like a madman in his cell, or a frantic hound in its leash, and before we reached Cape York he had fought his way through and into the van again. Early in the day's contest, the Aurora, with us in her wake, ran into a lead that closed upon her. Another lead showed to the southward, and we took this just as the Aurora passed safely through followed by the Wolf. Our lead looked narrow, and Schley slowed the Thetis. She stuck fast, and in backing out we broke our rudder against the floe. There was danger in proceeding, but it was time enough to halt when we could not help it; so on we plunged after the Aurora and Wolf. All that night the race continued, and at one o'clock in the morning of the 18th we entered the North Water. The Aurora was first through, five hundred yards in the lead; then came the Wolf, with our flying jib-boom over her rail, and the Bear directly astern of us. As we made the North Water, cheer after cheer arose from each ship, for we all rejoiced at our release from the icy grasp of Melville Bay. The whalemen, now fairly on their fishing-ground, were jubilant at their vic-

tory over the rest of the fleet, and as for ourselves we
were truly proud to have arrived simultaneously with
them.

The race was now renewed for Cape York in a clear
sweep of water. The Aurora and Wolf had proved themselves as fast as either the Thetis or Bear, which had
about the same boiler power; but the latter was a larger
and narrower ship and a little speedier than the Thetis.
So Captain Schley ordered her to advance if possible,
and reach Cape York ahead of the two whalers, and if
either or both went on to Littleton Island to proceed in
company with them, while we landed at Cape York and
searched the coast-line north. Accordingly she pushed
forward, passing the Aurora and Wolf, and followed a
lead in shore, dropping a boat to communicate with the
Cape. Then seeing no vent ahead she turned back; and
meanwhile the Thetis, accompanied by the Arctic, having approached, the whaling captains boarded our vessel
and bade us good-by, saying they would yield the chase
and proceed westward to their whaling-ground.

Standing in towards the Cape, we then picked up the
boat, sled, officer, and three men, dropped by the Bear,
and running our bows against the solid floe-edge spoke to
a native who had seen our ships in the offing and had
come down on a dog-sled to hail us; but we only gleaned
from him that nothing had been seen or heard of the
white men at or near Cape York. The Bear had been
instructed to attend the movements of the whalers, lest
they should play double by crossing the North Water to
the westward, and following the west water of the middle pack north to Littleton Island. So doubling Cape
York at once, we now stood along the land, passing the
Esquimaux village that was originally located on the
eastern side of the Cape, but which has been shifted half
an hour's ride to the westward. It comprised six huts,
tepees or *igloos*, but how many inhabitants we did not

learn, seeing but one. About three P. M. we arrived at
Conical Rock, having halted to put in several bolts that
had been broken during our struggle with the floes. It
was fortunate, indeed, for us that they had remained in
place while we rammed and raced across Melville Bay.

Thus far we had received no reliable intelligence of
Greely, for Captain Walker's Cape York native was
doubtless a liar. From the broken and scattered condition of the ice in Davis Strait, I was convinced that there
had been open water there for more than a month, and
that it was then navigable all the way up to Littleton
Island. Floes and rafts, it was evident, had been driven
hither and thither, and the ice had packed on either
shore according as the wind blew; but then, from the
easy manner in which the Thetis cut her way through
the rotten and friable floes, I felt confident, I repeat, that
the Strait had been open for many days. Bergs, of course,
were numerous, and out of every ravine there pushed a
glacier of greater or less magnitude. We were now in
plain sight of the Petowick Glacier, which reaches from
Cape Parker Snow to Cape Dudley Digges, a magnificent
extent of glistening white, the surface as rough and troubled as a great heaving sea, and from every facet and
angle of which sparkled the rays of the sun. Its drop-
pings studded the Strait for miles in every direction.

CHAPTER III.

THE RESCUE.

Saunders Island. — Cape Parry. — Littleton Island. — Finding the Records. — The Greely Party. — Scenes at Camp Clay. — Preserving the Dead.

WE all felt relieved now that the race for the $25,000 was over or abandoned, — not that we could claim the reward, but Congress by an unlimited appropriation had authorized the fitting of our expedition at a cost of $750,000, and then deliberately offered a prize of $25,000 for the whalemen to beat us ; and had Greely been at Cape York and the whale-fleet but a few hours ahead of us they would have earned the money, with us in sight at the time of rescue. It may be said to this that the prospect of reward hurried the whale-fleet to the front, and that to save our reputation we could not lag behind ; but then it must be remembered that the whalemen were as anxious for their whales as we were for Greely, and that, as previously stated, they have been known to be off Cape York as early as June 3d, while with all our exertion of energy we only arrived there on the morning of the 18th.

We finally after twenty-four hours tired of waiting for the Bear, and got under way from Conical Rock, where there was little game of any kind, birds, seal, or walrus. The Bear we supposed was working her way north on the west side of the pack toward the Cary Islands, where there was a large cache of provisions, and where, as Mrs. Greely

had told Lieutenant Emory, her husband would surely
be found. We were obliged to stand well off-shore to
keep in the open water, which extended to the north as
far as we could discern from aloft. Placing a cairn on
Woostenholm Island to inform the Bear of our movements,
we came to off Saunders Island on June 20th, and found a
large summer encampment of natives, probably one hundred and sixty in number, who had journeyed, they said,
from North Star Bay in spring time. They visited us on
eighteen dog-sleds, and men, women, and children roamed
all over the ship. They had nothing to barter, but took
anything that was offered them, and modestly asked for
everything they saw. They told us they had seen the
Proteus party on its retreat last summer, but had not
shown themselves because they were afraid; three or four
of their men and women having even remained hid on
Saunders Island while Garlington and his men camped
there. Since then, however, they had not seen or heard
of any white men from the north. One of our male visitors had lost a foot at the ankle, and with his leg encased
in a tight-fitting moccasin was stumping unconcernedly
around on the joint. Whether any one assisted Jack
Frost in amputating the foot we could not discover.
Another old fellow was possessed of an ancient and dilapidated gun of German manufacture. The cock would
not stand for him, so he held it up in order to snap a cap.
A second native owned an old United States Springfield
musket, stamped 1862, but like the first he had neither
powder, lead, nor caps, for which we were deaf to their
entreaties. I observed while in Siberia that the Yakut
and Tunguse dislike the cap gun and cling persistently to
the old-fashioned flint-lock, because the steel will practi-
cally last forever and they can always procure a flint.

Upon giving the natives a few scraps of iron they immediately indicated that they wished them fashioned into
spear-heads; and indeed a blacksmith would be altogether

a more useful and acceptable gift than the missionary that the Greenland governor promises to send them. Their sleds were, like those built by all Esquimaux, marvels of ingenious and patient workmanship; the runners and cross-bearer bars consisting of small bits of wood and walrus bone bound together with thongs, and shod their whole length with ivory, but so covered with grease, blood, and filth that it was almost impossible to see how they were knit or joined. Like the natives themselves, who were living on the fat of their land, to wit, on seal and walrus fat and blubber and water-fowl of all kinds, the dogs were in excellent condition, but as wild as wolves. The sight of our ship and people seemed to fill them with terror, and they tugged frantically away at their leashes and traces until some at length succeeded in freeing themselves and dashing across the floe. Hitherto these natives had never made use of boats, but they were now utilizing a small one abandoned by the Proteus party. and were constructing the frame of a *kyack* from small pieces of drift-wood lashed together. What a godsend to these people it would be to dump in their midst the condemned material of a cooper's yard! As an illustration of the rudeness of their work, I saw a harpoon staff six feet long composed of ten distinct pieces of wood lashed comparatively straight, and shod with walrus ivory.

After a halt of several hours at Saunders Island we pushed on to Cape Parry, where we landed and left a record marked by a flag for the Bear. Here there were fresh sled-tracks on the snow and ice, turning into Whale Sound; and on the low beach that stretches along the foot of the mountains were numerous Esquimaux graves, and many little cairns and spots covered with refuse, where the natives had made their summer encampments. Rabbits were seen in great numbers on the highlands and in the valleys; there were fresh evidences of the presence of deer; and indeed, from the abundance of walrus, seal, and bear bones scattered along the beach, it was evident that this is a decidedly favorite hunting-ground. While there I secured the skulls of five Esquimaux, all in a fair state of preservation, but covered with moss and lichens and very delicate to handle.

When leaving our anchorage close in-shore, where we had fastened to the narrow ice-foot, the Thetis slightly scraped the bottom on the starboard bilge, but no damage was incurred, and we proceeded without delay. The ice lay in a solid mass about three feet thick across the mouth of Whale Sound, but we skirted the edge in beautiful open water, seldom having occasion to struggle with the floe. As we ran along we noticed a fresh sled-track proceeding from Northumberland Island towards Cape Parry, but the round grooves in the snow and ice showed that the runners had been shod, Esquimau fashion, with walrus tusk ivory. The team had consisted of three dogs, and we could distinguish the mark of the whip trailing behind in the snow.

On the morning of June 21st we arrived at Littleton Island, the objective point of our cruise. It had been our intention to pass between Littleton and McGarry islands, but the wind blew shoreward, and upon slowing the ship she sagged down to leeward out of mid-channel, and we struck heavily on the Littleton Island shore. The engine was kept in motion, however, and the ship gradually forged ahead with helm hard a starboard, until, hanging for an instant amidships, she at length came off without injury, and we rounded McGarry Island and made fast to a grounded berg. Life-Boat Cove was in plain sight, but no vestige of the old Polaris house remained. It had been so strongly built of the deck-house and upper timbers of that vessel that I had thought it would surely last forever, but the Esquimaux have evidently torn it down for the wood.

After an early morning meal we landed to search the

island for cairns and records, confident that Greely had
done everything in his power to reach this point. We
found untouched and in good condition the Neptune cairn
and the one erected by Sir George Nares, together with
the coal and stores cached by the Neptune party; but
there was no record whatever of Greely or his men.
While lying here we shot a few ducks and gathered a
number of eggs; and on Sunday morning (June 22d) we
landed one thousand rations of bread, pemmican, tea,
sugar, and alcohol for fuel. The Bear had not yet shown
herself since we separated at Cape York; and now, after
waiting thirty hours for her, with fair and promising
weather, Commander Schley had become very anxious
to advance. But as it had been part of the general plan
of the expedition that our two ships should keep together
as constantly as possible so as to avoid the disaster that
overtook the Proteus during the absence of the Yantic, he
finally decided to run over and search Cape Sabine, the
water being open and favorable, and then return to Littleton Island to await the arrival of the Bear. This we set
out to do, and the men were actually on shore about to
cast off the lines, when greatly to our delight we descried
the Bear coming up the channel.

It appeared that in company with the whale-ships she
had proceeded but a short distance to the westward,
when it was discovered that the North Water did not extend any further. So they were all obliged to return,
and, being unable to find entrance into the land lead,
worked their way north through the middle pack, failing
to see us when we lay at Conical Rock, and to observe
our cairns or signals there or at Woostenholm Island and
Cape Parry. The whole fleet continued on to the Cary
Islands, where they halted in the hope of finding a passage to the west of the pack. The Bear, after examining
the cache of provisions there and finding it in a good
state of preservation, then pushed on to Littleton Island,

29

having learned no more of the fate of Greely and his men than did we. But those who had known the man, his orders, and that clause of his own final instructions, directing that a depot of supplies be made at Littleton Island, and that a sharp lookout be kept on Cape Sabine for his retreating column, were assured that he had used every means in his power to gain Littleton Island ; for, as a regular officer of twenty-three years' service, it was not likely that he would deal lightly with orders, or pursue a course different from the one he had marked out. Some persons there were, it is true, who professed to believe that he would not abandon Fort Conger ; and others again, with even less foundation for their belief, that after proceeding as far south as Cape Sabine he would attempt to return ; but it was our general and only warranted impression that we would find Greely as near to Littleton Island as he could possibly come. And therefore, since we had searched all the prominent points north from Cape York, the next places in order were Cape Sabine, Payer Harbor, and Brevoort Island, all within a radius of several miles of each other. So about a quarter past seven in the afternoon of the 22d both ships got under way standing across the bay, and soon brought up against the fast ice north of Payer Harbor between Brevoort Island and Cape Sabine.

As heretofore, every one who could be spared was sent on shore to search for cairns and records, Commander Schley ordered the steam-cutter of the Bear to be lowered and got in readiness for a visit to the cairn built to the westward of Cape Sabine by the Neptune, and the depot of five hundred and fifty rations made by Lieutenant Garlington at the same place. A party was also dispatched to examine a cairn left by the Neptune on a low neck of land connecting Brevoort Island with the main Stalknecht Island ; a party under Lieutenant Taunt of the Thetis was detailed to search Brevoort

Island ; another under Ensign Harlow, to visit the cache on Stalknecht Island ; and finally officers and men from both vessels were dispersed in all directions to find what they could.

Dr. Green of the Thetis and myself made straightway for Stalknecht, passing Dr. Ames, who was on his way to the main-land. We could distinguish the cairn on the island, but there was broken ice between us ; and presently we saw Ensign Harlow and command make a detour along the land for the solid floe. At this instant a cheer arose from Taunt's party on Brevoort Island. They had discovered the first record, and we next observed seaman Yewell, of our ship, waving it above his head and running like a deer towards the Thetis. Hastening on in the direction of the Stalknecht cairn, now in plain sight, we were suddenly startled by another cheer, issuing this time from Ensign Harlow's force, likewise of the Thetis. Joining him we found that he had tumbled down the cairn and come upon a box of choice photographs, and papers, instruments, and records of the Greely expedition ; together with a separate record left by Lieutenant Lockwood. It was dated September 22d, 1883, just nine months past, and stated that the party had gone "into camp four and a half miles west of Cape Sabine, or about midway between that point and Cocked Hat Island. Twenty-five men, all well."

Twenty-five men, all well ! At this good news my companions seemed overjoyed, but I reminded them that the record was written nine months before, and that ere now all hands may have starved to death. However, the Thetis and Bear were now blowing their steam-whistles for our return, and we had seven pieces of baggage to carry back over the rugged ice. Ensign Harlow signaled the Thetis for a sled, but five of us, without waiting, seized each a package of books or papers and started across the floe, meeting at length the sled-party,

and soon we were all again on board ship. Meanwhile
Commander Schley had ordered a party composed of
Colwell, Lowe, Ash, and Dr. Ames of the Bear, and Mr.
Norman, the ice-pilot of the Thetis, with a couple of men,
to proceed to the spot where, in the record found by Lieu-
tenant Taunt, Greely and his men were said to be.

By this time the screeching of our whistles had alarmed
the unfortunates at Camp Clay, and three of the strongest
of their number, Fredericks, Long, and Brainerd, tottered
down to the edge of the rocky promontory to look for
the relief they were sure had come. The whole party
had been lying under a portion of the fallen tent for
forty hours, some of the men being buried beneath it
and unable to move, for a gale had been steadily blowing
for fifty-six hours. The three had gazed long and anxiously to the eastward, whence they were expecting succor, but no sight or sound of the longed-for ship gladdened their eyes or ears. They were nearly palsied
with disappointment, and one of them declared that he
now despaired for the first time.

Returning then, to their starving comrades with the
sad intelligence that no vessel was in sight, a garrulous
discussion arose as to the cause of the prolonged and distinct blast which sounded so strangely like a steam-whis-
tle, for they had heard but one blast, albeit the whistles
of both ships were kept blowing for more than half an
hour. Lieutenant Greely at last told the men to cease
their quarreling and save their strength for a better purpose. Long then said he would go again to the low
promontory and take another look. He did so, and gazing eastward beheld our steam-cutter. With unspeakable joy he tried to raise a signal of distress — the loom
of an oar with three old rags nailed fast to it; but the
furious wind blew it down. The steam-cutter had now
observed him, however, and run in-shore to the ice-foot,
down upon which Long contrived to roll and scramble.

Clamoring meanwhile for food, he informed his eager questioners that his comrades were over the hill, and that seven still survived, one of them an officer, Major Greely. At this Mr. Norman, our ice-pilot, bounded out of the boat and up the hill, and was the first of our company to greet Greely, as he had also been the last to see him three years before, when the Proteus had carried the command up to Lady Franklin Bay.

" Greely, are ye there? How do ye get in?" Mr. Norman hailed from without the tent, the rear end only of which was standing.

"Is that you, Mr. Norman?" responded Greely at once.

" Yes, it is; you are all right now, succor has come; " shouted Norman, who was now joined by Lowe, and Ash, the ice-pilot of the Bear. Following then the directions of Lieutenant Greely, he cut the back out of the tent with his pocket-knife.

While this was transpiring Commander Schley had leaped on board the Bear and backed her out of her ice-bed, instructing Lieutenant Sebree of the Thetis to pick up the search parties from both ships and bring them around to Camp Clay. So only a few minutes elapsed ere the Bear had followed the cutter, and the Thetis the Bear, and soon every officer and man in our expedition who could be spared was doing his utmost to transport the survivors on board ship, and to gather together the wreckage of the camp. All the while the wind was blowing so fiercely that only a strong man could withstand it. Luckily it was off-shore, and the ships could safely steam straight against it; but so powerful was the gale that though going at full speed they were driven back, and when turned around by the force of wind under bare poles were placed in no little peril. Indeed, it seemed as if the evil fortune that had pursued the luckless band of heroic explorers was invoking the aid of the elements to prevent at this critical moment their rescue before death.

The scene itself was indescribable, and I shall not attempt to depict our pity and horror as we viewed it. A
cold barren plateau, between a small outlying promontory and a bleak weather-riven rock of red syenite reaching to the skies, on which even the mosses and lichens
would scarce grow. The raging of the wind and the
pitiless sea, and the roar of the black water of the bay
dashing over the ice-foot, made the lonesome picture
look colder and more appalling. Drifts of ice and snow
choked the ravines and hollows; but, saving ourselves
and the famished, skeleton-like survivors, not a living
thing appeared on the whitened landscape. The region
truly seemed to be the most desolate on the face of the
earth. It looked as though the curses of ten thousand
witches had descended upon and blasted it, and even the
birds would not dare to take their flight across the lifeless
land lest they too fall victims into the death-gap below.

Struggling up the valley of death, against the frantic
wind, from the low point to the westward of the camp,
where we managed with difficulty to effect a landing in
our whale-boats, we first came upon the remains of the
winter habitation, a parallelogram of four walls about
three feet high, built of loose stone, the inside dimensions
being perhaps 18 × 22 feet, with a tunnel or covered way
facing the mountain to the southward. This hut had
been roofed over with the whale-boats turned upside
down and covered with the sails and tent-cloths; the
smoke-flue, made of old tin-kettles bound with bits of
canvas, was thrown to one side; and water had risen
in and about the wretched dwelling-place to a height of
eight inches, concealing much of the foul evidence of
squalid misery in which its poor occupants had lived.
Cast-off fur and cloth clothing, empty tin cans, and the
sickening filth of twenty-five men for nine months, lay
heaped and scattered about — a veritable Augean scene.
Continuing up the valley toward a little rise of ground we

passed the dead body of a man laid out on a projecting plane of rock. A woollen cap was pulled down over his face, his hands were crossed on his breast, and his clothing and blankets were fastened around him with old straps and shreds of rope or yarns. Further up the hill lay the summer camp or tent, black with smoke and partly blown down, the flaps flying in the wind, which was blowing loose papers, leaves of books, and old clothing hither and thither; and on their backs within this half-open inclosure lay the poor creatures whom we had come to rescue, now more dead than alive.

Greely, in his sleeping-bag, and resting on his hands and knees, was peering out through the open door-way; his hair and beard black, long, and matted, his hands and face begrimed with the soot of months, and his eyes glittering with an intense excitement. For what terrible days of agony had been swept into oblivion by this supreme moment of joy. Succor had come at last! And yet he scarcely seemed to realize it. Mr. Norman told him who I was, and he said he was glad to see one of the people of the Jeannette, for he had learned a great deal of the history of our expedition from scraps of newspapers that had been wrapped around some lemons left by the Garlington party. Alongside of him lay a man on his back, Sergeant Ellison, to whom he introduced me, and who said he would like to shake hands with me, but his hands and feet were both frozen off. I looked down and saw that his nose was likewise gone. Yet he seemed cheerful and bright, and coolly discussed his sorrowful plight, thrusting out one of his arm-stumps, which I shook in lieu of a hand. Higher up and beyond the tent was the burial-ground, where ten bodies lay in a row, some barely covered with loose earth and stones. The first grave, or one nearest to the northern crest of the hill, had been very carefully made, for it was that of Sergeant Ross, the first man to die, and the survivors were then still strong enough to endure a little exertion. To the southward, or toward the face of the mountain, the graves became more and more shallow, just as the strength of the party was waning. All the faces were covered with woollen hoods and cloths or handkerchiefs; and each body was stretched out on its back, with the hands crossed on the breast and the clothing bound round. Only one corpse was found unburied, that of Private Henry; but the six that had been interred in the ice-foot were of course beyond recovery.

In the camp all was bustle and confusion. One man, Connell, was to all appearance lifeless; his face was fixed in death; he was cold from the hips down; and he scarcely breathed. Three days before he had eaten his last ration of seal-skin, and, abandoning all hope, had calmly determined to die. Doctors Green and Ames had their hands full of work. Water-kettles were heated, and the clothes being stripped from the half-dead Connell, he was wrapped in a blanket dipped in hot water. A little brandy was then poured down his throat, but it ran out at the side of his mouth until, catching his breath, he drew in sufficient to choke him and blew out the rest. Yet the few drops he retained sufficed to revive him, and rolling his head to one side he said wearily, "Let me die in peace." Not realizing that succor had arrived, he thought his comrades were still laboring with him. However, he survived, and still lives. He was a vivacious sort of man, and when on board the Thetis a few days, remarked, — "Well, boys, it was a pretty close squeeze for me. Death had me by the heels, and you pulled me out by the back of the neck."

Stretchers were brought from the ship, and the survivors carried to the steam-cutter and then transferred to the Thetis; all save Fredericks and Long, who, as hunters for the party, had been allowed additional rations from the game procured, to maintain their strength for the

1 The Devil's Thumb. 2. Lieutenant Greely 3. Finding the Greely party.
4. Carrying them down to the boats

extra exertion demanded of them. The camp was devoid of all food except a few pounds of boiled seal-skin strips, contained in tin cans. The final division of this food had been made some days before, and each man had charge of his own meagre supply. Considerable wood, including about four feet of the bows of the light-boat, still remained as fuel; and the bodies of the two ducks just killed, and one as yet untouched, were found at the old winter hut. Here, let me again observe, that this seems to be the most desolate, inhospitable spot on the face of the earth; while only twenty-one and a half miles across the Sound, sea-fowl and the eggs thereof are as plentiful as mosquitoes in Siberia. We killed two or three hundred braces of eider ducks at Littleton Island, and our people would not touch gull eggs, so bountiful was the supply of duck eggs. At Cape Sabine the famishing camp seldom saw a bird of any kind, nor any walrus, and but few seal; while on the opposite coast there was an abundance of game a month or six weeks earlier in the season.

The faces of two of the men were so swollen that they could scarcely see, and the rheum and slime had gathered in their eyes and half blinded them. They were too weak to help themselves, and dipping an old woollen sack in warm water I cleansed the eyes of one who lay upon his back gazing dimly in the direction where our mast-heads could be seen across the rocks.

Commander Schley stood by and said, —

"My man, don't you see the ship's masts? Don't you see the flags?" For we had mast-headed our colors.

"Please lift me up a little," he urged huskily, "that I may see." Then catching sight of the colors, he cried, — "Hooray! There is the old flag again; now, boys, we'll get some mush." And he did his best to raise a feeble cheer, while tears of joy ran down his cheeks as we supported him in his sleeping-bag.

When I shook poor Ellison by the stump, he said, —

" So you are one of the officers from the Jeannette, and poor De Long is dead. You must have had a terrible time." Here was sympathy sure enough. A man with nose, feet, and hands frozen off, who for months had been helplessly stretched upon his back enduring every agony and horror but death itself, could nevertheless find room in his bleeding heart to pity the past sufferings of others. A noble nature, indeed. He it was who sacrificed his life on the expedition to Cape Isabella for the English beef, when Sergeant Rice likewise perished.

And these are the great souls who die for their companions; who, with their lives in their hands, crawling on their very knees, go bravely forth to meet an heroic death, while their comrades are in their sleeping-bags, or their cruel critics away off in comfortable pot-houses are penning their uncharitable and infamous obloquies. Yes, when the cold waves extinguished the life of that poor Esquimau whose frail *kyack* was cut through by the treacherous ice while he so bravely strove to catch the seal his white friends were dying for,—there perished one of the noblest of souls.

And wolves, and ghouls, and would-be critics of Arctic toil and suffering, halt and know that the men whom you traduce or whose memories you would blast forever, perhaps for a penny a line, are made of finer clay than you; men who were and are yet ready to sacrifice everything on earth save honor for the sake of science and the benefit of mankind. Men who did their best; and that best is so far ahead of the conception of their malicious judges that it is a nation's shame that it permits its heroes, living and dead, to be dragged through the slime of public courts and press for the gratification of the prurient multitude of scandal-mongers, gloating over the silly effusions of the Arctic critic who never ventures his dear life nearer to the Arctic Circle than can be seen from the window of some tall printing-house south of 50° N. latitute.

It was after midnight of June 22d before we finished our sad duty of removing all the dead and living. together with the books and papers and certain relics, from Camp Clay to our two vessels; and we then sought shelter from the gale under the lee of Brevoort Island. The next morning saw both ships moored together at Payer Harbor; but when the fury of the wind had abated, Captain Schley sent back in the Bear a party of officers and men selected from both companies to go over the ground more carefully at Camp Clay and gather up all overlooked articles that might be of value either as mementoes or a part of the history of the expedition.

The pendulum and a case of photograph negatives which we had left the previous day on Stalknecht Island were taken on board; and after a search of several hours the shrill whistle of the Bear recalled us in haste, and we left behind many articles, of no value, however, except as relics. The ice that had been driven up the Sound by the gale was now returning with a dangerous rush; so we steamed across to Littleton Island and made fast to a small berg.

All the dead except Private Henry had been laid out on the Thetis and covered with ice readily hoisted over the ship's side for that purpose, and now the question arose as to their care and preservation; for, albeit the temperature was far below freezing point, the sickening odor from the bodies pervaded the whole ship. It was at first proposed to build an ice-box, but then we bethought ourselves of the alcohol we had on board and the oil-tanks in the engine-rooms. So these were called into requisition, and five of the dead being transferred to the Bear for preparation, those on the Thetis were stripped of their clothing, and bandaged after the manner of an Egyptian mummy. They were then sewed up in sheets, chocked tightly in the tanks with billets of wood, and covered with alcohol, sixty per cent. pure, which not only prevented further decay but rendered the bodies hard and solid to the touch. This last and melancholy office was performed by the surgeon of each ship assisted by several of the officers who volunteered their services, and the disagreeable duty was done behind a canvas screen on the forecastle, away from the idle gaze of the crew. To avoid any possible error or difficulty in the future identification of the bodies, a piece of numbered canvas was sewed on each one, beginning with Sergeant Cross, the first man to die, who was consequently marked number one, and so on down to the last body in the row of graves, number ten; then came Private Snyder, whom we had found on the projecting rock; and finally — number twelve — the remains of Private Henry, recovered from the ice-foot.

CHAPTER IV.

THE RETURN VOYAGE.

Reminiscences. — Foulke Fiord. — The Inconstant Esquimaux. — The Burial of Frederic Christiansen. — St. John's. — Portsmouth. — New York.

As will be supposed, there were many incidents of absorbing interest related to us by the rescued party. At one time, they said, their hunter killed a dovekie weighing about a pound. Greely assigned the whole of it to the hunter, to encourage him in his good work and keep up his strength. But one poor fellow clamored for his share of the food. In vain Greely tried to show him that his quota would only amount to an ounce, and that it would be far better for him in the end to yield it to the hunter. He claimed his ounce, and said he would pay $300 of the money due him from the government for the bird. He burst into tears, and finally, to quiet him, he was given an ounce of the raw flesh. He ate it and was satisfied, although his comrades ridiculed him and called him coward. Three days later he died of starvation.

The skins and bones of the birds procured were used for shrimp-bait. These little shrimps or sea-fleas are about the size of a grain of millet seed, and were caught in tin cans punched full of small holes, which at intervals were drawn up, cleaned of their contents, and then reset. Of these diminutive crustacea it requires 2,300 to fill a gill measure! Ye fishermen who fish for a living, think of this! And still on these they contrived to eke out an existence for weeks.

We left Littleton Island early in the morning of June 24th, and had gotten well out in the Strait, when the Bear signaled us to wait. We afterwards learned that as she was about to get under way, an *oomiak* full of natives came alongside of her. They had been living at Life-Boat Cove, just to the north of where we were lying, but had escaped our notice. Steaming along the coast, now quite free of ice, we at length ran into the mouth of Foulke Fiord, or Port Foulke.

The harbor was still locked by the fast bay ice, but we could see far up the Fiord, beyond the little island which afforded Dr. Hayes such a snug anchorage. Cold and bleak though it be, it is a beautiful spot, and, compared with Camp Clay, or the everlasting ice-pack that surrounded the Jeannette, looked a perfect paradise. In the dim mist the distant mountains could be seen stretching far inland, and but for the raw chilly air of the day one of a dreaming nature might have idly stood, and, surveying the peaceful scene, have lost himself in reveries of the *dolce far niente*. Small wonder that Hayes selected this superb place for his winter harbor; it is enchanting enough to tempt any one to winter there and drink in its Arctic glories, from the great rocks fading away in the fog from black to a hazy purple, to the dazzling purity of the crystal glacier. There is nothing so grand in nature; and, as I said before, I cannot help marveling why our millionaire yachtsmen do not cruise there and enjoy these matchless sights. The voyage can be made in two months, July and August, with entire safety to the frailest of their steam-yachts.

Towards noon we again put off and sailed down as far as Northumberland and Hackluyt islands, where we lay all night. In the morning we found that the ice had drawn us down on the land, and it was with no little difficulty that we released ourselves. About ten P. M. of that day we were off Cape Parry, and making fast to the ice sent a party on shore to remove our old record and replace it with a new one for the Alert, in case we should pass her on our way south.

Proceeding, we spoke the whalers Jan Mayen, Esquimaux, and Narwhal, which were working their way across the North Water, and on the morning of the 26th we espied two more under the land at Woostenholm Island. We then ran into the mouth of North Star Bay and moored to the floe. Here a number of Esquimaux, seven dog-sled loads of men, women, and children, paid us a visit and were photographed for their trouble, though in addition we gave them bread and plenty of wood and needles. Resuming our voyage we were off Conical Rock before morning, and lifting the record we had left for the Bear, deposited in its stead a fresh one for the Alert.

Now that our duty of rescuing the perishing party was happily performed, we had plenty of leisure to jog easily along on our return home and do the artistic Arctic thing of tying up to a berg, or, as the Arctic poet would more prettily put it, " lie out a gale under the lee of a friendly berg." We tried to do this that night, but the berg we selected was too high and not at all friendly ; it carried away our flying jib-boom and head-gear, and knocked the cheek, breasts, and arms from " Mrs." Thetis. So as heretofore we ran the ship into the pack, where we lay quietly enough, drifting gently along.

While we were threading our way through the loose ice, one of our two Esquimaux interpreters was seen to leap suddenly over the ship's side and dash towards the land like a deer. Instantly both vessels were put about and driven through the pack to cut him off, and several men from the Bear took up the chase on foot. As the fellow ran, he would occasionally halt, roll over on his back, and elevate his legs in the air to let the water out of his moccasins, which were evidently weighing him down ; but finally, after a hot pursuit of a mile or two, he was captured and returned to the ship. He would offer no explanation of his conduct, and becoming insolent was at last put in irons for safe keeping. We could assign no reason for his wishing to desert, unless it were to escape marriage to some dusky maiden at Disco to whom he had become engaged, for the governor sometimes forces the gay young Lotharios into matrimony. Perhaps he longed for the freedom of the Etah Esquimaux whom we had just left, or he may have found a new love among their number. However, he fled the ship utterly unprepared for a journey; having neither provisions of any kind about him, nor weapons with which to procure game. And previously he had been entirely happy and contented among us, and had grown as fat as a porker. It may be that he desired to convert himself into a mountain spirit, for these natives have a superstition that all who stray away into the mountains and starve to death straightway become powerful spirits for good or evil to friends or enemies; and possessed of this belief many of them have been known to wander away and voluntarily perish.

Desertion among the Esquimaux is by no means uncommon. Hans left both Kane and Hayes for the Etahs; Joe Iberbing, growing weary of the restraint on the Tigress, asked to be put on shore with his people at Niantilisk Harbor; Greely told us that his two natives, taking a notion at one time to go home, set boldly forth and had to be forcibly arrested, one of them starting off in midwinter, without his mittens or any means of procuring food, to make the long journey back to Proven or Disco; and I remember that poor Alexia often spoke of securing permission to return home from the ice-bound Jeannette.

As we advanced to the southward our patients grew stronger — all save Sergeant Ellison, whose mind gave way. The shock was too great, and then as we fed him the blood coursed more freely and there was danger of the old wounds at his ankles breaking out anew. Dr. Green told us that second amputations would be necessary when we reached warmer weather. Poor fellow, he had such queer fancies; and it was so sadly plain that he must die.

Rounding Cape York we were now fairly on our way home. On the evening of June 30th, while still in sight of the Devil's Thumb, we met the Alert and Loch Garry. Our mail was brought on board, and the night passed merrily. But what a constant surprise and curiosity that "iron tank," the Loch Garry, was to us; for verily I do not believe there was a single individual in St. John's of any ice experience whatever, or who gave the subject any thought, who for one instant imagined that the Loch Garry would ever cross Melville Bay. Nevertheless, she got half way over, and the worst half at that; though there can be no doubt that had she been nipped in the tenderest fashion, she would have collapsed like a blacksmith's bellows. This simply shows what kind of vessels the almighty dollar can induce to venture into the Arctic regions; then why not our strongly-built gunboat types? Where there's a will, there's a way, and it was really astonishing to watch the Loch Garry pushing along with the rest of us.

Early in the morning of July 2d, while running briskly ahead, the Thetis just cleared a rock, but the Bear, which seemed to possess a *penchant* for such things, ran hard and fast aground. However, our combined efforts shortly set her free; and now the time had come for the Thetis and Bear to part company for a while with the Alert and Loch Garry, they to proceed directly to Disco, while we halted at Upernavik to take up the coal that had been deposited there by the Loch Garry. We did not enter the Danish harbor to the north of the village, for it was too small; but both ships anchored in the fiord off the town. As a matter of course we got adrift, and the

30

Thetis slipped her cable and lost an anchor and sixty fathoms of chain to avoid a collision with the Bear. It is always preferable, if there be but one or two ships, to anchor in the Danish harbor; but in the fiord there is also a ten-fathom bar which affords fair holding ground. To prevent dragging, the anchor should be dropped to the southward and westward of the bar, which is located about half a mile up the fiord from the town.

Landing our crazy Esquimau, we departed from Upernavik at six P. M., and stood out towards Disco, scraping one of the islands as we passed. The Bear kept in mid-channel, where there is water enough, and did not touch. The better passage going south is between the islands and the main toward the south or southwest; and contrariwise, going north, the proper way is along shore towards the north and westward, avoiding the channel between the little islands just off the harbor, unless the vessel is kept straight in mid-channel.

The water was now entirely open and clear, and we did not encounter any ice from Upernavik to St. John's, saving the bergs, to be sure, which were numerous; but as it was constant daylight, these occasioned us no trouble. We celebrated the Fourth of July at sea with a capital dinner, but in a miserable snow and hail-storm. The next morning we arrived at Lievely, God-Haven. Disco Island was clad in its summer garb; the ice had left its harbor, the snow had melted from its hills, and the mountain sides were rich with purple flowers and green mosses. All the inhabitants were out sunning themselves, and the *kyack* men flocked like ducks around the ship trucking their articles of trade. I purchased a full-size *kyack* and some eider quilts and clothing.

The body of Frederic Christiansen, the courageous Esquimau of Greely's party, was here placed in a coffin covered with blue cloth and bearing a brass name-plate. At two P. M. the officers and crews of the other vessels

came alongside of the Bear, and the coffin, guarded by
four pall-bearers, was towed ashore in one of the cutters
to the little landing-place, where the inspector, the governor, and most of the natives were congregated to receive it. A salute having been fired from the three-gun
battery, the coffin, now covered with the American Jack,
was laid upon a bier that had been made on board the
Bear; and the poor Esquimaux buried it beneath crosses
and wreaths of wild-flowers and blue heather gathered
from their native hills. The *cortége* then advanced in
military order to the little church, where the body was
placed in front of the chancel, and the inspector made a
clever address in English. He received from the hands
of the American Government the body of the faithful
Esquimau, who, he said, had nobly died doing his duty
by the people whom he had engaged to serve, and whose
memory, he knew, would live on and be kept green by
the good feeling of the two governments. The governor
next addressed the natives in a similar vein in their own
tongue, after which the Lutheran Church service was
conducted by a native preacher, the Esquimaux congregation singing a psalm. The body was then carried to
the cemetery and interred with military honors.

That same morning poor Ellison died, and his remains
were bandaged and placed like the others in an alcohol
tank for conveyance to the United States.

Here the machinery of the Alert met with an accident,
slight, but in one of its vital parts, which detained us several days, but meanwhile we coaled our ships and got
ready for the passage to St. John's. We found excellent
cod-fishing in the harbor, and from the number caught I
should think that the villagers, if diligent, could capture
a sufficient food supply for their support from the sea
alone. The *kyack* men brought many fine large salmon-
trout for sale, that had been taken from the cool water
at the glaciers. They were a splendid fish, weighing

from two to four pounds, spotted and beautifully marked, with flesh of a bright-red salmon color. The head is sharp, with a protruding lower jaw, and altogether the fish has the appearance of being very choice game to take.

We cleared from Disco on the morning of July 9th; and there were doubtless some among us who with pleasure bade a final farewell to " Greenland's icy mountains." I did not; but, quite otherwise, felt a strong reluctance at leaving this land of wonders, which, long before the time of Columbus, was visited by the Norsemen, and which ever since has continued a land of adventure and conjecture ; and I hope sincerely it will not be the last time I shall see its ice-crowned peaks or green and purple valleys. There is still a great and important work to be done in studying the glaciers of Melville Bay and North Greenland, — the former having never been charted or even surveyed except from a distance — and exploring the Terra Incognita from Lockwood's farthest down the east coast of Greenland to the highest point attained by the German expedition under Koldeway.

There was nothing eventful in our passage to St. John's. We experienced no extraordinary wind or sea — nothing but thick foggy weather, which rendered our entrance to the harbor a difficult one. There we arrived, however, on the morning of July 17th, and instantly the telegraph announced to the civilized world the return of our expedition with the dead and living of Greely's command.

It now became necessary for the proper preservation of the remains that they be placed in hermetically-sealed metallic coffins, but such as we wished could not be procured at St. John's. This contingency had been foreseen, and drawings for the casting and proper fitting of cast-iron caskets had been prepared on board ship; but as the foundry and machine facilities at St. John's were in-

adequate to perform the work in time, caskets were made
of one-tenth inch boiler-iron riveted to angle-iron frame-
work with lids bolted on. They were thus neat, light,
serviceable, and perfectly air-tight, and to each one six
polished handles were attached for convenience in transportation, and a silver plate was fixed on each casket
inscribed with the name of the corpse and the date of
death. When the bodies were removed from the alcohol
tanks, particular care was taken to identify them, and
each number, which was stamped on a tin tag and fastened to the remains, was buried with it. Hence no mistake could possibly have been made.

And here it may be of interest to record the names of
the party, with the death dates of the poor victims : —

Sergeant Wm. H. Cross, general service; died Jan. 18, 1884.
Frederic Christiansen, Esquimau,   "   April 4,   "
Sergeant David Lynn, general service,   "   "   6,   "
1st Lieut. Jas. B. Lockwood, 23d Infantry,   "   "   9,   "
Sergeant Geo. W. Rice, special service,   "   "   9,   "
Jans Edwards, Esquimau (drowned),   "   "   10,   "
Sergeant W. F. Jewell, special service,   "   "   12,   "
Private Wm. A. Ellis, 2d Cavalry,   "   May 19,   "
   "   Wm. Whistler, 9th Infantry,   "   "   21,   "
Sergeant David Ralston, special service,   "   "   22,   "
   "   Edward Israel,   "   "   "   27,   "
1st Lieut. Fred. F. Kislingbury, 11th Infan-
   try,   "   June 1,   "
Corporal Nicholas Sailer, 2d Cavalry,   "   "   2,   "
Surgeon Octave Pavy,   "   "   6,   "
Private Chas. B. Henry, 5th Cavalry,   "   "   6,   "
   "   Jacob Bender, 9th Infantry,   "   "   6,   "
Sergeant H. S. Gardiner, special service,   "   "   12,   "
Private R. K. Snyder, 1st Artillery,   "   "   19,   "
Sergeant Joseph Ellison, 10th Infantry,   "   July 7,   "
   (On board the Bear at Disco.)

The survivors are : —

1st Lieut. A. W. Greely, 5th Cavalry;
Sergeants Brainerd, Fredericks, and Long;
Hospital Steward Biderbeck;
Private Maurice Connell.

The scientific value of the work accomplished by these men, living and dead, can only be estimated after their observations have been compiled and computed, compared and applied — all of which will involve years of patient toil.

We left St. John's July 26th for Portsmouth, N. H., where we arrived August 1st, and where the cruise of our rescue ships virtually ended.

We were received by the Secretary of the Navy and the entire North Atlantic fleet and practice squadron; and the inhabitants of Portsmouth tendered us a royal welcome. Here we transferred Greely and his command to the care of the Navy Yard, but afterward the bodies of the dead were returned to us for delivery to General W. S. Hancock at Governor's Island, New York Harbor.

Sailing then from Portsmouth on the morning of August 5th, we reached New York on the morning of the 9th and discharged our last expeditionary duty in surrendering the remains of the heroic dead into the tender custody of their comrades of the army.

A PROPOSED METHOD

FOR

REACHING THE NORTH POLE.

Let the stately Polar bears
Waltz around the Pole in pairs,
And the walrus, in his glee,
Bare his tusk of ivory;
While the bold sea-unicorn
Calmly takes an extra horn,
All ye Polar skies, reveal your
Very rarest of parhelia;
Trip it all ye merry dancers
In the airest of 'Lancers;'
Slide, ye solemn glaciers, slide,
One inch further to the tide,
Nor in rash precipitation
Upset Tyndall's calculation.
Know you not what fate awaits you,
Or to whom the future mates you?
All ye icebergs make salaam, —
You belong to Uncle Sam!"

BRET HARTE.

I.

WITHOUT entering into the question of the utility of
Polar exploration, which has been so thoroughly discussed by the scientific societies of the world, and universally passed upon by press and people with widely
different results, I will state at once my theory of the
proper means and way of reaching the North Pole.

Many modes of travel have been proposed, — by ship,
balloon, dog and deer-sleds, and by boat, not to dwell
upon a variety of submarine vessels, which, in fancy, are

made to dive beneath the ice in an easy-going manner, and reappear in open places to blow like whales or walrus, — but the detailed plans of which I have never had the pleasure of seeing. With all of our modern appliances, we have been able to make no further " northing " than did the old Dutch navigators three hundred years ago, for the approach to the Pole has been steadily barred in about the same latitude. This is not because our means or men are faulty or less fearless, but simply because that impenetrable ice-barrier, against which so many stout hearts have forced their ships in vain, floats, as it will forever, between the Arctic explorer and his goal. Vessels cannot go through it or around it, and were the power of all the steamships in the world concentrated in one ship, it could not push the ice aside and make the passage to the Pole. Nor is it probable that any ship or other locomotive machine will be devised to cross *over* the ice.

I have heard considerable talk of the proper shape, strength, and material of which Polar cruisers should be built, chiefly from people who have never been in the ice of the Arctic seas, and who have a very faint idea of the flotative capacity of bodies. A moment's reflection, and it will occur to every one that, for a body to float at all, its specific gravity must be less than that of water ; and that, to resist the crushing strain of ice, it must be more tenacious and solid than ice. And here the problem presents itself, — could a vessel of such construction (lighter than ice) withstand the enormous pressure ? I think not, even though it were " built in solid."

II.

Suppose a ship constructed in the shape of a parabolic spindle, its greatest transverse diameter thirty feet, and its length, say two hundred feet. This would give a body of fine lines, good rising power, if nipped below its

greatest diameter, and for speed and strength be an acknowledged model. Now build this spindle solid, — that
is, without an inner hold wherein to store men or provisions, — and of buoyant material, hooping it like a mast
with iron or steel bands so arranged with reference to
number and weight that the spindle will float in the manner of ice, or about one eighth part above water. Yet
even this pattern of strength would be an egg-shell in the
power of the mighty moving masses of ice. never at rest,
but always grinding like the everlasting gods, and grinding exceeding fine even the granite hills and islands.

From various causes there is a constant drift of all free
floating bodies from the Poles towards the Equator. In
the Antartic Ocean, we find a constant procession of majestic bergs drifting north, until they are dissolved by
the warm current that sets to the southeast from the
Equator, the counterpart, though in the opposite direction, of the Gulf Stream, which forces its way to the
northward and eastward.

Whether the cause may be the inflowing warm currents forcing the ice from the Poles towards the Equator,
or the centrifugal force that influences loose objects on
the surface of a sphere rotating on its axis, to which was
originally attributed the flattening of the earth at the
Poles, I will not attempt to decide. It is sufficient for us
to know that such is the fact; that the ice of the Polar
regions is continually being carried towards the Equator,
winter and summer, though much more rapidly in summer time, and that it moves much faster on the outskirts
of the Polar ice-fields than at the Poles. Within the
latitudes of 70° and 80° N. we find the southerly drift
swifter, because of the loose condition of the ice to the
southward, than we might be led to expect it would be
at 85° or 90° N. ; and here arises a question on which, to
a great extent, will depend the practicability of reaching
the Pole. If my premises are right, there can be no
doubt of my ability to do so.

The flattening of the earth at the Poles is admitted by all scientists, and from the formation of the earth, and by meteorological observations taken near the Poles, a lower barometric pressure is universally conceded. Whether this partial vacuum is caused by the rotation of the earth on its axis causing the currents of air to rise from the shoulders of this earth, tangent to the earth's surface (which, in truth, is an ellipsoid instead of a sphere), still remains a mooted question among the scientists of the world, and is one of the problems of which we hope to find a solution by reaching the Pole. Enough for my premises to know that these axioms are fully established.

The ice-cap, then, that covers the earth's surface at the Pole is held in place by the projecting islands which doubtless extend directly to the Pole, for in the history of Polar exploration each successive advance to the north has revealed new islands extending in small groups or chains towards the Pole; and the evidence of all Arctic explorers has been that they saw sea-birds and water-fowl still winging their northward flight, presumably towards the yet undiscovered islands dotting the path to the Pole, there to breed in quietness and safety on the land as yet untrodden by the foot of man.   This nucleus of pointed island peaks, if nothing more, will hold the ice fast at the Pole; and if we have the partial vacuum covering the flattened portion of the earth's surface around the Pole, and the air currents swirling in space above it, we should consequently have all the air motion *above* the earth's surface, and a comparative calm on the surface itself.   Or, in other words, we would not have the gales necessary to drive the ice towards either outlet, and as the centrifugal influence is acting equally in all directions, and tending to pull the ice-cap towards the Equator, it can only carry away those detached portions of ice broken near the outskirts of the ice-cap; or,

say, that portion that lies to the southward of about 85°
N. latitude, where we find the southerly drift almost too
rapid to march upon with any prospect of reaching the
Pole. But after crossing the eighty-fifth degree of latitude, if my premises are right, the traveler will come to
that immovable ice-cap which will, in all probability,
prove to be a paleocrystic sea of ice and snow. If so,
instead of having the terrible chaotic mass of ice described by Commander Markham and Sir George Nares,
we should have a clear unbroken surface to travel upon,
subject, of course, to fissures and shrinkage cracks. Indeed, the very fact that the sea of ice traversed by Markham was broken and chaotic is conclusive proof, to my
mind, that it was in motion and moving out, and by no
means paleocrystic.

Having reached the firm ice-cap which covers the earth
to the north of 85°, the travel will be smooth and easy,
and the traveler will not be carried south by the current
faster than he can travel north.

I therefore consider it impossible to construct a floating body which will be able to resist the tremendous
strain of the Polar ice-packs. It might not be crushed
for weeks or months, but then the contingency might
arise, on the first day it entered the pack, that two floes
would close upon and overwhelm it like an almond in
the jaws of a nut-cracker. For the wonderful potency
of these floes is incredible, and can only be calculated in
millions of tons, or rather square miles, of ice, averaging
twenty-five feet in thickness, or forty feet, where the
usual winter's growth of ten or twelve feet is rolled up
into hummocks — and I need not mention the colossal
floe-bergs one hundred or more feet in height. Telescoping and piling up, these vast masses form the great gorges
which only the hydraulic power of nature can move.
And this is forever occurring, in all seasons of the year,
though faster in some than in others; and the countless

million square miles of ice annually expelled from the Arctic Ocean through the three great outlets, — between Nova Zembla and Spitzbergen, between Spitzbergen and the east coast of Greenland, and the course through Baffin's Bay, — alone prove the fallacy of a " paleocrystic sea of ice."

III.

Did such a thing as a " paleocrystic sea of ice " actually exist, the task of reaching the North Pole would be one of comparatively easy accomplishment; for in winter a smooth, hard-beaten surface of snow would invite the traveler, and in summer a glassy surface of ice, and at either time depots of supplies could be laid out at convenient distances, as proposed by many clear-thinking persons, or Howgate's colony system be adopted. But, as it is, when we leave the land, both plans are impracticable ; for if we make a depot of supplies on the ice to-day, it is gone to-morrow — snowed under, overrun, or drifted out of position. The whole appearance of a pack may be changed in a day, or it may be so uniform that a definite location cannot be made. During the month of October, 1879, the Jeannette broke out of her ice-bed, and was whirled along an open lead for a few hours. We left a canvas structure on the floe, under the lee of a hummock, and for nearly a year and a half we looked for it unavailingly, albeit the ship had moved but a few miles. Finally it was discovered by one of our Indians, who returned to tell us, one day, in a state of trepidation, that he had found a " two-man house," for he really failed to recognize the spot alongside of which we had lain for a month. Similarly, on the evening of the same day the ship broke loose, Alexia shot and killed a bear, which, despite our constant endeavors, was never found ; for in all this while the entire ice-pack had drifted nearly five hundred miles into the northwest, and had swung around in a zigzag course more than one thousand miles.

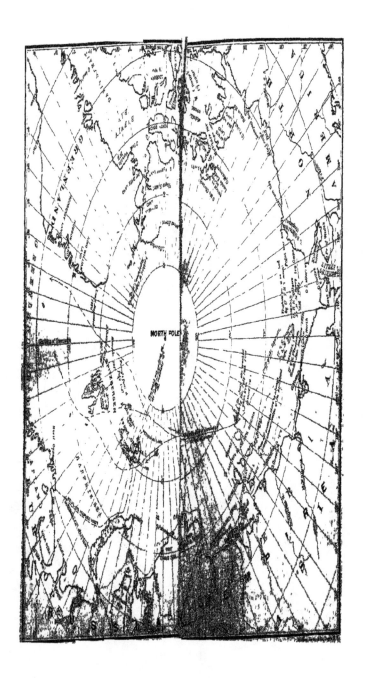

And so I iterate that depots made upon the ice are fugitive and utterly unreliable.

IV.

By observing, however, the drift and discharge of the ice, it may be possible to make good use of it on the retreat if not on the advance. The southerly ice-drift is well known, — down through Baffin's Bay and between Spitzbergen and Greenland: It was against this latter current that the gallant Parry marched so persistently for weeks, only to be thrown back upon his ship at the northern point of Spitzbergen.

The drift between Nova Zembla and Spitzbergen, although not so well traced as the others, has yet been satisfactorily observed along the east coast of the latter island, which is almost inaccessible because of the huge masses of ice heaped upon its shores, and the drift-wood which for ages has been accumulating from the great Siberian rivers. This drift continues down between the southern end of Spitzbergen and Bear Island, where a shoal has grown from the dropping of the stones and dirt from the floes as they jam and grind along. A study of the current charts of the Arctic Ocean, and the course in which the Jeannette drifted for twenty-two months, as well as the last drift of the Thegetoff, when she brought up on the island of Wilczek, to the southward of Franz Josef Land (all of which territory was discovered and charted by the two most determined and heroic explorers, Weyprecht and Payer), — a study of these charts will show that had the Jeannette held together, she doubtless would have drifted out into the Atlantic Ocean, either to the southward of Franz Josef Land and along the east coast of Spitzbergen, passing safely out, or else being crushed and cast upon the shore along with other drift-wood; or, rounding the north end of Franz Josef Land, she would have drifted down with the familiar current between Spitzbergen and Greenland.

V.

The northeast coast of Greenland has never been explored, and but little is known of the currents along its face. Nor has any considerable amount of drift-wood ever been seen floating from the coast of Siberia on the current which runs south along the shores of Greenland, or on the current which impinges against the northern coast of Spitzbergen. Here, then, is negative evidence that there is no passage around the northern end of Franz Josef Land, which must consequently be a large continent, or chain of islands reaching to, or well up towards, the Pole.

In either case this is now the way to the North Pole. It has been fully demonstrated by Weyprecht and Payer in their ships, the Isbiörn (Ice-bear) and Thegetoff, likewise by their escape in boats, and by Leigh Smith's subsequent voyages in his summer yacht, that the southern coast of Franz Josef Land is yearly accessible by steamer, and that, in case of emergency, retreat is easy and sure.

Now if Franz Josef Land extends to 85° north latitude, the Pole is within our reach, — if it extends farther into the north, so much the better. A series of depots can be established on the land as far as it goes, and a march of ten degrees, that is, five to the Pole and five back, is by no means beyond the power of human endurance. Let the state of the ice be as it may, it certainly can be no worse than the broken path over which the Jeannette's crew marched from the point where the ship sank to the mouth of the Lena, a distance of five hundred miles, — only one hundred miles less than the proposed journey from 85° north latitude to the Pole and return.

All this, to be sure, is upon the supposition that Franz Josef Land extends as far north as 85°. We have no

positive assurance that this is so, but it has been explored
to about 83°, and still stretched northward as far as the
eye could reach, which, on high ground and a clear day,
was fully sixty miles, or one degree — say to 84°.

So the first work will be to explore this land to its
northern limit, and if depots of supplies can be advanced
as far as 85° north, the feat of marching to the Pole and
back will, I repeat, be easily practicable.

VI.

But what would it avail a man to reach the Pole and
never return to discover to the world his success? Hence
I say "to the Pole and *back*," and now emphasize that
word; for thinking men no longer, if they ever did, act
upon that reckless dictum, — "Cut yourself off from all
that is behind you; let your retreat take care of itself."
And if there are no thinking men to undertake this hazardous enterprise, it had better be abandoned altogether,
else it must surely come to grief.

Now, I clearly see my retreat on this route. There
can be no doubt of the escape of the Austrian expedition, or of Leigh Smith's party, from the south side of
Franz Josef Land; and, to make doubly sure this line
of retreat, I would place supplies and additional boats on
the Pankratieff Islands to the southward of Cape Nassau,
Nova Zembla, and at Silver Bay in the western side of
Matotchkin Strait; and no boats, but food-supplies, at
South Goose Cape, where fishing vessels go and return
every year from June to September, — often as late as
October.

The other route that I would propose for the retreat
of the party which will essay to reach the Pole is by
way of Spitzbergen, — should the ice be drifting in a
proper direction, namely, to the westward, or southward
and westward. For, in such an event, the same current
which baffled Parry's efforts will assist and advance the

party, — floating them homeward when they lie down to sleep.

I would station boats and supplies at Parry's Harbor, the northernmost of the Seven Islands, which is almost annually accessible, and can be retreated from in boats, since the current here is continually setting to the southward, though more swiftly in summer than in winter. However, I would be entirely governed by the state of my provisions whether to await the coming of summer, or take advantage of the winter drift, as did the crew of the Hansa, the Germania's tender, and thus be ready to take to the boats in spring, or as soon as I reached open water. If the ice moved too fast or too slow to the westward, with no " southing," I would return to the point whence I started, or, should Franz Josef Land stretch far into the north, say to 87°, I would retreat upon it.

VII.

Concerning the depots of supplies at the points designated, I intend that they should be actually placed there — not the mere promise of establishing them. For I would intrust this duty to intelligent and intrepid officers who would execute their orders in the face of all difficulties.

Each officer, with a small party, should remain to guard his depot from bears, and also to secure a regular series of meteorological observations in those high latitudes. Only the two main positions, however, would require guards, namely, the Pankratieff Islands, off the coast of Nova Zembla, and the Seven Islands to the north of Spitzbergen. The ship will return home after landing the parties, which will consist each of one officer and three men, provisioned for four years, and equipped with house, sleds, boats, instruments, etc. When three years have elapsed these parties are to abandon their posts and set out for home in boats provided for that purpose, leav-

ing the balance of their supplies and boats for the use of
the main marching party, in case it should retreat that
way.

And likewise I would land this main band, consisting
of one surgeon, two officers, and twenty men, on the
southern coast of Franz Josef Land; with house, sleds,
boats, instruments, and four years' supplies, including a
sufficient quantity of dried fish for dog-food; the ship to
proceed home immediately after discharging its cargo.

VIII.

Three good teams of seven or nine dogs each would be
of great service in advancing supplies to the northern
border of Franz Josef Land. but the picked company of
one officer and ten men are to make the grand march to
the Pole without the aid of dogs; for though it is true
that when of no other use they are at least good to eat,
and may be regarded as so much "meat on the hoof,"
still, in my opinion, they are not economical draught animals for a party cut off from its base of supplies.

It was our experience, on the Jeannette retreat, that a
dog exacted about one half as much food as a man, and
performed about one quarter as much work. If a constant supply of game were assured, the offal would suffice for the dogs, and yet on a long march I question even
then if they would be an aid.

From the pack of twenty-four dogs which left the Jeannette we mustered two good teams of seven dogs each,
and these for the first couple of weeks were of little or
no use. Later on, under the guidance of two seamen
and two natives, they worked fairly well; but from the
time the ship was crushed until the dogs were shot or
lost, the labor performed by them in proportion to the
labor of the men, expressed in pounds of food (pemmican) consumed, was less than one tenth, — bearing in
mind, too, that four of our largest men were steadily employed in keeping the sleds upright.

So, I insist, well-drilled dogs, worked by competent drivers such as I had in Siberia, would doubtless be of considerable assistance in the transportation of supplies along the coast of Franz Josef Land, but on the long journey across the ice they would simply embarrass the party's progress.

IX.

This is my theory of reaching the North Pole — as far as I can see, a certain and safe one, attended by no sacrifice of life or property, for it will be remembered that the ships are to return as soon as they have unloaded, and that each party is provided with every possible means of escape. To recapitulate, then, I would have —

Depots established at designated points, and guarded by parties which are to abandon their positions and take care of themselves at a specified time; and an advance party on Franz Josef Land furnished with two feasible lines of retreat.

The details of general equipment, sleds, boats, food, and the season of year for traveling, I have fully examined into, and satisfactorily.

X.

And, finally, I propose to prove this theory of reaching the North Pole *by going there myself.*

GEO. W. MELVILLE,
Chief Engineer U. S. N.

When the report first reached this country that Baron Nordenskjöld would in the summer of 1885 command a South Polar expedition under the auspices of the Swedish Government, certain of Chief Engineer Melville's friends, interested in the success of his proposed expedition to the North Pole, conceived the idea of *simultaneously* dispatching the two explorers on their respective and kindred missions.

With this idea in view the Editor opened a correspondence with the Baron, inviting him to briefly outline the plan of his rumored enterprise for the pages of this work. The following reply was received : —

"STOCKHOLM, *April* 4, 1884.

" HONORED SIR, — The rumor that I propose to start on a South Polar expedition in the summer of 1885 is not exact.

" I have the greatest interest in Mr. Geo. W. Melville's (U. S. N.) new expedition, but it is, unfortunately for me, impossible to find the time necessary for directly contributing to your new book. Yours most truly,

"A. E. NORDENSKJOLD.

"MELVILLE PHILIPS, ESQ.,
    " Philadelphia, N. America."

APPENDIX.

———————

THE main advantages of the Melville sled over all others for
Arctic travel are its great strength attained by so little additional weight, and the facility with which it can be packed and
will hold its load securely.

It is impossible to build a sled combining all the advantages
that a critical novice might exact for Arctic use; for although
weight will not necessarily give the maximum of strength, yet
the minimum of weight can without doubt give the minimum of
strength, and one element of weakness may destroy its total effi-
ciency. It is only when his sled breaks down that the Arctic
traveler realizes how fatal it is to rely upon this argument of a
minimum weight.

On the retreat from the scene of the sinking of the Jeannette,
every sled save the three solid oaken ones was broken in less
than a week, and even these oaken sleds were many times temporarily disabled by the turning under of the runners. This
was the first indication of weakness in the best sled ever used
for Arctic service (the McClintock); and when the cross-bars
were firmly lashed in order to prevent the turning under of the
runners, then the top-rails, where they were pierced by the
tenons of the uprights, were wrenched off or split open from
mortise to mortise. And this, indeed, proved to be the weak
point of all the McClintock sleds on our retreat, and it was
the repairing of them, so that they could perform the work under which they had broken down, that suggested the idea of the
"double-bow" and "bow-string" runner of the Melville sled.
Once, when the top-rail of a sled had split open, a new rail

made from part of an oar was lashed next the broken one, and
vertical posts were fastened alongside of the old posts, which in
turn were lashed together across the top of the load, and it at
first seemed curious indeed that a broken sled should thus be
made stronger than a whole one.

But the reason simply was that we used the vertical post for
a cantilever, the bow-string of the sled acting as a centre about
which the lever turned. We likewise observed that the St.
Michael's native sleds were furnished with a top-rail to support
the uprights, and that these sleds, though old and dilapidated
when received on board, nevertheless continued to do a vast
deal of work, although, it is true, the ship carpenter skillfully
strengthened them. After our arrival on the Siberian coast I
rode many thousand miles on native sleds built upon the same
plan, but of far inferior workmanship, and their stoutness under
the roughest kind of usage was most extraordinary, for they
only became disabled by the wearing through of the lashings or
the unshod runners.

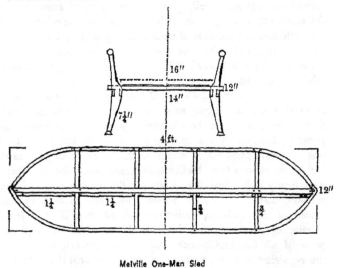

Melville One-Man Sled

Double bow and string of hickory. Shod with light hoop iron. Iron extends four
inches on top-rail. One-quarter inch iron rod passes through rear end posts. Cross-
bearers lashed to runners. Rails mortised and fastened with copper rivets All super-
fluous wood to be finished out and the sled varnished.

EQUIPMENT OF ARCTIC EXPEDITIONS.

IN answer to a request from the Navy Department, Chief
Engineer Melville submitted a series of papers on the fitting
out of the Greely Relief Expedition, wherein he gave at length
his views upon the proper equipment of Arctic cruisers.[1]

I.

Replying to the first question, — as to what provisions were
best to lay in store, — he says substantially as follows : —

It is well to recollect that the natural taste of all prepared
food should be retained as much as possible, and that goods put
up in tin cases or canisters should be avoided, not only because
of the poison imparted to them by the lead in the solder and
adulterated tin, but also on account of the nauseating gases generated in the blackened cans.

Soups should be of the richest kind, — turtle, mock-turtle,
ox-tail, vegetable, Philadelphia pepper-pot, and mutton-broth.
None of these, however, should be the soup pure and simple as
prepared for the market, but should be the "stock" of which
the soup is made, stored in small wooden casks. Fresh meats
and poultry of all kinds should be cut in good sizes and roasted,
and then packed in fifty-pound casks and covered with hot lard
or refined gelatine. The poultry, roasted in the usual manner,
should be split open longitudinally, to admit of closer stowage.
Oysters (fried) can be barreled in the same way ; while stewed
oysters for pies (a favorite dish with the Jeannette people) can
be put up in kegs or stone jars.

Fried potatoes — not "Saratoga chips," which have all the
nutriment cooked out of them, but potatoes quartered, and
underdone — can likewise be stored and covered with hot lard,
and it will be seen that this is a convenient way of carrying the
lard, which can all be used again. Boiled potatoes with their
"jackets" on, fresh carrots, parsnips, turnips, etc., can be
similarly preserved, and eggs done up in small packages and
covered with scalding lard will keep indefinitely. Salted or

[1] The plans of a model Arctic cruiser designed by Mr. Melville
while on the Jeannette have already been published.

spiced meats are a wholesome relief from fresh canned food, and under this head I would recommend sausage meat, and what is known to Philadelphians and Baltimoreans as "scrappel," which is simply head-cheese mixed with a quantity of well-boiled corn meal and buckwheat, and which, fried like sausage meat, is equally palatable and nutritious, having all the good qualities of the *erbswurst*, — a staple food in the German army, and an article abundantly supplied to the Austrian Arctic expedition. Bologna sausages, boiled eggs, boiled hams, raw hams, and sides of bacon, should all be packed in lard and placed in store.

The chief reason for cooking many of these foods before packing them is to economize the fuel account of the galley. It would be well, for variety's sake, to lay in a moderate supply of cooked corned beef (not salt beef), in tubs. Clean pork, in half barrels, to be cooked with beans or sauer kraut, and pig's jowl, in one quarter casks, would likewise be a most welcome accession to the ship's larder. Oatmeal, maccaroni, and corn starch should not be forgotten, nor — in the list of compressed vegetables — beans, lima beans, green peas, whole hominy, split peas, rice, and corn meal. But desiccated potatoes should be shunned, for on both the Arctic ships in which I have served, it was impossible for the ingenuity of the cooks to render them relishable.

Fresh bread is a necessity once or twice a day, and hence first class flour and good yeast are requisite ; but as here again the economy of fuel comes into consideration, a variety of dry baked sugar-cakes should be supplied in lieu of hard bread, thus affording the crew a change of diet, and an antiscorbutic in the shape of the sugar. Pails of peach and apple butter, prepared pumpkins for pies, mince-meat in tubs, plum-pudding in jars or cans, — all these are not only toothsome but antiscorbutic. Pickles of every kind, including cabbage, and whole onions in casks filled up with equal quantities of water and vinegar, are an excellent provision, as we of the Jeannette had cause to know. Nor should I forget to enumerate butter, cheese, and syrups.

Tea, coffee, and chocolate might be supplemented at dinner by a half pint ration of claret, both as an antiscorbutic and a

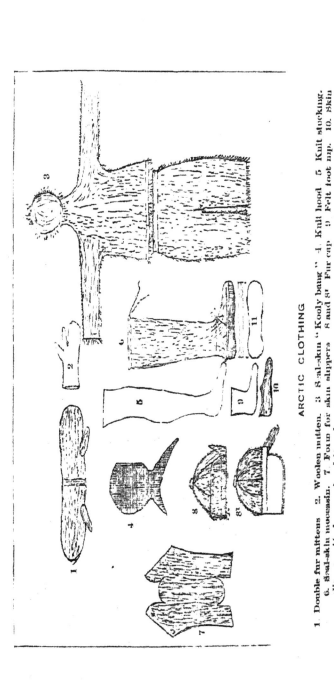

ARCTIC CLOTHING

1. Double fur mittens 2. Woolen mitten. 3. Seal-skin "Kooly bang" 4. Knit hood 5. Knit stocking. 6. Seal-skin moccasin. 7. Form for skin slippers 8 and 8' Fur cap 9 Felt foot nap. 10. Skin slipper. 11 Inner soles, felt and canvas

most agreeable addition to the bill of fare. Lime juice comes properly under the head of medicines, but when mixed with sugar and water it also becomes a pleasant beverage. I cannot advise spirits or wines of high alcoholic standard, yet I do know that, on the Jeannette, when each officer and man was given about three ounces of sherry wine with his dinner, it seemed more palatable, and the company became brighter, conversation was more sprightly, and general good-humor prevailed in the place of silence and gloom. On Wednesday evenings two ounces of American whiskey, with sugar and candied lemon peel for punch, were handed round to officers and men, enlivening all, and that day became the red-letter one of the week. There was not enough alcohol in the allowance to enervate us, while the capital effect in manner, speech, and good-fellowship would be noticeable for days.

Sauces of every description should be supplied in liberal quantities, to make as tempting as possible the food, whose sameness, though daily changed, will surely and quickly pall upon the crew.

It will thus be seen that substantial and nutritious food is the kind required, and that persons whose habits have been such that they cannot digest it are unfit for Arctic work. So that while preparing food for the Arctic voyager, the voyager himself should be selected to suit the food. A diet of cream-puffs and marmalade would not yield the stamina and bodily heat required for life in high latitudes.

II.

And as to the query, — " What quantity of each kind of food will be needed per day for each man ? "

It will be a variable quantity, dependent upon the quality. Then, too, the stomach, for comfort, exacts distention as well as nourishment.

While on the retreat after the Jeannette went down, our rations were one pound and a quarter of pemmican, half a pound of hard bread, beef tea, sugar, and tea or coffee, making in all about two pounds of food per man ; and we found the allowance ample. Some of us experienced at times a sense of repletion, others could have eaten more, but the question is, could they have assimilated it? I think not; for though we trained down to rather light weights, it must be borne in mind that the work we were performing demanded an exhaustive expenditure of strength. Therefore, it is my opinion that on the march a ration of two pounds would be plenty, providing it consists of hard bread and pemmican. On board ship, however, each man should be given about four pounds of all kinds of food. De Long's provision book shows as large rations as seven and a quarter pounds; but then the greater part of this food was not eaten, because of its disagreeable taste and odor.

By four pounds of food I mean *solid* food, as served on the table, including condiments, — not as it is issued from the storeroom.

III.

" As to variety." — The greater the better. It cannot reach too wide a range, and should only be limited by the storage capacity of the ship.

IV.

" As to the best manner of packing the provisions." — I have proposed wooden casks of different sizes bound with iron bands, and they should of course be air-tight, thoroughly cleaned, and coated with refined glue.

Glass and stone jars are too fragile and heavy; whereas the wooden casks, when emptied of their contents, may be used for fuel. Such provisions as are free of acids or alkalies can for convenience be packed in tin cans, — which, however, should be made of and sealed with pure tin, since it is a well-known fact that a single duck-shot dissolved in a bottle of claret may have as deadly an effect upon the human body as when fired from a gun.

All casks and packages should be made with an air-space, so that they will float if cast into the sea. This can be accomplished by fitting a double head upon one end of each cask, with an empty space between; and for ease in handling, a ring should also be attached, and no package should exceed fifty pounds in weight.

v.

In answer to the question, — "What improvements can be made in the usual cooking apparatus?" — it will be easier for me to state what would not do than what would.

Any cooking-stove with a capacity for thirty or forty men would be better than the galley (small navy pattern) which we had on the Jeannette. Two cooks in succession, whom we shipped at San Francisco, deserted their posts because of it, at least this was their excuse; and only by employing the patience of a Chinaman were we enabled to use it to the end.

As I do not know of any particular make of galley which meets with my approval, I would propose two large kitchen ranges set back to back, and discharging their gases into one common flue. This arrangement will economize the consumption of fuel, and at the same time offer two fire-places, either one of which can be used, or both when necessary. Water-backs can moreover be utilized to great advantage to melt ice or snow for general purposes.

These ranges would dispense with the necessity for a sheet-iron baking oven; yet if one be desired I would recommend an oven of just sufficient size to bake the bread for the ship's company; with one flue or a pair of flues set on top of the galley over one or two of the fire-holes, so as to permit the heated gases to pass up and around the oven, and be delivered into the galley pipe through an adjustable flue. The bread oven, again, may be used on the berth deck as a heater.

vi.

In the matter of clothing, I would advise for summer travel, —

A suit of red flannel underclothes, woolen stockings (first-class) to reach above the knees; heavy cloth trousers, either fitting tight from the knees down, or knee-breeches, so that the moccasin legs will come over them; and they should not have fly-fronts, but button up squarely to keep out the cold, and be upheld by a waist-belt, not too wide.

The moccasins should be oil-tanned without hair, — what are known as "water-boots," with one canvas and six inner soles. A pair of blanket foot-nips is necessary, — hay is sometimes used, but not always procurable ; and a pair of rubber sandals, with toe-and heel guards and strap across the instep, will save the moccasin soles in summer. These sandals should have a large diamond mesh or rough surface, similar to that of a rubber door-mat.

A blue flannel overshirt with neckerchief; a fur cap with ear-laps, a guard for the back of the neck, and an extensive visor to protect the face; and a " lammie," a short close-fitting cloth and lined coat coming to the hips, closed in front, and furnished with two breast-pockets, — complete the outfit.

To prevent taking cold when halting at night, each man should be provided with a fur jacket, knee-trousers, and a sleeping-bag, winter or summer. A sheath knife is indispensable; and in a small rubber bag he should also carry one pair of socks, one pair of foot-nips,[1] one undershirt, one pair of drawers, some patches and sewing material. Moreover, for general service, one jacket and a pair of trousers should be kept in the boat; and likewise an abundance of navy flannel for underclothing.

On board ship each person should sleep on a double blanket and have two pairs over him; on the march a half-blanket and a sleeping-bag will suffice. A tent is necessary for winter travel, but in summer it may be dispensed with, and the boat-cover and sail used for shelter instead.

[1] The foot-nips are made of blanket stuff, in the manner of ankle-buskins.

INDEX.